Elementary Excitations in Solids, Molecules, and Atoms

Part A

NATO ADVANCED STUDY INSTITUTES SERIES

A series of edited volumes comprising multifaceted studies of contemporary scientific issues by some of the best scientific minds in the world, assembled in cooperation with NATO Scientific Affairs Division.

Series B: Physics

Volume 1—Superconducting Machines and Devices
edited by S. Foner and B. B. Schwartz

Volume 2—Elementary Excitations in Solids, Molecules, and Atoms
Part A edited by J. T. Devreese, A. B. Kunz and T. C. Collins
Part B edited by J. T. Devreese, A. B. Kunz and T. C. Collins

Volume 3—Photon Correlation and Light Beating Spectroscopy
edited by H. Z. Cummins and E. R. Pike

A Continuation Order Plan may be opened with Plenum for Series B: Physics. Subscribers to this scheme receive the same advantages that apply to all our other series: delivery of each new volume immediately upon publication; elimination of unnecessary paper work; and billing only upon actual shipment of the book.

This series is published by an international board of publishers in conjunction with NATO Scientific Affairs Division

A Life Sciences **B Physics**	Plenum Publishing Corporation London and New York
C Mathematical and Physical Sciences	D. Reidel Publishing Company Dordrecht and Boston
D Behavioral and Social Sciences	Sijthoff International Publishing Company Leiden
E Applied Sciences	Noordhoff International Publishing Leiden

Elementary Excitations in Solids, Molecules, and Atoms

Part A

Edited by

J. T. Devreese

Department of Physics and
Institute for Applied Mathematics
University of Antwerp
Antwerp, Belgium

A. B. Kunz

Department of Physics
University of Illinois
Urbana, Illinois

and

T. C. Collins

Solid State
Physics Research
Laboratories
A.R.L.
Dayton, Ohio

PLENUM PRESS • LONDON AND NEW YORK
Published in cooperation with NATO Scientific Affairs Division

Lectures presented at the NATO Advanced Study Institute on Elementary Excitations in Solids, Molecules, and Atoms, Antwerp, Belgium, June 18–30, 1973

Library of Congress Catalog Card Number 74-1247
ISBN 0–306–35791–7

A Division of Plenum Publishing Company, Ltd.
4a Lower John Street, London W1R 3PD, England
Telephone 01-437 1408

U.S. edition published by Plenum Press, New York,
A Division of Plenum Publishing Corporation
227 West 17th Street, New York, N.Y. 10011

Set in cold type by Academic Industrial Epistemology, London
Printed in Great Britain by The Whitefriars Press Ltd., London and Tonbridge

PART A

LIST OF LECTURERS

P.W. Anderson
University of Cambridge, Cavendish Laboratory, Cambridge CB2 3RQ, England

T.C. Collins
Aerospace Research Laboratories, Wright-Patterson Air Force Base, Dayton, Ohio 45433, U.S.A.

L. Hedin
University of Lund, Department of Physics, Sölvegatan 14A, 223 62 Lund, Sweden

J. Heinrichs
Institut de Physique Théorique, Université de Liège, Sart Tilman par Liège I, Belgium

A.B. Kunz
University of Illinois, Department of Physics, Urbana, Illinois 61801, U.S.A.

P. Longe
Institut de Physique Théorique, Université de Liège, Sart Tilman par Liège I, Belgium

A. Lucas
Faculté Universitaire de Namur, Département de Physique II, rue de Bruxelles, B-8000 Namur, Belgium

S. Lundqvist
Chalmers University of Technology, Institute of Theoretical Physics, Fack, S-402 Göteborg 5, Sweden

N.H. March
Imperial College of Science and Technology, Department of Physics, London SW7 2BZ, England

P.M. Platzman
Bell Laboratories, Murray Hill, New Jersey 07974, U.S.A.

S. Rodriguez
Department of Physics, Purdue University, West Lafayette, Indiana 47907, U.S.A.

D.F. Scofield
Aerospace Research Laboratories, Wright-Patterson Air Force Base, Dayton, Ohio 45433, U.S.A.

PREFACE

The Advanced Study Institute on 'Elementary Excitations in Solids, Molecules, and Atoms' was held at the University of Antwerp (U.I.A.) from June 18th till June 30th 1973. The Institute was sponsored by NATO. Co-sponsors were: Agfa-Gevaert N.V. (Mortsel – Belgium), Bell Telephone Mfg. Co. (Antwerp – Belgium), the National Science Foundation (Washington D.C. – U.S.A.) and the University of Antwerp (U.I.A.). A total of 120 lecturers and participants attended the Institute.

Over the last few years, substantial progress has been made in the description of the elementary excitations of the electronic and vibrational systems and their interactions. Parallel with this, the experimentalists have obtained outstanding results, partly as a result of availability of coherent light sources from the far infrared through the visible region, and partly because of the availability of synchrotron radiation sources in the soft X-ray region. The results of today will lead to further progress over the next years. It was the purpose of this NATO Advanced Study Institute to present a state of the art, namely a survey of experiment and theory.

Although the title of the Institute incorporated Solids, Molecules, and Atoms, the emphasis was on Solids. However, the study of elementary excitations in Solids, Molecules, and Atoms involves many common techniques and insights. E.g., bandstructure calculation starts from atomic wave functions. The methods of studying correlations are often similar in Solids, Molecules and Atoms. Therefore it is hoped that molecular and atomic physicists as well as solid state physicists will benefit from these lectures.

The material in these lectures is presented in two parts. Part A deals with the excitation energies of the electronic system. Band theory, study of exchange, correlation and relaxation phenomena are included here as well as the collective excitations. Part B is devoted to the vibrational excitations and their interactions with radiation fields and electrons.

I would like to thank the lecturers for their collaboration and for preparing a number of manuscripts. Thanks are also due to the members of the International Advisory Committee: Professors, M. Balkanski, F. Bassini, F.C. Brown, M.H. Cohen, W. Hayes, S. Lundqvist, G.D. Mahan, N.H. March, and D. Schoemaker. Further

I am much obliged to Professor A.B. Kunz and Dr. T.C. Collins, who acted as co-directors.

The secretarial tasks of the Institute were extremely heavy, partly because the Institute was organized on a new campus. Dr. V. Van Doren, Drs. W. Huybrechts and Miss R.M. Cuyvers receive special thanks for their outstanding contributions to the practical organization, and to the editing of the lectures. Dr. V. Van Doren, Drs. W. Huybrechts, Dr. F. Brosens and Dr. P. Cardon de Lichtbuer also contributed to the preparation of the lectures and the subject index. The author index was prepared by Miss R.M. Cuyvers. To these co-workers I express here my sincere gratitude.

All those who further contributed to the success of the school by way of practical organizational help, copying lecture notes, organization of social programs etc., all the members of the physics department of the U.I.A. and my co-workers of R.U.C.A. are sincerely thanked.

I gratefully acknowledge the financial support of the NATO Scientific Affairs Division and the co-sponsors. This support made it possible to invite a panel of distinguished lecturers and to provide a substantial number of grants to students.

My sincere thanks also to the Board — especially Prof. F. Nedée, president, and Prof. L. Vandendriessche, rector, — and the administrative directors of the University of Antwerp (U.I.A.) who provided accomodation and very valuable organizational support.

Jozef T. Devreese

Professor of Theoretical Physics,
University of Antwerp, (U.I.A. and R.U.C.A.)
Director of the Advanced Study Institute

February 1974

CONTENTS OF PART A

ATOMS MOLECULES AND LIQUIDS

SEMINARS

CONTENTS OF PART B

MANY-BODY EFFECTS AT SURFACES

P.W. ANDERSON

Bell Laboratories, Murray Hill, New Jersey

and

Cavendish Laboratory, Cambridge, England

The lectures I will give here will be based in the first instance on the thesis of one of my Cambridge students, Dr. John C. Inkson [*1*], which has been the source of a number of articles in the literature [*2*]. John's work in turn leans heavily as far as formalism is concerned on some papers by D.M. Newns [*3*], and on Denis Newns' advice and help during the period when he was at Cambridge. All of his work is concerned with many-electron effects at interfaces; the main intention was to study the metal-semiconductor interface with a view to understanding the so-called surface state phenomenon, but in the end we found ourselves concerned with all kinds of surfaces and with the whole range of electronic excitations which may occur at surfaces, whether metal-vacuum, metal-insulator or metal-semiconductor. I will also draw on work on surface states by various people at Bell and Cambridge: Heine, Pendry, Appelbaum and Hamann.

In fact, the ideas go beyond surfaces to the very fundamentals of the electronic theory of metals and insulators, which is the main reason I chose it as a topic to talk about here. There is in fact so much physics to talk about that I will continually dodge the formal mathematics of the problem, which in some cases is rather formidable. This formal mathematics is to be found in Inkson's papers, which may be looked up in the literature or obtained by writing him.

The genesis of this problem which concerned me lies far back, in the ideas of Bardeen [*4*] and Shockley [*5*] which played an important role in the invention of the transistor. What they were trying to do was to create what is now known as the 'field effect transistor': to apply sufficient voltage to the surface of a piece of semiconductor to move the Fermi level relative to the energy bands and thus to control the flow of current inside it

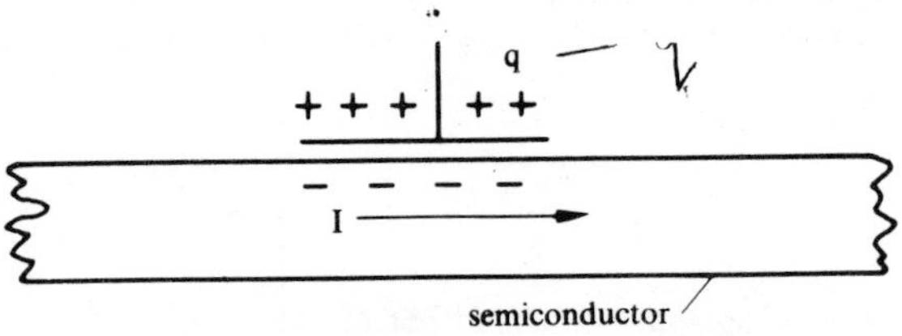

Why should have this device worked as, incidentally, they now do — it is amusing, and I think very instructive in understanding the nature of science, to realize that the work which earned Bardeen his first Nobel Prize was essentially completely out of date by the time he got his second. (This absolutely is not to say that he didn't deserve it, any more than Einstein detracts from Newton).

To see this we have to understand the nature of rectification and the Schottky barrier layer on a semiconductor. In general, it *is* true that the Fermi level at the surface of a semiconductor is not the same relative to the valence and conduction bands as it is in the bulk. In the bulk it is controlled by the nature and density of impurity levels within the band gap (P in Si, for example),

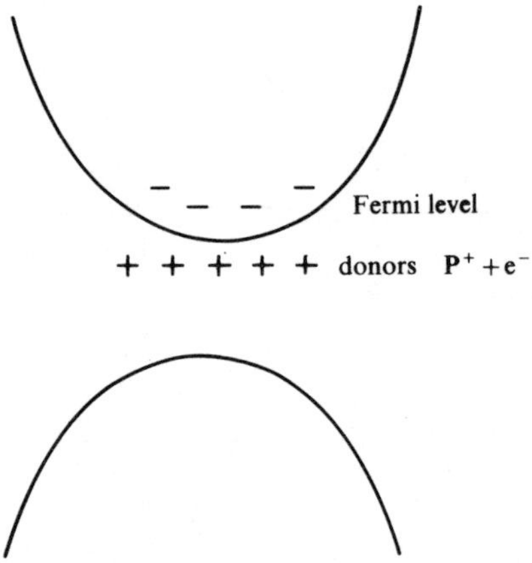

but at the surface it is determined by surface properties, by any metal we may put on, etc.

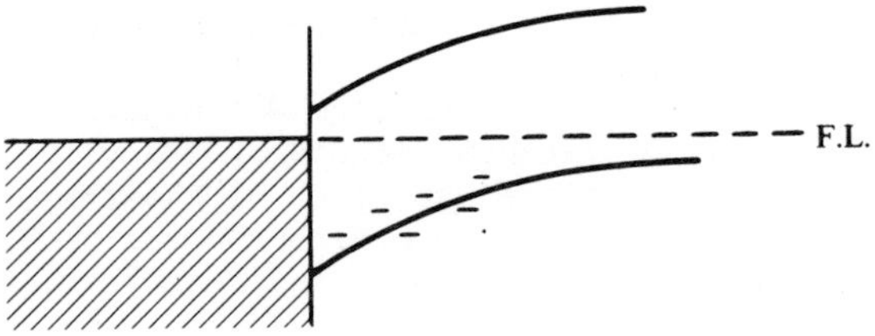

The solution to this problem is the Schottky barrier: the bands bend according to $\nabla^2 V = 4\pi\rho$, the charge being provided by charged impurities: in the above case, extra electrons captured by the acceptors; B^- bends the bands down, and we get a barrier

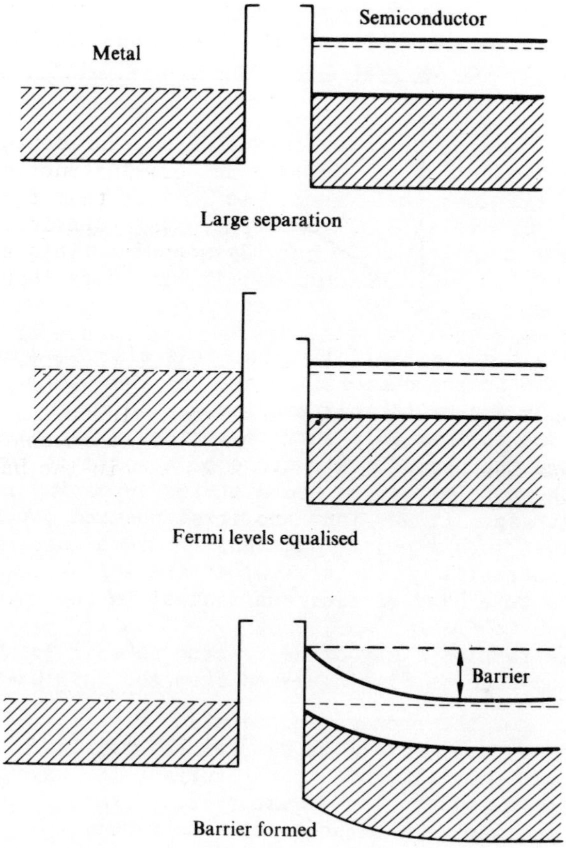

Classical Schottky barrier formation

Figure 1

region of space charge. (See also figure 1)

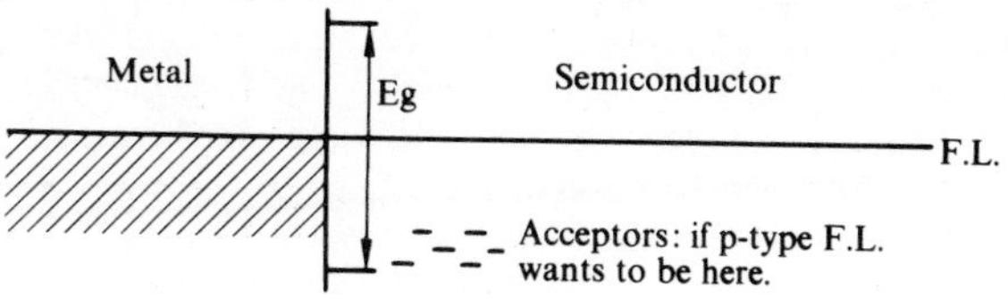

The thickness of this barrier region is quite large if the material is pure (screening length $\simeq (kT/4\pi ne^2)^{\frac{1}{2}}$) and it therefore rectifies current, for reasons easily available in solid state physics texts.

What was expected was that extra charge applied to the surface, say by applying an electric field to a metal electrode or by changing the work function of the metal, would change the barrier very considerably. Since this extra charge would have to be compensated by just that many more charged impurities.

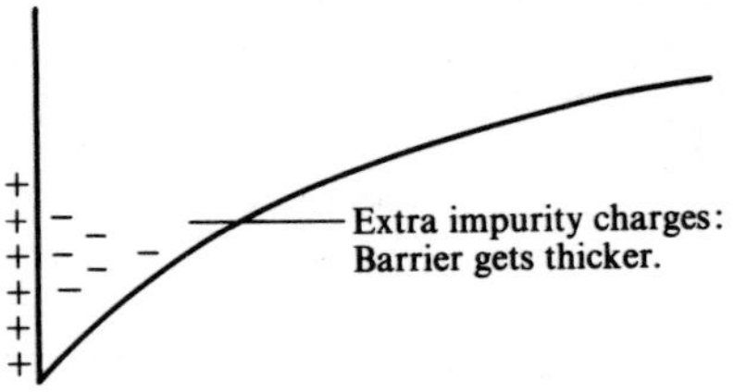
Extra impurity charges:
Barrier gets thicker.

But they found essentially that this effect never works, at least on the semiconductors (Si and Ge) they had available then.

Bardeen's answer to this problem was to invoke a relatively old idea due to Tamm [6]: surface states. In general, at any point where the periodic structure of a solid is interrupted, one has the possibility of bound states appearing in the forbidden energy gap. It was Tamm who first pointed out that a surface is just such a point, and that the mere interruption of the periodic potential of the crystal at the surface would, in general, lead to a band of 'surface states' in the forbidden energy gap.

The old-fashioned way of describing this is to observe that the allowed band is distinguished from the forbidden one by whether or not the propagation constant k in $\Psi = u(r)e^{i\vec{k}\cdot\vec{r}}$ is real or imaginary. If it is complex no normalizable states in the bulk are possible, but at the surface the wave-function may decay into the bulk. The oldest surface state is probably the Rayleigh surface wave on an elastic continuum.

Shockley in particular gave good arguments for the idea that there would always be surface states in the gap of the usual covalent semiconductor — arguments which seem to be correct. Therefore, they argued, there will be a very high density of the equivalent of impurity states at the semiconductor surface. A band of surface states will have 2 states/surface atom, or $\sim 10^{15}$-$10^{16}/\text{cm}^2$, in the 1-2 eV energy gap, which is capable of absorbing any amount of charge without allowing any appreciable excursion of the Fermi level. Any great excursion of the Fermi level will charge up the surface states and it is thus pinned at a point in the gap. If I have time, I would like to talk to you about surface states on clean semiconductors as they have been calculated by Appelbaum and Hamann, and very probably measured by Rowe at Bell; the old ideas of Shockley are confirmed and extended by their results, so all of this probably *is* actually true of a clean semiconductor surface in perfect vacuum. However, the experiments until recently have all referred to an entirely different physical situation. Twenty-five years ago it is likely that all surfaces were heavily oxidized or otherwise chemically contaminated; and it is now known that an oxide layer has an enormous effect. But even more important, most of the interesting experiments are not at semiconductor-vacuum interfaces at all, but at semiconductor-metal contacts. What seems not to have bothered people at all until Volker Heine wrote his 1965 paper [7] is that at a semiconductor-metal contact the original Tamm and Shockley arguments for surface states are wholly meaningless, since whatever the electronic state behaves like on the *semiconductor* side of the barrier, on the metal side we must have a propagating state with a real k value, so that there can be no question of true bound states in the energy gap. What Heine suggested, and made a start at discussing, is that whatever electronic states exist in the semiconductor are the decaying *tails* of the wave functions from the metallic side, and that the really correct point of view is to ask what the metal electrons see when they get to the semiconductor surface, and whether they can be expected to have short or long experimental tails. This is what will determine the amount of charge on the semiconductor surface, and above its Fermi level.

Since Heine's paper, it has become clear that there are actually two distinct types of semiconductor-metal interface. This is the result of a series of investigations by Mead, Kurtin et al., [8] which are summarized in figure 2. The experiment Mead et al. do is simply to study the type of surface barrier which

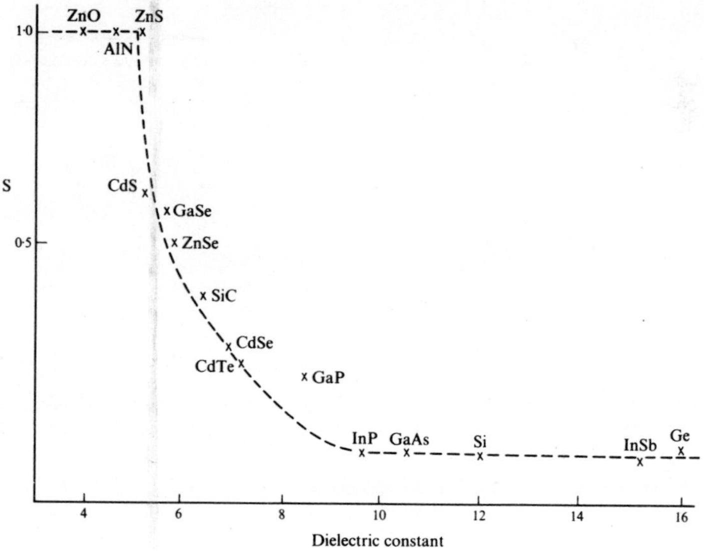

Figure 2

results when metals with a wide variety of work-functions are put on the surface of a given semiconductor. It turns out that there is rather a sharp distinction. The 'classical' covalent semiconductors Si, Ge and many others all behave as though the surface were covered with a dense layer of Bardeen surface states — Phillips has called this a 'Bardeen' Barrier. That is, independently of where the Fermi level in the metal is, an amount of charge adequate to bring the Fermi level at the surface of the semiconductor to a fixed level about 1/3 up in the gap always forms:

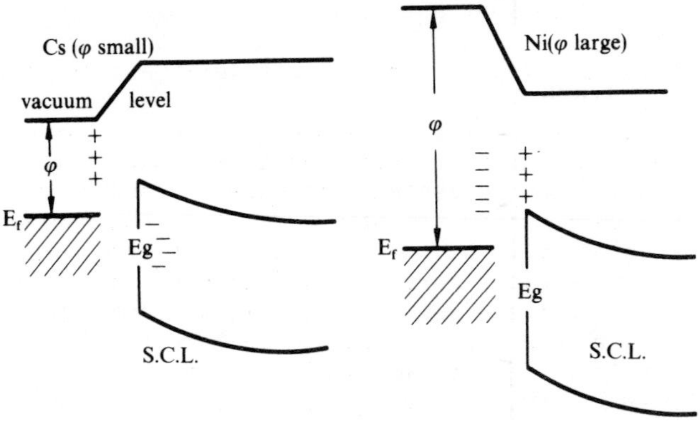

On the other hand, great numbers of other semiconductors, especially the ionic ones, with wide band gaps, behave in essentially the classic, Schottky-barrier way: E_F at the surface just equalizes to that in the metal, with no large surface charge effects.

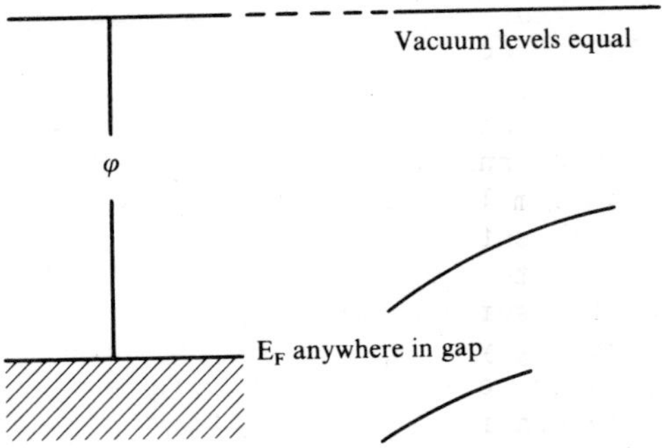

It was this dichotomy in behavior that I called to Inkson's attention, and that we feel we found an explanation for. Our explanation envisages a qualitative difference between the two types of semiconductors near a metal: the ones which form 'Bardeen' barriers essentially become metals themselves in the first layers next to a metal, losing their energy gaps entirely, while the others remain insulators, with a well-expressed gap at the metal surface (see figure 3).

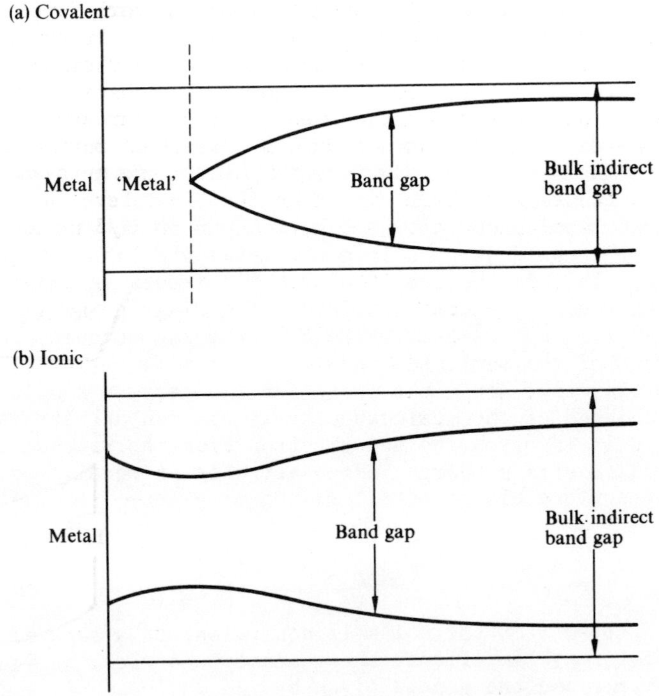

Indirect band gap distortion (schematic)

Figure 3

Like one of the seven blind men confronted with an elephant, then, I see what interests me most when confronted with nature: a many-body effect of the nature of a *surface* metal-to-insulator transition. I should emphasize that I have now changed the distance scale and on these figures the distances are not tens or hundreds of angstroms as they are for Schottky barriers, but one or two atom layers — very short indeed. But, of course, if the gap is absent even in one atom layer the metal states will spill right into the semiconductor, and charge neutrality of the semiconductor surface will require that the Fermi level be pinned very strongly at a position near the middle of the energy gap — actually we even find that it should be a bit below the middle. In other words, no matter what the relationship between the Fermi level and the vacuum potential in the metal, a dipole layer must build up at the surface which sets the semiconductor Fermi level somewhere near the middle of its gap, the electrostatic potential level floating to the necessary height. In the other case, the Fermi level may be almost anywhere in the semiconductor gap, since the charge in the semiconductor is rather independent of where the Fermi level is, and it is then controlled just by the electrostatic potential barrier at the metal surface, just as though the two were separated by a vacuum.

It is probably equally valid to look at this elephant from a chemical point of view and to say that the one group of semiconductors forms a *metallic bond* between the last atomic layer and the metal, while the second does not form any strong chemical bonds, or only electrostatic ones. One of our points is to be that one can predict whether or not this essentially chemical process takes place by studying the dielectric properties of the two substances and their effect on the electrons.

In any semiconductor, as I have shown on the figure, at relatively large distances from the metal the band gap will decrease. This is the result of the electrostatic image force for an electron in the metal surface. The image potential is of course given by $-e^2/2\varepsilon a$, where ε is the macroscopic dielectric constant of the semiconductor.

On the other hand, the image force effect on a hole in the valence band of the semiconductor is also an attractive potential: that is, removing one electron from the valence band leaves behind a *positive* charge, which attracts an equal *negative* charge to the surface of the metal, giving

$$V_{\text{hole}} = -\frac{e^2}{2\varepsilon a} .$$

But an *attraction* for holes is equivalent to a *repulsive* force for electrons: in effect, the valence band *rises* as we approach the metal, and the gap is given by

$$(E_g)_{a>>a_0} = E_g(\infty) - \frac{e^2}{\varepsilon a} .$$

With the dielectric constant of Si, 12, the image force term reaches the energy gap of about 1 eV at 1.2 Å from the surface, and the gap is reduced by half at 2.4 Å, which is certainly a distance at which macroscopic electrostatics should be valid.

Now you can see the reason why we went off into many-body theory. This very peculiar effect which is *opposite* in sign for valence and conduction bands does not fit into any one-electron theory — it simply cannot be handled as an ordinary potential, or even as an exchange effect, since the electron 2.5 Å from the metal certainly is not exchanging with any of the metal electrons. So in order to keep this simple-sounding and very big effect in the problem we have got to go to a full many-body treatment of the interactions of the electrons.

In order to keep in the many-body effects, we have to hope that it is not too much of a simplification to use relatively schematic models for the metal and the semiconductor. It is probably quite accurate to treat the metal as a free electron gas with a plane surface — as far as low-energy and relatively long-range effects are concerned, that should be a good approximation. We shall be considering metals with screening lengths of order ½ Å, and that should be some kind of indication of the thickness of the metal surface.

For the semiconductor we shall also take a homogeneous model — the so-called Penn model [9], actually due to Phillips and Cohen, which is a good approximation to covalent semiconductors and a rough one to most others. The idea of the Penn model is to ignore the detailed crystal structure and treat the Jones zone at which the energy gap appears as a sphere with a uniform gap across it. The gap is caused not by simply displacing the free electron states by a fixed amount but by mixing the states k and $k_F - k$ by a potential V_g, just as though it were a true gap due to scattering by a periodic potential. We assume that this gappy free electron gas also has a plane surface at the interface. I will shortly give reasons why this ignoring of the periodic structure of the two substances is a pretty good approximation.

What we now wish to do is to study the propagation of the electron in the presence of these two substances and this interface. To do so we must calculate the *self-energy* of the electron,

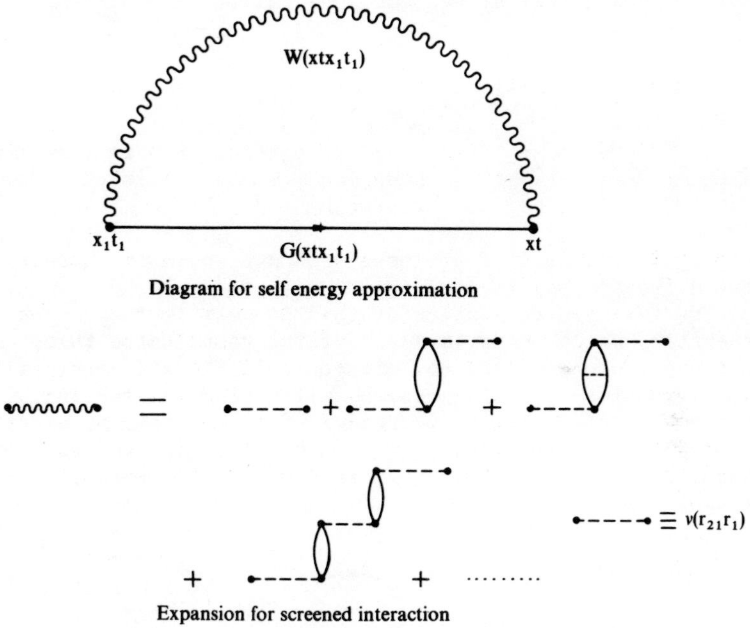

Figure 4

i.e. we must learn to deal with the exchange and correlation energy. It has already been pointed out by Bennett and Duke [10] that a great fraction of the work function of most metals is caused by the attraction of the electron to its exchange-correlation hole, of which, again, the image potential is the longest-range piece. Every electron in a metal is accompanied — because of the exclusion principle — by a positive hole of exactly the same charge, called the exchange-correlation hole. When the electron is outside the metal, its hole is the image charge: as it merges into the metal, the image charge becomes its exchange hole, and the attraction to this hole is the sum of its exchange and correlation energies. Our problem is to see how this process goes when the vacuum is replaced by a semiconductor.

What we cannot at first do is to draw the kind of diagram we have been drawing so far, in which we simply draw out the band structure as a function of position. That is because in many-body theory the energy of a particle is not a local operator, as it is in one-electron quantum mechanics. In one-electron theory, we write

$$\mathcal{H}\Psi = \left(\frac{p^2}{2m} + V(r,t)\right)\Psi = i\hbar\,\frac{\partial\Psi}{\partial t}$$

$$= E\Psi,$$

but as you no doubt know, in Hartree-Fock theory the exchange part of the potential becomes nonlocal:

$$V(\vec{r})\Psi(\vec{r}) \rightarrow \Psi(r)\int\rho(\vec{r}\,')V(\vec{r} - \vec{r}\,')d^3r'$$

$$- \int d^3r' \sum_{\substack{occupied \\ states\ n}} \phi_n{}^*(r)\phi_n(r')V(\vec{r} - \vec{r}\,')\Psi(r'),$$

the last term being a nonlocal exchange operator which relates two different points $\vec{r}$ and $\vec{r}\,'$.

The full generalization of this concept is the *self-energy* operator, which summarizes all of the complicated things which can happen to an electron. In general, the self-energy is nonlocal in time as well as space. It represents the sum of all the possible scattering processes which can take an electron from point r,t to point r',t': in diagrams (which you do not need to understand thoroughly in order to follow what I am going to say),

where •——→—• is an electron propagator $G_0(\vec{r},\vec{r}\,',t - t') = i\langle\Psi(\vec{r}t)\Psi^\dagger(\vec{r}\,'t')\rangle_0$ and - - - - is the interaction $V(\vec{r} - \vec{r}\,')$. Σ is the correction to the energy due to the interactions; we express this by doing the diagram sum $G = G_0 + G_0\Sigma G_0 + G_0\Sigma G_0\Sigma G_0 + \ldots.$ which represents the 'Dyson equation'

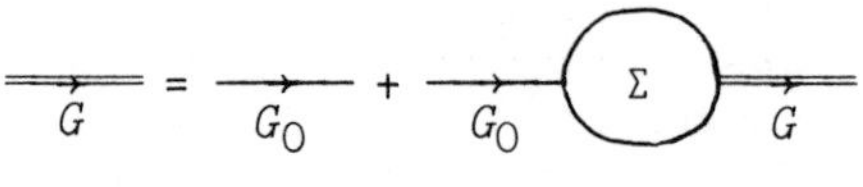

where G_0 is the propagator in the absence of any interactions

$$\left(E - \frac{p^2}{2m}\right) G_0 = 1,$$

$$G_0 = (E - \mathcal{H}_0)^{-1} \quad \text{and} \quad G = (E - \mathcal{H}_0 - \Sigma)^{-1}.$$

We will not by any means use the full complication of this kind of formalism; what we will use is called, in the case of uniform systems, the *screened interaction approximation* of Quinn and Ferrell [11], and Hedin and Lundqvist [12]. This approximation may be described diagrammatically as *neglecting all 'vertex corrections'*. It is basically just a Hartree-Fock theory in which we use not the Coulomb interaction $V = e^2/r$ but the fully dynamically screened interaction W, which in a uniform system would be the Fourier transform of $V(q,\omega)/\varepsilon(q,\omega) = 4\pi e^2/q^2 e(q,\omega)$. The errors made in this kind of theory are harmless for our purposes; the short-range parts of the correlation energy are slightly overestimated. What is neglected are repeated strong scatterings: paramagnons, hard cores, and the like. The RPA, as well as Migdal's theory of electron-phonon interactions, are included within this approximation. The content of the approximation is to set

$$\Sigma = G \cdot W,$$

where W is the fully screened interaction. Thus once we understand how our semiconductor-metal junction acts to screen the Coulomb interaction, we will easily be able to work out the self-energy of an electron in the vicinity of the junction.

Again following rather similarly to what Hedin and Lundqvist chose to do, Inkson decided that any possible simplification for such a complicated problem was more than justified. Thus he chose to use certain analytic approximations for the two dielectric constants $\varepsilon_{\text{metal}}(q,\omega)$ and $\varepsilon_{\text{sc}}(q,\omega)$ which determine the screened interaction. What he uses for the metal already has a distinguished history; it was used by Lundqvist as an approximation to the Lindhard dielectric function of the free energy gas.

$$\varepsilon_M = 1 + \frac{k_s^2}{q^2 - \dfrac{\omega^2}{\omega_p^2} k_s^2},$$

where

$$k_s^2 = \left(\frac{6\pi n e^2}{E_F}\right) \qquad \omega_p^2 = \frac{4\pi n e^2}{m}$$

are the Thomas-Fermi screening length and plasma frequency. As you see, it behaves correctly both for q and $\omega \to 0$:

$$\varepsilon_M(q = 0) = 1 - \frac{\omega_P^2}{\omega^2} ,$$

$$\varepsilon_M(\omega = 0) = 1 + \frac{k_s^2}{k^2} ,$$

and is chosen on this basis plus two simplifying conditions: (1) the only pole of $1/\varepsilon$ is at $\omega_P^2(q)$ where

$$\omega_P(q) = \omega_P\left(1 + \frac{q^2}{k_s^2}\right)^{\frac{1}{2}} ,$$

and (2) this pole satisfies the sum rule.

It is interesting to contrast the behavior of ε — which determines optical properties, for instance — with that of $1/\varepsilon$, which determines the interaction and the self-energy. The former has its poles at a median free-electron excitation energy $\omega_e = qv_F/\sqrt{3}$, the latter has these poles completely suppressed and one sees only the plasma frequency, as in the electron energy losses. We will see the same phenomenon in the semiconductor, and it is a very important effect.

This dielectric constant is not new, but the one John invented to fit the semiconductor is, and I think has a lot of very useful and transparent properties. He uses,

$$\varepsilon_s(q,\omega) = 1 + \frac{\varepsilon_0 - 1}{1 + \frac{q^2}{\gamma^2}\,\varepsilon_0 - \left(\frac{\omega^2}{\omega_R^2}\right)\varepsilon_0} .$$

Here ε_0 is the observed electronic dielectric constant, given in terms of the 'average (or isotropic) gap' ω_g of the Penn model by

$$\varepsilon_0 = 1 + \frac{\omega_P^2}{\omega_g^2} .$$

The other two parameters are

$$\omega_R^2 = \omega_p^2\left(\frac{\varepsilon_0}{\varepsilon_0 - 1}\right) = \omega_g^2 + \omega_p^2 ,$$

$$\left(\omega_p^2 = \frac{4\pi ne^2}{m}, \qquad \text{again}\right),$$

and

$$\gamma^2 = \left(\frac{\varepsilon_0}{\varepsilon_0 - 1}\right)k_s^2.$$

Again we have a dichotomy between the poles of ε, which occur at the 'excitonic' frequency

$$\omega_e^2 = \omega_g^2 + \frac{q^2 v_F^2}{3},$$

and the plasmon poles of $1/\varepsilon$, at

$$\omega_p^2(q) = \omega_R^2 + \frac{q^2 v_F^2}{3}.$$

ω_R^2 in fact, in the case of Si, is a slightly better fit to the observed bulk plasma frequency than is ω_p^2.

I have a couple of figures showing how remarkably good this Ansatz for the dielectric constant is (see figures 5,6), both in terms of $\varepsilon(0,\omega)$, which can be measured optically, and of $\varepsilon(q,0)$ which can be calculated from the Penn model. I think this dielectric constant is one of the nice achievements of Inkson's thesis; it makes a lot of results in semiconductor physics rather transparent - such as the plasma frequency. He has recently been extending this to study the nonuniform parts of the dielectric constant — that is, the off-diagonal matrix elements in reciprocal space, $\varepsilon_{q\to q+G}$ where G is a reciprocal lattice vector. These tell us how nonuniform the dielectric response is because of the pileup of electronic charge in the covalent bonds. It is part of the Phillips mythology of dielectric effects that the 'bond charge' is given by the missing screening of the atomic charge, i.e.

$$\text{bond charge} = \frac{4}{\varepsilon_0} \div 2 \text{ bonds/atom} = \frac{2}{\varepsilon_0}.$$

This myth is in fact supported by some arguments due to Pick, Cohen and Martin [*13*] based on sum rules connecting the diagonal and off-diagonal parts of the dielectric constant; it appears that $1/\varepsilon_0$ is in fact a good measure of the relative importance of nonuniform terms. Thus in the good covalent semiconductors like Si and Ge with ε_0 of 12 or 16 these terms are indeed less than 10% of the uniform ones, which justifies, to some extent, our rough continuum model.

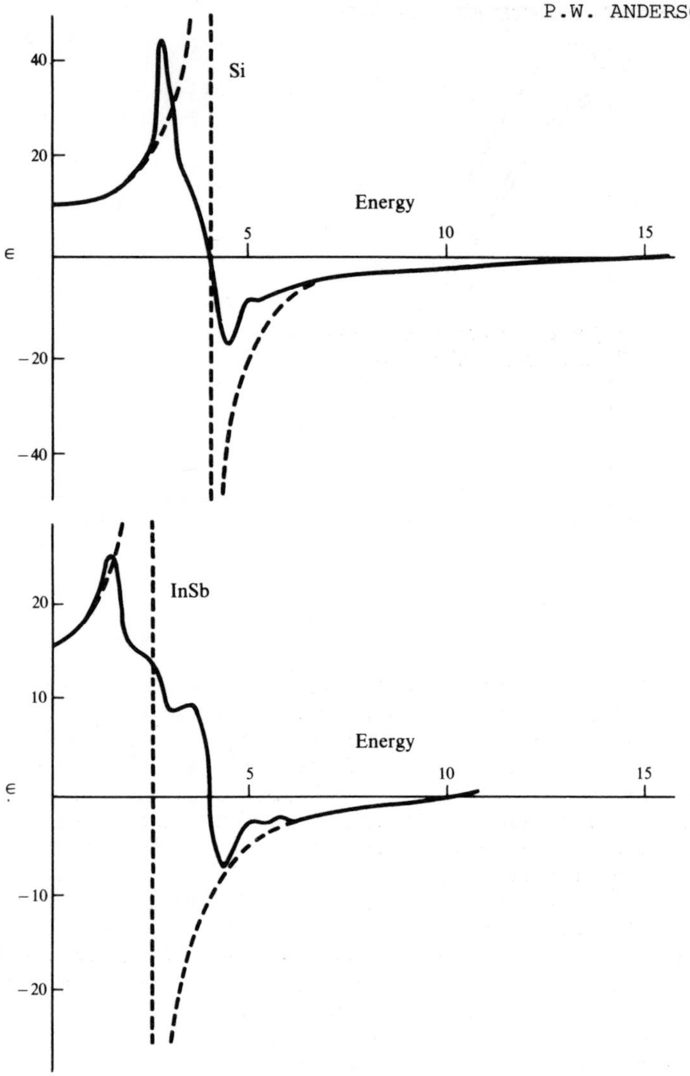

Comparison of model and experimental dielectric functions

Figure 5

It is these simple, and yet at the same time very realistic, approximations to the dielectric functions which make it possible to get anywhere with the problem at all. Already with these dielectric functions it is possible to get good estimates of something which was not previously available in the literature — surface plasmons for semiconductor surfaces.

The long-wavelength, observable surface plasmons follow directly from a simple argument due to Stern [*14*]. We want to know at what frequency a charge fluctuation of infinite wavelength at the surface will give an infinite E-field, i.e. we can

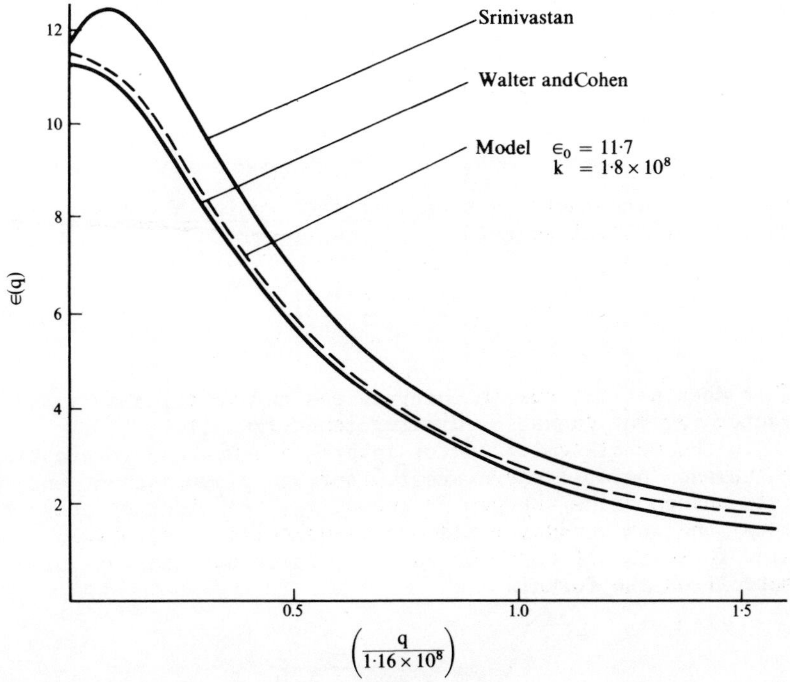

Theoretical static dielectric function
Figure 6

have a spontaneous oscillation of the E-field. We have two media of dielectric constants ε_1 and ε_2,

$$D_1 = \varepsilon_1 E_1, \qquad D_2 = \varepsilon_2 E_2.$$

From Poisson's equation

$$\nabla \cdot D = 4\pi\rho$$

we get

$$\varepsilon_1 E_1 - \varepsilon_2 E_2 = 4\pi\rho_{surface}$$

and thus if $E_1 = - E_2 = E$ (no external sources for the fields, or, equally, V constant at ∞),

$$(\varepsilon_1 + \varepsilon_2)E = 4\pi\rho_{surface}.$$

Thus

$$\varepsilon_1 + \varepsilon_2 = 0$$

gives us a finite E for zero ρ.

This gives us the famous $\sqrt{2}$ factor for a metal-vacuum surface:

$$1 - \frac{\omega_p^2}{\omega_s^2} + 1 = 0,$$

$$\omega_s = \frac{\omega_p}{\sqrt{2}} .$$

For the semiconductor-vacuum interface, using the Inkson dielectric constant we get

$$\omega_s = \omega_R \left(\frac{\varepsilon_0 + 1}{2\varepsilon_0} \right) .$$

This does not fit Rowe's observations any better than a simple factor $\sqrt{2}$, but cannot be distinguished from it.

In the metal-semiconductor interface a new and interesting phenomenon occurs: there are *two* surface plasmons, not one; one gets pushed below the gap frequency ω_g, the other up above the metal and semiconductor plasmon frequencies. The physics behind these is not complicated — the lower one can just be estimated from the formula

$$\omega_{M-s} = \frac{\omega_p}{(1 + \varepsilon)^{\frac{1}{2}}} ,$$

where ε, the semiconductor dielectric constant, is somewhat greater than ε_0 in this range. Since ω_R and ω_p are about the same, and $\varepsilon_0 = \omega_p^2/\omega_g^2$, ω_{M-s} is about $\frac{1}{2}$ to 2/3 the gap at long wavelengths.

To do the dispersion of the surface plasmons it is not adequate to apply the Stern-Ferrell relationship, since we cannot guess that the field is uniform in the z direction, and in fact it presumably decays exponentially. This result, however, is one which falls naturally out of our central problem; the calculation of the potential of interaction of electrons in the surface region - i.e. we put a charge at a distance a from the surface and ask for the potential $\Psi(\rho,z)$ at a point z from the surface with the radial coordinate ρ. (We will always work in frequency space rather than time; V and Ψ are Fourier components with frequency ω).

Since this calculation is central to all the further work, I will do it out, at least in the notes, in a bit of detail. Of course it is important to the self-energy problem, because our approximation is just to take the self-energy as that of the appropriate charge distribution in the given medium - basically, the straightforward generalization of the image potential problem. So what we need most to know is $\Psi(\rho \to 0, z \to a)$ which gives us the self-potential. Second, the *poles* of this interaction in frequency space are the plasma frequencies — they give a potential response in the absence of charge. In general, they will

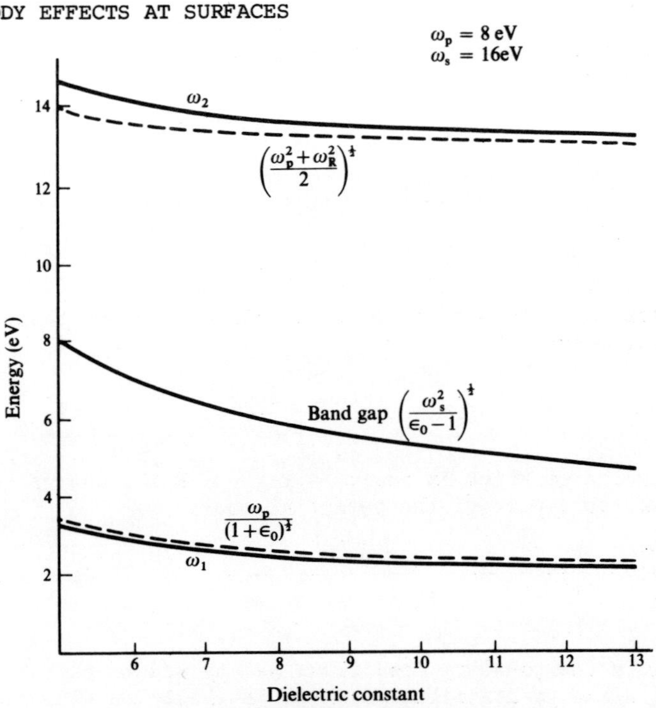

Long wavelength surface plasmon energies

Figure 7

be complex — decaying — but usually only slightly so. These poles will form a continuum for a point charge, but if we Fourier transform in the transverse direction there will be one or two discrete ones for each transverse momentum m. It is convenient to use circular coordinates in this space and thus we transform using Bessel rather than Fourier series,

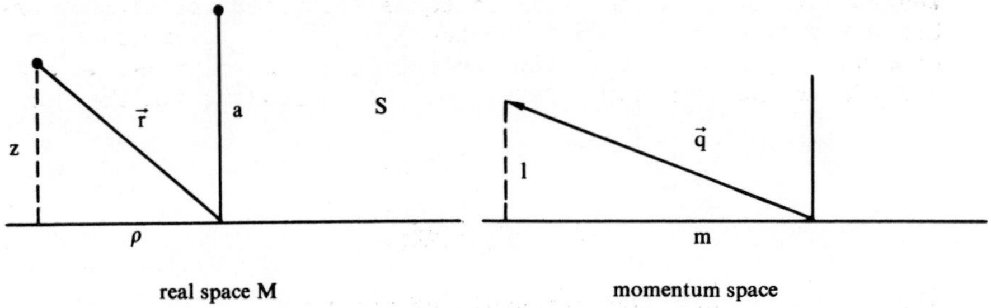

$$V(\rho,z) = \frac{1}{(2\pi)^2}\int m\,dm\; V(m,z)J_0(m\rho)$$

and the problem separates in the transverse momentum m (there is no angular dependence, of course).

The solution is carried out using the continuity conditions at the surface,

$$\Psi_1(\rho, z = 0) = \Psi_2(\rho, z = 0)$$

and

$$\frac{\partial \Psi_1}{\partial z} = \frac{\partial \Psi_2}{\partial z} ,$$

and a general method of images. Within either substance the potential obeys

$$\nabla^2 \Psi(q,\omega) = \frac{4\pi\rho}{\varepsilon(q,\omega)} .$$

The only charge which is *inside* a region is the charge source a, 0, and so a part of the potential obeys

$$V = \frac{4\pi e}{(2\pi)^2} \int \frac{m\,d\ell\,dm}{\ell^2 + m^2 \varepsilon_1((\ell^2 + m^2)^{\frac{1}{2}}, \omega)} e^{-i(z-a)\ell} J_0(m,\rho).$$

To satisfy the boundary conditions we may add to this the potential of a charge distribution which is wholly outside region 1, but computed as though region 1 filled all space; this might as well be on the surface

$$\Psi = V + \Phi,$$

$$\Phi = \frac{4\pi e}{(2\pi)^2} \int \frac{m\,d\ell\,dm}{(\ell^2 + m^2)\varepsilon_1} f_1(m,\omega) e^{-i\ell z} J_0(m,\rho),$$

where f_1 is the charge distribution. Correspondingly, the potential in region 2 can be taken to be the potential of some arbitrary surface charge distribution f_2. f_1 and f_2 are functions of m which are related by the two continuity conditions, which thus determine them completely. We get two equations of the form

$$\alpha + \beta f_1 = \gamma f_2,$$

$$\Delta - \nu f_1 = \mu f_2,$$

where α, β, γ, etc. are integrals of the type

$$\int \frac{d\ell\, e^{ia\ell}}{\varepsilon_{1,2}(\ell^2 + m^2)} \quad \text{or} \quad \int \frac{\ell\, d\ell\, e^{ia\ell}}{\varepsilon_{1,2}(\ell^2 + m^2)} .$$

These integrals would be hopeless with real dielectric functions but with the simple analytic approximation we have chosen they are not very difficult by contour integration; there are two contributions, one coming from the pole at $\ell = im$ and the other from the zero of the dielectric constant at

$$m^2 + \ell^2 = q^2 = k_s^2\left(\frac{\omega^2}{\omega_p^2} - 1\right) .$$

But of course the resulting equations for f are complicated and while one can indeed make a great deal of sense of the result physically, it is so complicated as not to be worth our writing it down.

One thing which *is* worth noting is that in general f_1 and f_2 represent the response of the surface to the charge. Thus the point at which they blow up is clearly a surface plasma, and it is to be obtained by the singular case for solution of the linear equations for f_1 and f_2:

$$\text{surface plasmon:} \quad \begin{vmatrix} \beta(m,\omega) & -\gamma \\ \nu & \mu \end{vmatrix} = 0.$$

The integrals are in this case quite simple and John has worked out the two cases. In the metal-vacuum case,

$$\omega(m) = \frac{\omega_p}{2k_s}\,(m + (m^2 + 2k_s^2)^{\frac{1}{2}}),$$

which is a bit surprising in that it gives a linear dispersion law, but perfectly permissible.

For the metal-semiconductor, the solutions had to be obtained numerically and are given in the figures (see figure 8). We note that the dispersion is quite complicated. One noteworthy feature which we have *not* analyzed, but would like to think further about, is whether there is any appreciable electron-electron resonant attractive coupling via the low-frequency surface plasmon. The strength of this pole in the interaction is rather low, and as you can see its dispersion is steeply varying, but one suspects it is far more strongly coupled to the electrons than any so-called 'excitonic' excitation.

So this completes one of the results which is my excuse for entering a school on 'elementary excitations' — surface plasmons. The second result is what we will now take up — how elementary electron and hole excitations behave near the surface. Here we keep in mind the surface state problem; although actually several other applications of these ideas have been made or suggested — one very important one, for instance, being the calculation of potentials and energy losses for medium-energy electrons impinging on a metal or semiconductor surface, which is a key question for LEED.

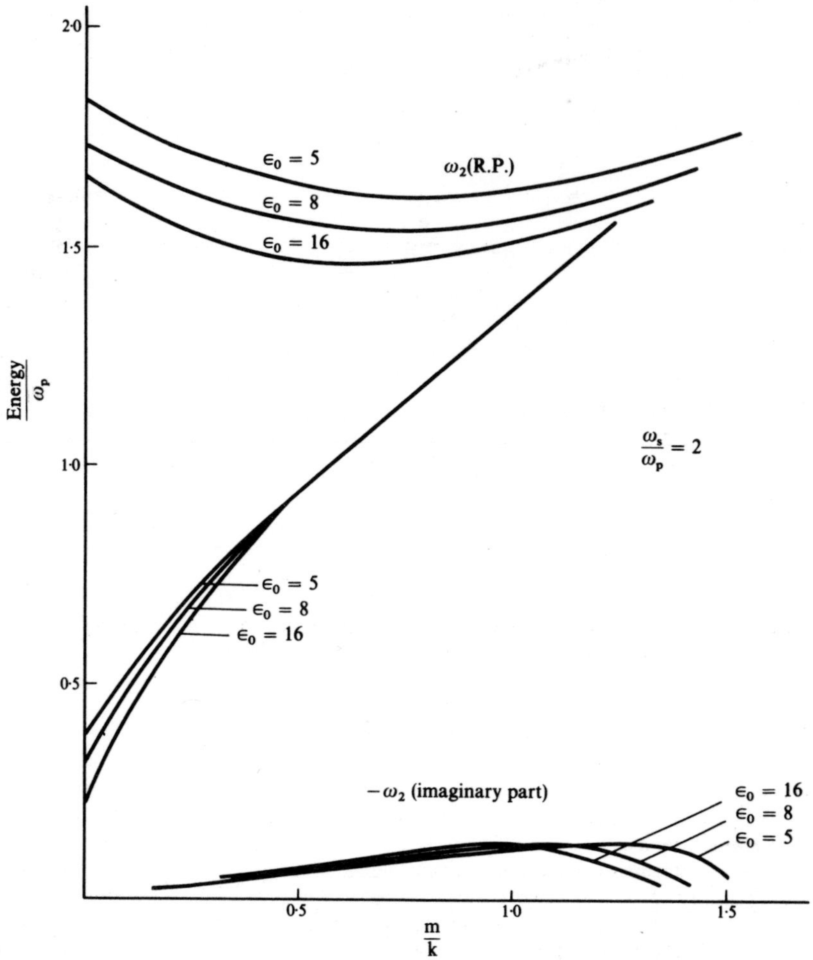

Surface plasmon dispersion relation (dielectric constant varying)

Figure 8

Next I would like to bring out a very over-simplified approximation which nonetheless does bring out a great deal of the physics we are after. This is what Inkson calls a 'Thomas-Fermi' approach but it is actually even oversimplified relative to that: we could call it linearized Thomas-Fermi. The basic approximation is to neglect the frequency-dependence of ε entirely, and just to treat the two media as dispersive classical state dielectrics, and the electron and hole as static point charges. In other words, what if we generalize the image potential idea by taking the short-range dielectric behavior of metal and semiconductor correctly into account?

This had already been done by Newns [3] for the metal-vacuum interface, where one can use the static metal dielectric constant

$$\varepsilon = 1 + \frac{k_s{}^2}{q^2},$$

or equally the equivalent linearized Thomas-Fermi equation

$$\left(\nabla^2 - k_s{}^2\right)\Phi(r) = 0,$$

can be integrated to give the potential in the metal. The same thing can be done in the semiconductor for our dielectric constant

$$\varepsilon = 1 + \frac{\varepsilon_0 - 1}{1 + \frac{q^2\varepsilon_0}{\gamma^2}}.$$

If we introduce the vacuum potential U which satisfies $\nabla^2 U = 4\pi\rho$, the real potential obeys

$$(\nabla^2 - \gamma^2)\Phi = -\frac{\gamma^2}{\varepsilon_0}U(r).$$

By using this differential equation, or just returning to our image techniques of before, one may arrive at the image potential for a classical point charge approaching the metal, which is shown in the figure 9.
One sees that it is very heavily dependent on the ratio of the screening constant γ in the semiconductor to that (k_s) in the metal. When the *short-range* screening in the semiconductor is better, the image potential reverses near to the metal and can even become positive. That is, where the electron density — at least where the electrons are — is higher in the semiconductor, it will in fact have a higher exchange-correlation energy and thus the image attraction will be reversed near the metal. In general, this *is* the direction of the trend which is seen; and while this is not the whole story — in particular (a) the potential is nonlocal, in general, where this is local; (b) the hole is not a classical point charge, by any means — I think this reversal gives one a clear understanding of why so many semiconductors do *not* lose their band gaps near a metal.

While this picture, if oversimplified, gives a good account, it is enlightening to go on and try to achieve some degree of realism in the same calculation. This is unfortunately the point at which one must really leave the world of the living for the complex plane, but I think one can discuss the physics, if not the math, with some realism.

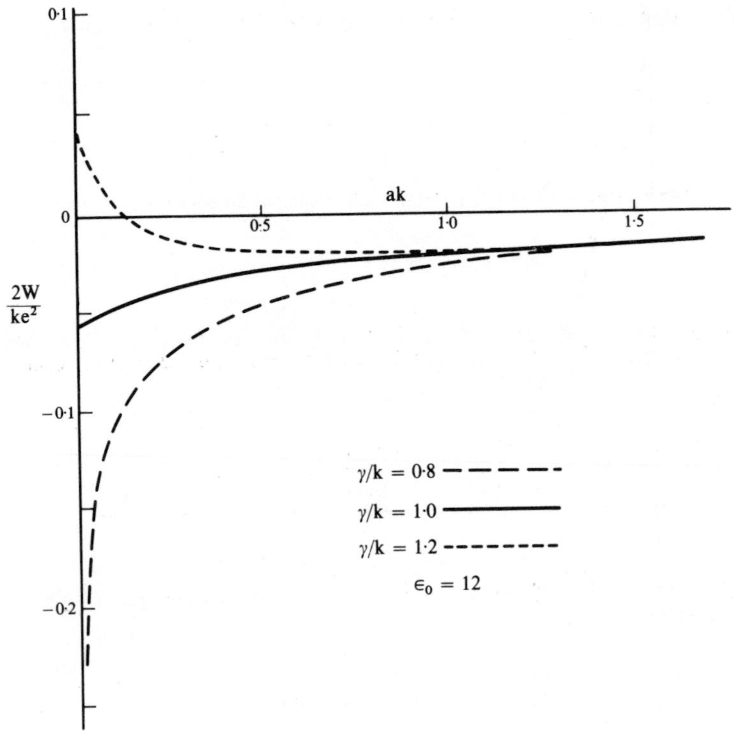

Variation of image potential with distance

Figure 9

The self-energy in space-time terms is the product

$$\Sigma(x,x',t - t') = G(x,x',t - t')W(x,x',t - t')$$

of G, the amplitude for the electron to get from x to x', and W, the interaction. In Fourier space then it is a convolution integral

$$\Sigma(x,k,\omega) = \frac{i}{(2\pi)^4}\int W(x,q,\omega)G(x,k - q,\omega - \omega')e^{-i\delta\omega'}\,dq\,d\omega',$$

where we have Fourier transformed in time and in $x - x'$ space. One has left over the x-dependence, which is necessary here because we do not have a homogeneous problem. This divides naturally into two parts, which we can describe as the 'correlation' and the 'screened exchange'. If W were an ordinary time-dependent potential, the exchange term would be the only contribution to the ω' contour integration, and would arise from the poles of G at $\omega - \omega' = E_n$, the occupied state energies; so the screened exchange term is taken to be just that term from the poles of G. There is also a 'correlation' term, which comes from the poles

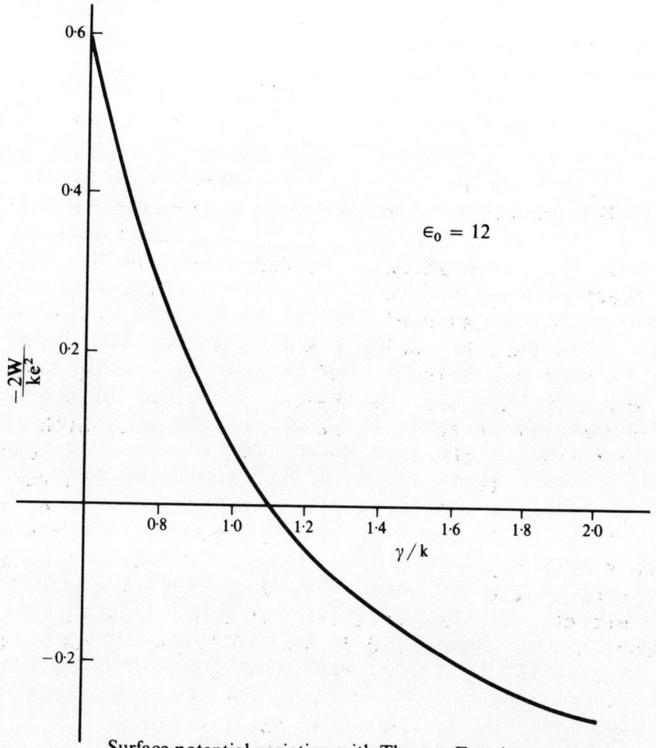

Surface potential variation with Thomas-Fermi wavevector

Figure 10

of W: the various bulk and surface plasmons we have been discussing. Of course there is correlation energy, as normally defined, in each term.

The model for G which is used is the kind of two-band model we have already described — incidentally, one of the surest approximations we can make is to ignore the effect of the boundary on G, and use the unperturbed semiconductor G;

$$G(x,x',\omega) = \sum_n \frac{\Phi_n(x)\Phi_n^*(x')}{\omega - E_n + i\delta\,\mathrm{sgn}(E_n - \mu)}$$

is its definiton. In the two-band model

$$\Phi_{\pm k}(x) = \frac{1}{(1 + \alpha^2(k))^{\frac{1}{2}}}\,(e^{ik\cdot x} + \alpha e^{i(k-G)x}),$$

$$E_\pm(k) = \tfrac{1}{2}[E^0(k) + E^0(k - G) \pm (E_g{}^2 + (\Delta E^0)^2)^{\frac{1}{2}}],$$

$$\alpha^\pm = \frac{1}{2}\,\frac{E_g}{E^\pm - E^0}\ .$$

This gives

$$G(x,p,\omega) = \frac{1}{(1+\alpha^2)^{\frac{1}{2}}}\left\{\frac{\Phi_{-p}(x)}{\omega - E_{-}(p) + i\delta} - \frac{\alpha^{\dagger}(p)\Phi_{-p}(x)}{\omega - E_{+}(p) - i\delta}\right\} .$$

Only the first term has a pole on the right side of the axis — corresponding to occupied states — to contribute to the integral in Σ.

Given G, then, and given W, we can calculate the self-energy of the electron near the surface. Before doing a final realistic calculation it was useful to go to the long-range limit $a,z \gg 1$. This was, for quite a while, a stumbling block for us because we were not able to find the physical image potential effect which we knew must be there. There was no way to calculate the *correlation* term, as we understood it, which did not give almost exactly the same answer for valence and conduction bands. For these terms the main difference between valence and conduction bands is the value of ω — positive for one, negative for the other. Only for the low-frequency surface plasmon term could this possibly make a difference, and for that term there is no mixing of the two terms involving k and $k - G$ and therefore no effect. So the 'correlation' term is practically identical for the two bands, and in fact follows our main 'Thomas-Fermi' calculation to a good approximation. Thus at long distances it goes simply to $- e^2/2a\varepsilon_0$: the classical image potential *for an electron* — no dichotomy of the two bands.

The screened exchange term, on the other hand, behaves quite differently. At long range this term is actually zero for the conduction band, and goes as $+ e^2/a\varepsilon_0$ for the valence band, thus giving us back our original semiclassical result. There is a good physical reason for this. The conduction band wave-function is orthogonal to all of the occupied states, so insofar as $W(x,x') \simeq$ constant, when $x \to \infty$,

$$\int dx' W(x,x') \sum_{occ} \phi_n(x')\phi_{cond}(x') \simeq 0.$$

On the other hand, the valence band wave functions do *not* have this orthogonality property and will see a full exchange term.

There is a very perspicuous way of seeing that this difference in screened exchange exists, which also brings us back to a rather fundamental point of semiconductor physics. First we note the fundamental Wannier theorem on full bands: that the wave functions of a filled valence band with a true band gap may always be written as linear combinations of exponentially localized Wannier functions:

$$\phi_k(x) = \sum_n W_n(x) e^{ik\cdot R_n}.$$

For a multiple set of bands as in a covalent semiconductor there may be several W_n's per atom — in that case actually the best set are of the symmetry of sp^3 hybrid bond orbitals.

The second remark is that the exchange energy of an arbitrary wave-function is the sum of the *Coulomb self-energies of its overlap charges with all occupied* wave-functions. That is,

$$\sum_{ex}(\phi) = \sum_{\substack{n \\ occ}} \int dx \int dx' \phi(x)\phi_n^*(x)V(x - x')\phi^*(x')\phi_n(x').$$

So we can define the *overlap charge*

$$\rho(\phi,\phi_n) = \phi(x)\phi_n(x).$$

The fact that one can always describe a filled band in terms of a Wannier representation means that it is equally true (for static V, at least) that the exchange self-energy is the sum of the self-energies of the overlap charges with the Wannier functions. For a bulk crystal all *these* terms are the same and we get

$$V_{ex}(\phi) = W\int dx \int dx' \phi(x)W_n(x)V(x - x')\phi(x')W_n(x').$$

Here we see a sharp dichotomy between conduction and valence bonds: namely

$$\int \phi(x')W_n(x')dx' = \frac{1}{\sqrt{N}} \qquad \text{(valence band)}$$

$$= 0 \qquad \text{(conduction band)}$$

i.e. the conduction band wave functions have no *net* overlap charge. What our difference in image effect is, then, becomes obvious: the exchange charge in the valence band induces an image and is therefore *screened* by the presence of the metal, leading to a net reduction of the exchange energy, which overcompensates the ordinary image effect. In the conduction band, on the other hand, there is no net overlap (or exchange) charge and the metal screening has no great effect.

This point relates at a number of places to physical facts. Firstly, it tells us a limitation on the effectiveness of the band-gap closure idea. It is a very reasonable one: simply, we cannot treat the image force on a more local basis than the size of a Wannier function. For average semiconductors this is not a severe limitation; while the exponential tails may extend out to neighboring atoms and bands, the bulk of the Wannier function can be rather sharply localized on its bond site. Thus I cannot see at all why this exchange screening difference cannot be big enough to close the gap entirely, and in fact as shown in the figures 11-14 actual calculations with the two-band model do show band closure in the suitable cases.

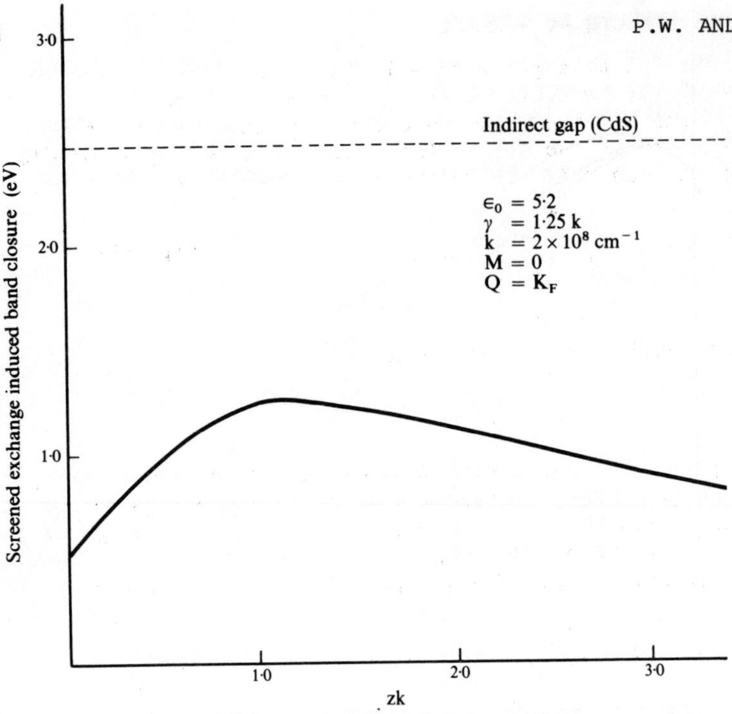

Band closure in wide gap ionic semiconductor.

Figure 11

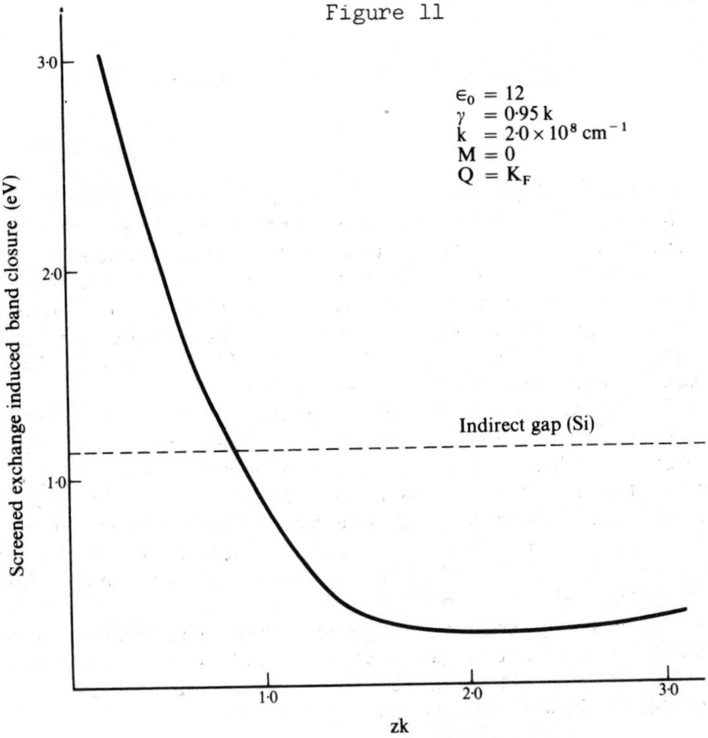

Band closure in narrow gap covalent semiconductor

Figure 12

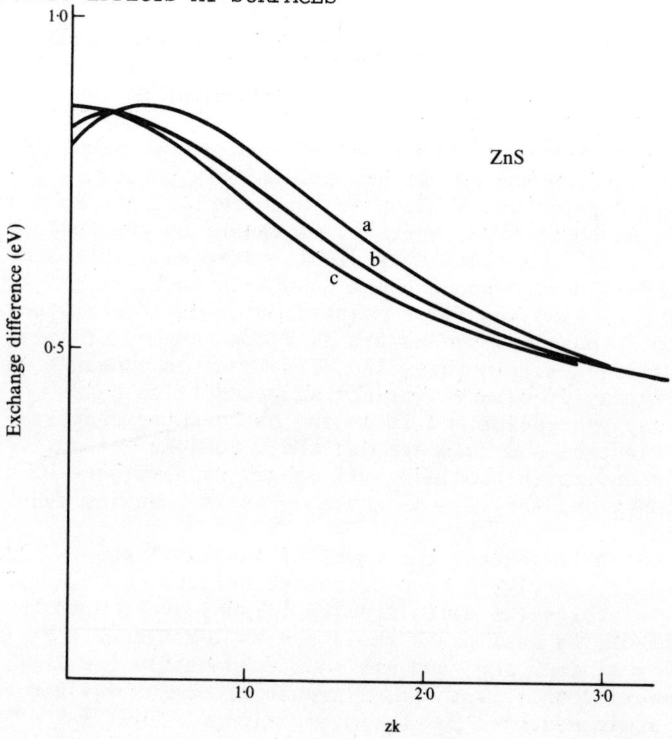

Figure 13

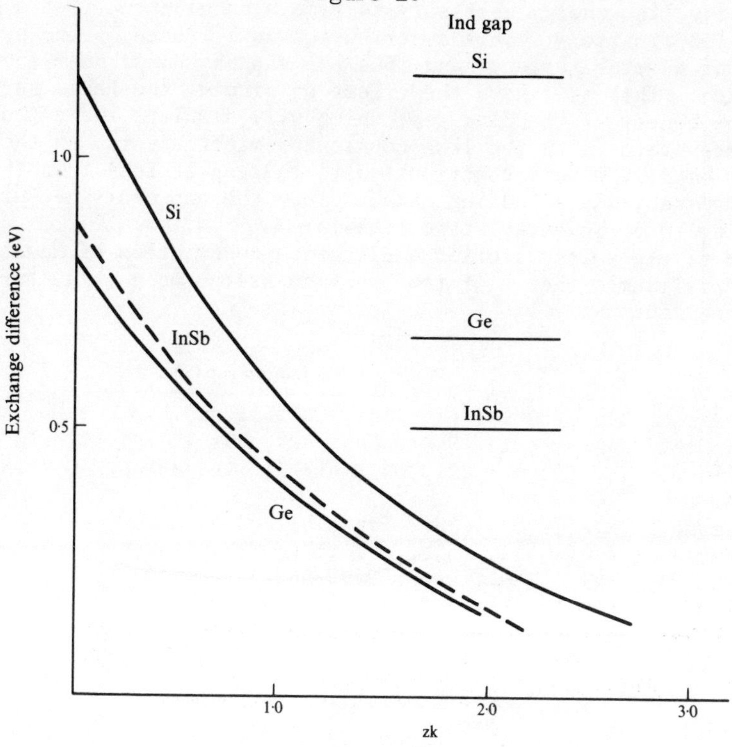

Figure 14

Second is a tie-in to a rather old but still inadequately understood bit of physics: the Mott or Mott-Hubbard gap and its relation to the energy gap in true semiconductors. One is taught — at least in reasonably good solid-state physics courses — that there are two kinds of insulators: simple band theory ones, where the energy gap is entirely caused by the strong pseudo-potential leading to large gaps at the zone boundaries; and the Mott type, where it is caused by the Coulomb correlation effect in a relatively dilute material. This simple distinction is here seen to break down — in fact, it was to my knowledge E.O. Kane who first pointed out that the existence of a true gap in most semiconductors is a consequence of large exchange differences. In fact, the difference in exchange and correlation energy between valence and conduction bands *is* just the Mott gap — he described it as the extra energy necessary to separate electron and hole against their Coulomb attraction, which is equal to the Coulomb self-energy of a valence band Wannier function, i.e. $e^2/\varepsilon a_0$ where a_0 is the Wannier function mean radius.

Thus, while in general the total or 'average' gap V_g in semiconductors is, mostly, a true potential effect, contributions of order 1-2 volts (as calculated by Inkson) are due to the exchange effect; in most cases these are at least as big as the indirect or minimum gap, and are thus responsible for it.

To summarize this last bit of argument: the energy gap in covalent semiconductors has a contribution of order 1-2 eV from the screened exchange term, which may be interpreted as the Mott gap, i.e. the energy necessary to remove an electron from a Wannier function to ∞. This contribution is screened by the presence of a metal surface, and thus the gap may close near the surface. This will have the effect of pinning the Fermi surface to the center of the band gap, which will simulate the effect of surface states. In the less covalent semiconductors, on the other hand, the Mott contribution to the gap is less than the minimum gap, and screening cannot close the gap near the surface, leading to a 'Schottky' type behavior.

If time permits, a brief qualitative description of Hamann and Appelbaum's theory of true surface states on a (111) surface will be given as well.

REFERENCES

1. Inkson, J.C. (1971). (Thesis), (Cambridge).
2. Inkson, J.C. (1971). *Surf. Sci.*, **28**, 69; (1972). (Surface Plasmon), *J. Phys.*, **C5**, 2599; (1973). (Band Gap Reduction), *J. Phys.*, **C.**
3. Newns, D.M. (1970). *Phys. Rev.*, **B1**, 3304.
4. Bardeen, J. (1947). *Phys. Rev.*, **71**, 717.
5. Shockley, W. (1939). *Phys. Rev.* **56**, 317.
6. Tamm, I. (1932). *Phys. Z. Sowjetu.*, **1**, 732.
7. Heine, V. (1965). *Phys. Rev.*, **A138**, 1689.
8. Kurtin. S., McGill, T.C. and Mead, C.A. (1969). *Phys. Rev. Lett.*, **22**, 1433.

9. Penn, D.R. (1962). *Phys. Rev.*, **128,** 2093.
10. Bennett, A.J. and Duke, C.B. (1967). *Phys. Rev.*, **160,** 541; *Phys. Rev.*, **162,** 578; (1969). *Phys. Rev.*, **188,** 1060.
11. Quinn, J.J. and Ferrell, R.A. (1958). *Phys. Rev.*, **112,** 812.
12. Hedin, L. and Lundqvist, S. (1969). *Solid State Physics, Vol. 23*, (Academic Press, New York).
13. Pick, R.M. Cohen, M.H. and Martin, R.M. (1970). *Phys. Rev.*, **B1,** 910.
14. Stern, E.A. and Ferrell, R.A. (1960). *Phys. Rev.*, **120,** 130.

INELASTIC X-RAY SCATTERING FROM ELECTRONS IN MATTER

P.M. PLATZMAN

Bell Laboratories, Murray Hill, New Jersey 07974

Scattering experiments have given a great deal of important information about the microscopic behavior of an enormous variety of physically interesting systems. Neutrons, [*1*], electron beams [*2*] and a range of electromagnetic radiation [*3,4*] have all been successfully utilized by several groups to elucidate many of the interesting phenomena in condensed matter. The exact information which can be extracted from a given scattering experiment ultimately depends on the characteristics of the probe (typically its energy and momentum) and on the nature of its coupling to the medium. Neutrons have energy in the millivolt range with DeBroglie wave vectors approximating 10 cm^{-1}. They couple primarily to mass (phonons) and magnetic moments (magnons, etc). X-rays are in a real sense the high-frequency limit of visible light. They extend manyfold the range of frequencies and wavelength one can hope to examine. They have energies in the 10 kilovolt range with wavevectors in the 10^8 cm^{-1} range. They couple primarily to charge and to light masses (i.e. electrons).

A typical scattering experiment for all probes is shown schematically in figure 1. The shaded region is the target. If the probe is scattered weakly by the target then the scattering may be treated in the lowest-order Born approximation, i.e. weak coupling between probe and system. In this case the spectrum of the scattered radiation reflects the properties of the medium (a correlation function) in the absence of the perturbing probe. For incident (scattered) photons of frequency ω_1 (ω_2) and wave vector $\vec{k}_1$ ($\vec{k}_2$), the scattering cross section contains both momentum $\vec{k} = \vec{k}_1 - \vec{k}_2$ and energy $\omega = \omega_1 - \omega_2$ information . (It may also, in the case of light, contain polarization information.)

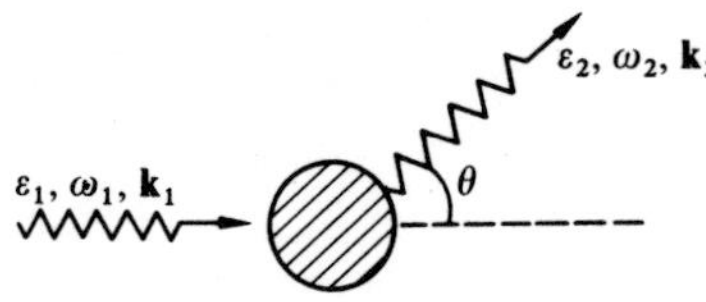

$\omega = \omega_1 - \omega_2$ (energy transfer)

$\mathbf{k} = \mathbf{k}_1 - \mathbf{k}_2$ (momentum transfer)

$$\frac{d\sigma}{d\omega\, d\Omega} = \left(\frac{d\sigma}{d\Omega}\right)_0 S(\mathbf{k},\omega)$$

Figure 1 - Schematic of a typical scattering experiment.

Since we expect the energy losses suffered by X-rays in their passage through atomic systems to be small, i.e. of the order of magnitude of a few electron volts the momentum transfer will be determined to a high degree of accuracy by the scattering angle,

$$k \cong 2k_1 \sin(\tfrac{1}{2}\theta). \tag{1}$$

Thus, a typical scattering experiment consists of collimating the characteristic lines from an X-ray source, impinging the radiation on the target, fixing the scattering angle with another set of slits and measuring the spectrum of the outgoing X-rays.

Because of the small energy shifts (a few parts in 10^4) the range of experiments one can perform depends to a great extent on the intensity of the source which is available. In the work I will tell you about how we have used a new high intensity X-ray source (60 Kev, 1 A). The source itself is a rotating anode type tube typically utilizing copper as the target material. A schematic of the experimental setup is shown in figure 2. It consists of the X-ray source, several sets of slits, and a double crystal monochrometer which is used to analyze the frequency distribution of the scattered X-rays. The analyzing crystals are typically Ge with resolution as good as 0.5 eV. This type of configuration allows us to span a wide range of energy and momentum space. (See figure 3). The region sketched allows one to investigate most of the interesting physical properties of solid state and atomic systems. However, it is clear that energy loss experiments involving small energy transfers (of the order of 0.01 eV) are out of reach of current X-ray technology.

The inelastic scattering of X-rays has, over the last five years, been used to investigate a wide variety of systems including atomic, liquid and solid [*5*,*6*,*7*]. For the purpose of these lectures we will concentrate on work in the three simple metals, Li, Be, and Na. The experimental data in these systems encompass, as one will see, almost all of the important physical phenomenon accessible to these types of studies.

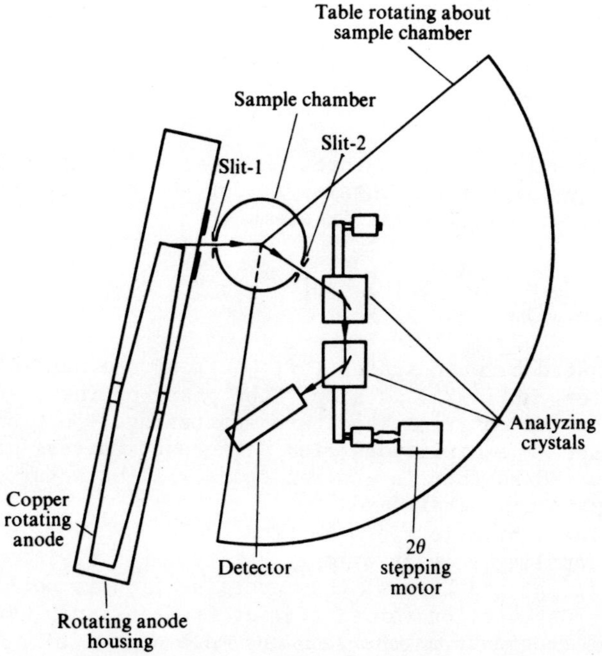

Figure 2 - Schematic of a typical scattering experiment.

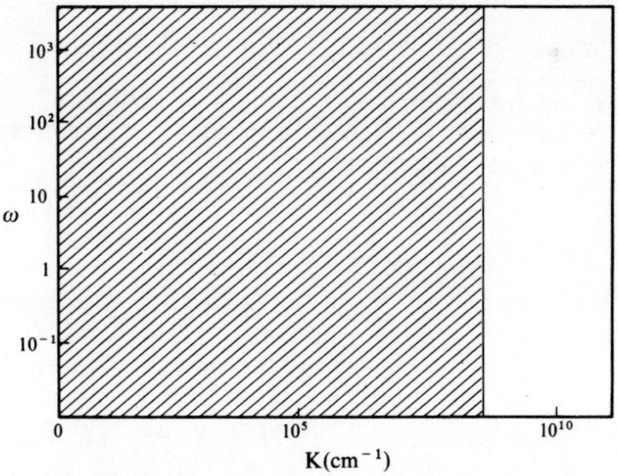

Figure 3 - Energy momentum space spanned in X-ray experiment.

To write down an expression for the cross section, one starts from the many-body Hamiltonian puts in the coupling to the radiation field and then computes to second order à la the golden rule. The cross section is almost always (light, neutrons, X-rays, electron beam) given by a formula of the form

$$\frac{d\sigma}{d\Omega d\omega} = \left(\frac{d\sigma}{d\Omega}\right)_0 \sum_f \bar{\sum_i} \left|\langle f|\sum_j \exp(i\vec{k}\cdot\vec{r}_j)|i\rangle\right|^2 \delta(E_f - E_i - \hbar\omega). \qquad (2)$$

Equation (2) contains three physically distinct pieces. (1) The energy δ function contains all of the relevant frequency information. (2) The matrix element

$$M = \langle f| \sum_j \exp[i(\vec{k}_2 - \vec{k}_1)\cdot\vec{r}_j]|i\rangle \qquad (3)$$

is evaluated between states ($\langle f|$, $\langle i|$) of the many-body system. The factor $\exp(i\vec{k}\cdot\vec{r}_j)$ is simply the phase of the scattering amplitude from the entities which are scattering (electrons, protons, neutrons). The amplitudes from different scatters are added and then squared so that in general there will be interference between scattering amplitudes.

(3) The intrinsic cross section $(d\sigma/d\Omega)_0$ describes the strength of the coupling to each primary entity. For electromagnetic radiation (light and X-rays) this cross section is well represented by the cross section for scattering from a simple harmonic oscillator of frequency ω_0 charge e and mass m, i.e.,

$$\left(\frac{d\sigma}{d\Omega}\right)_0 = \left(\frac{e^2}{mc^2}\right)^2 \left(\frac{\omega_1^2}{\omega_1^2 - \omega_0^2}\right)^2 (\varepsilon_1\cdot\varepsilon_2)^2. \qquad (4)$$

The frequency ω_0 can be thought of as a typical atomic or interband energy so that for visible light $(d\sigma/d\Omega)_0$, is often strongly frequency dependent. For X-rays $\omega_1 \gg \omega_0$ and the strength of the scattering is completely characterized by the non-relativistic Thompson cross sections, i.e.,

$$\left(\frac{d\sigma}{d\Omega}\right)_{Th} = \left(\frac{e^2}{mc^2}\right)^2 (\varepsilon_1\cdot\varepsilon_2)^2$$

$$\cong 10^{-26}\ \text{cm}^2\ (\varepsilon_1\cdot\varepsilon_2)^2. \qquad (5)$$

The scattering formula, equation (2) may be broken down still further by specifying the nature of the final states $\langle f|$ which are involved.

(A) If $\langle f| = \langle i|$, then the scattering is elastic and it is called Rayleigh scattering. In a solid this elastic scattering from a periodic array of scatters gives rise to Bragg peaks in the angular distribution. It is this type of scattering where almost all of the work on X-rays has been concentrated.

(B) If $\langle f| = \langle n|$, a discrete excited electronic state, then the scattering is called Raman scattering. No real X-ray Raman experiments have as yet been performed.

(C) If $\langle f|$ = continuum, then the scattering has been called Compton. We will be concerned exclusively with this type of scattering and the kinds of information it can yield about electrons in metals.

For all types of scattering it can be seen, by examining equation (2), that the process may be divided into two physically distinct regimes depending on the size of the momentum transfer k. If the length scale for the medium (interparticle spacing, screening length, etc.) is crudely characterized by a length (λ_{charac}), then for $k\lambda_{\text{charac}} \ll 1$ there is interference between the scattering amplitudes from many electrons, i.e. many r_j are important, and we are in the so-called collective regime. This is almost always the physically accessible region for light, neutrons and to lesser extent electron beams. The spectrum in this case displays the properties of the collective modes of the system (phonons, magnons, plasmons).

For $k\lambda_{\text{charac}} \gg 1$, the scattering amplitudes do not interfere with one another, and we are in the so-called single particle regime. The wave vector k determines the spatial resolution with which we look at scatter. For large k we are looking on a very fine distance scale and the scattering must take place from a single constituent of the many particle system.

It is easy to understand what we see in this regime by considering the kinematics of the scattering from a single particle moving with some momentum p. This scattering is diagrammed in figure 4.

Single particle regime

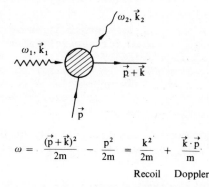

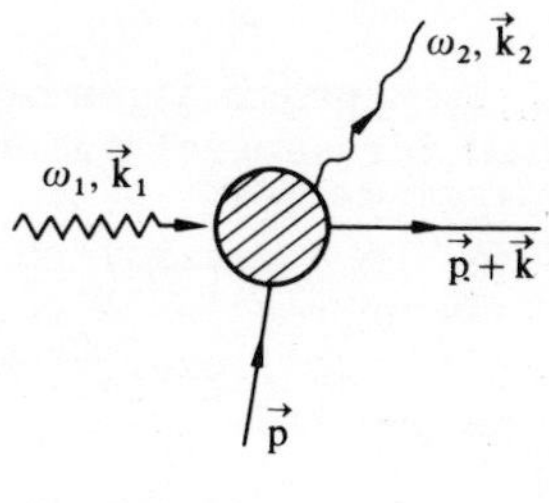

$$\omega = \frac{(\vec{p}+\vec{k})^2}{2m} - \frac{p^2}{2m} = \underset{\text{Recoil}}{\frac{k^2}{2m}} + \underset{\text{Doppler}}{\frac{\vec{k}\cdot\vec{p}}{m}}$$

Figure 4 - The kinematics of free particle electron photon "Compton" scattering

Energy conservation requires that for a fixed momentum transfer ($\hbar = 1$)

$$\omega = \frac{(\vec{p} + \vec{k})^2}{2m} - \frac{p^2}{2m} = \frac{k^2}{2m} + \frac{\vec{k}\cdot\vec{p}}{m} . \quad (6)$$

The first term on the right side of equation (6) is the recoil or Compton term and the second term, the Doppler shift associated with the electrons motion.

In a solid or a molecule the electrons momentum is smeared out, i.e., there is some probability $n_{\vec{p}}$ of finding an electron with momentum $\vec{p}$. For a single electron in an atom with a ground state wave function Ψ_0

$$n_{\vec{p}} \equiv \left| \int \exp(i\vec{p}\cdot\vec{r})\Psi_0(\vec{r})d^3\vec{r} \right|^2 . \quad (7)$$

For many-electron system

$$n_{\vec{p}} \equiv \langle 0 | a_{\vec{p}}^{\dagger} a_{\vec{p}} | 0 \rangle . \quad (8)$$

We can guess now that the cross section may (in this single particle regime) be written as

$$\frac{d\sigma}{d\Omega d\omega} = \left(\frac{d\sigma}{d\Omega}\right)\left(\frac{\omega_1}{\omega_2}\right)_0 \int \frac{d^3\vec{p}}{(2\pi)^3} n_{\vec{p}}\delta\left(\omega - \frac{k^2}{2m} - \frac{\vec{k}\cdot\vec{p}}{m}\right) \quad (9)$$

$$\equiv \left(\frac{d\sigma}{d\Omega}\right)_0 \left(\frac{\omega_1}{\omega_2}\right) J(\vec{q}),$$

with $\vec{q} \equiv \vec{k}\cdot\vec{p}$.

Equation (9) is the Impulse Approximation to equation (2). It has been shown to be valid for several interesting many-electron systems to terms of order $(E_C/E_R)^2$ [5,8]. The quantity E_C can either be a binding energy to a central potential for an atomic system or a correlation energy for electrons in a metal. The energy $E_R = \hbar^2 k^2/2m$ is the recoil energy involved in the scattering. The width of the spectrum is of order

$$\Delta\omega = 2\,\frac{p^2}{2m}\left(\frac{k}{p}\right) \cong 2E_C\left(\frac{k}{p}\right) , \quad (10)$$

i.e., it is much wider than a typical solid state or atomic energy E_C. For a 10 KeV X-ray back scattering from an electron gas in a metal and $p = p_F$ (the Fermi momentum) then $\Delta\omega = 100$ eV.

Typically then in this regime, we expect to see a spectrum which is qualitatively sketched in figure 5. The incident X-ray frequency appears as the narrow line centered at the origin and at

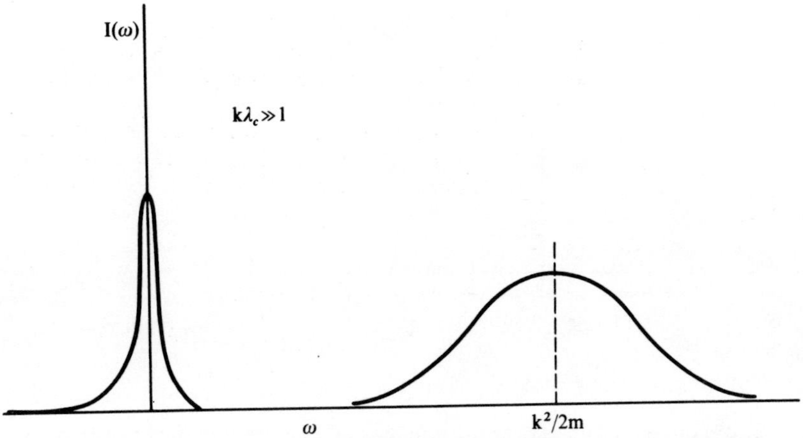

Figure 5 - Schematic Compton Spectrum.

lower energy appears a rather broad structure centered about the recoil energy $k^2/2m$ with shape and spread determined by the details of the electronic wave functions. In showing you some experimental data we will frequently shift our origin and plot the spectrum as a function of the momentum component q_z. Since this work is rather old I will only describe it briefly and refer the interested student to already published results.

In a metal like Li, Be or Na there are really two distinct types of electrons, core and valence. The energy level diagram of such a simple metal is shown schematically in figure 6. The Fermi energy is of the order of 10 electron volts while the binding of the core levels is of the order of 100 electron volts. Because of the order of magnitude energy difference separating these two types of carriers we can in fact think of them as separate entities. In a typical experiment we of course obtain contributions from both sets of carriers which may then be separated in a very straightforward manner. The core electrons gave us a momentum distribution characteristic of (for Li) a tightly bound He like system. Generally, we are not very interested in the details of this core spectrum since it contains very little solid state information. We almost always will subtract it from the data utilizing for example experimental measured values for He itself.

To give you some feel for the kinds of data one gets in simple atomic systems like He where there is really only one kind

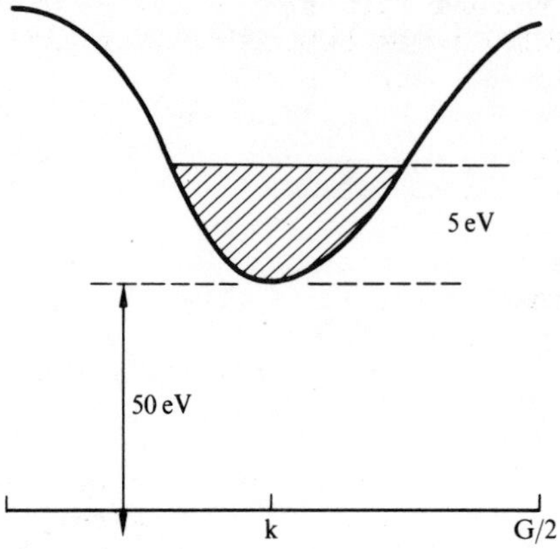

Figure 6 - Energy level diagram for a typical light alkali metal.

of electron we display the results of a Compton scattering experiment on gaseous He (figure 7) [5]. The solid curve is the experimental result and the points are a theoretical fit obtained using equation (9) along with a simple Clemente Hartree-Fock

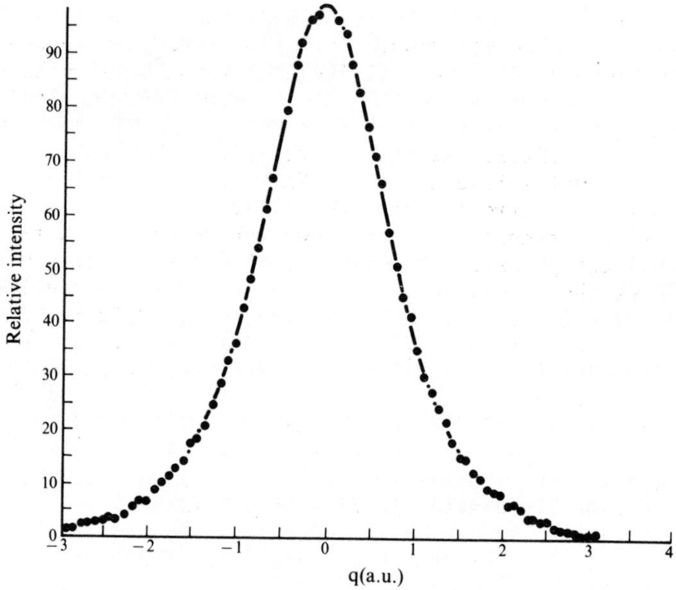

Figure 7 - Compton Profile of He (theory — dots, experiment — solid curve).

wave function for the ground state of the two He elctrons. The agreement between the two curves as expected is almost perfect.

The momentum distribution of the outer or conduction electrons in Alkali metals is not such a simple or an uninteresting problem. As we shall see the effects of both band structure and inter-electron Coulomb interactions are very non-trivial. The simplest picture of a metal assumes that the electrons are non-interacting and occupy a set of states in momentum space uniformly filling the three-dimensional Fermi sphere (see figure 8). The Compton profile resulting from such a simple distribution is obtained by integrating over two components of the momentum p. This procedure yields a parabolic shape which is sketched in figure 9.

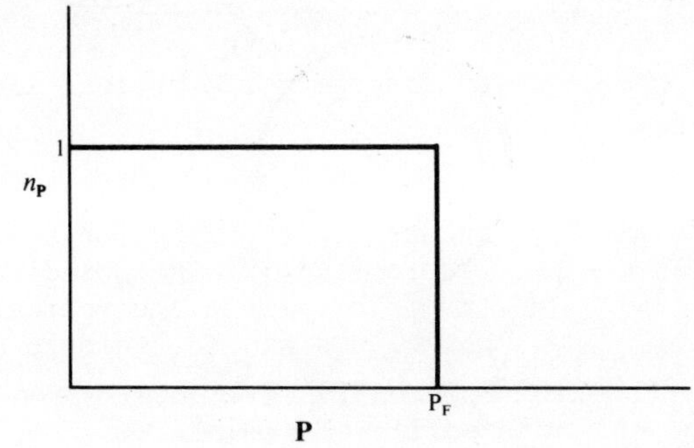

Figure 8 - Three-dimensional Fermi distribution for non-interacting electrons.

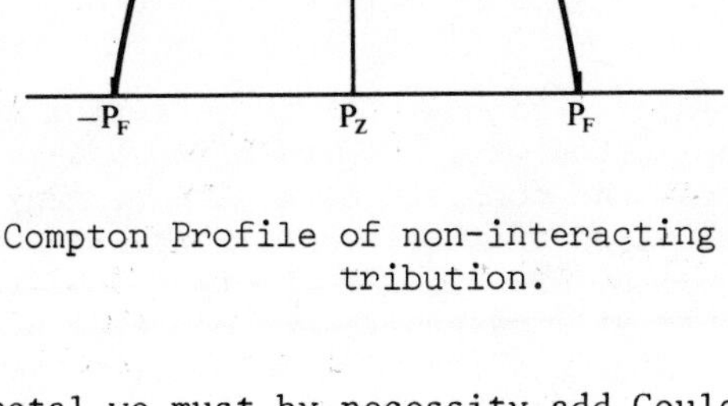

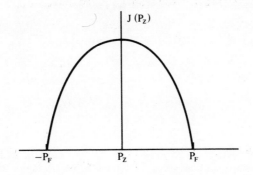

Figure 9 - Compton Profile of non-interacting Fermi distribution.

In a real metal we must by necessity add Coulomb and band structure effects. In a world where we can neglect Coulomb interactions except in so far as they determine a simple periodic one electron potential we may simply describe the new momentum distribution. The old sphere in momentum space is replaced by many spheres each centered at a particular reciprocal lattice point (G_n). This situation is shown schematically in figure 10.

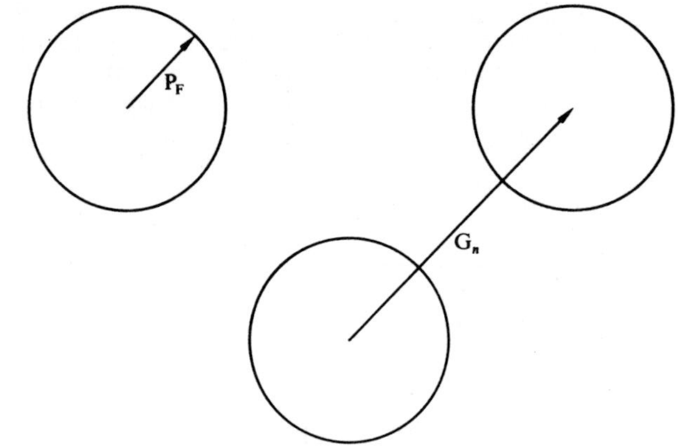

Figure 10 - Momentum spheres in reciprocal space for non-interacting electrons.

The amount of momentum in these higher spheres is directly proportional to the appropriate Fourier coefficient of the Bloch part of the conduction electron wave function $a_{n\underset{\sim}{p}}$, i.e.

$$\Psi_{n\vec{p}}(\vec{r}) = e^{i\vec{p}\cdot\vec{r}} u_{n\vec{p}}(\vec{r}), \tag{11}$$

with

$$u_{n\vec{p}}(\vec{r}) = \sum_{G_n} a_{n\vec{p}} e^{i\vec{G}_n\cdot\vec{r}}. \tag{12}$$

The Coulomb interactions affect the momentum distribution in a much more complicated way. In order to analyze the problem properly one needs the complete machinery of many-body physics. However, an evaluation of these formal expressions usually require a weak coupling RPA type of theory. Qualitatively, however, we know what will happen to the momentum distribution in the presence of inter-electron Coulomb interactions [9]. The distribution becomes somewhat rounded below the Fermi momentum. The electrons removed out of this region are placed into states above the Fermi momentum. In addition, the unit discontinuity in the original distribution is weakened but remains finite (see figure 11a). This modified distribution will give a Compton profile with a smaller break in slope at the Fermi momentum and with a tail existing to several multiples of the Fermi momentum (see figure 11b) [*10*,*11*].

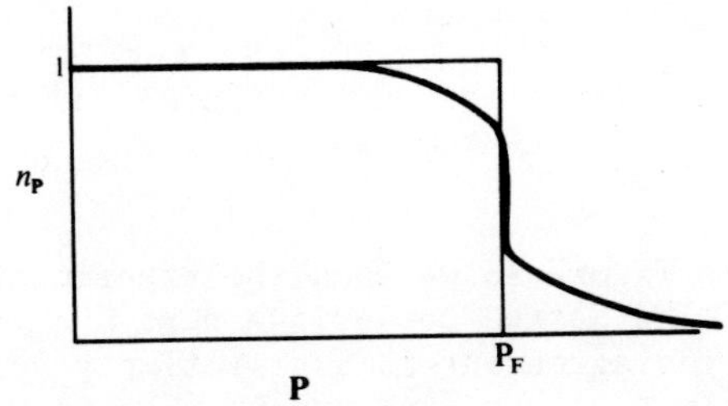

(b)

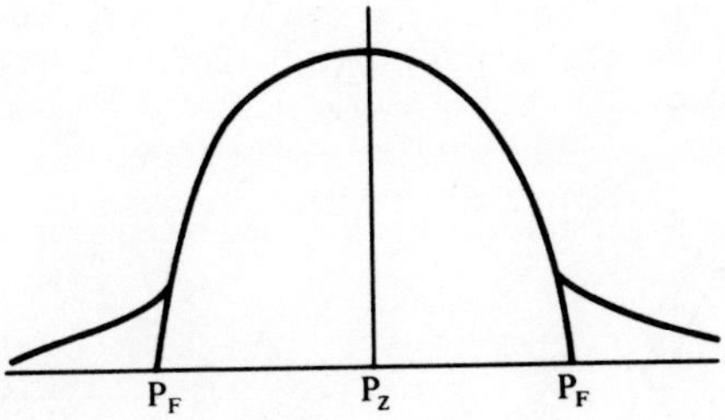

Figure 11 - Coulomb modifications of the momentum distribution.

In order to compare some theoretical results with the experimental spectra which we will show you shortly we should understand at least qualitatively the nature of the RPA calculation of the momentum distribution. In the usual scheme we compute, via standard techniques the single particle Green's function $G(\vec{p},\omega)$ by summing the simple geometric set of diagrams shown in figure 12 and relating $n_{\vec{p}}$ to $G(\vec{p},\omega)$ by the general formula

$$n(\vec{p}) = \int_{-\infty}^{+\infty} G(\vec{p},\omega)d\omega. \qquad (13)$$

Fortunately, in Na 'real'band structure effects are small. The Compton profile primarily reflects the effects of Coulomb

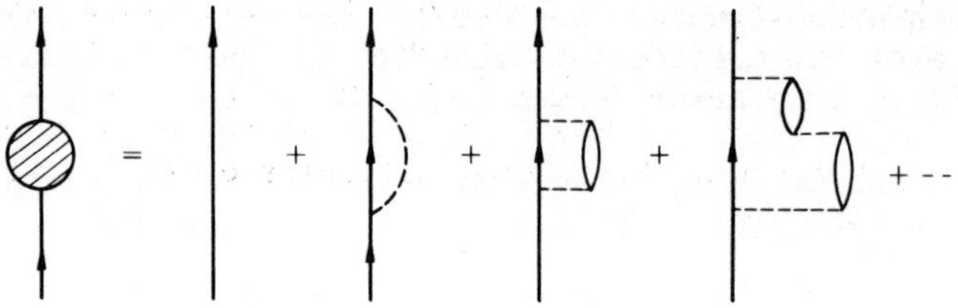

Figure 12 - The ring diagrams for the RPA calculation of momentum density.

interactions. In figure 13 we show the experimental Compton profile for Na. The dotted curve is a pure RPA calculation with r_s = 3.6. The agreement is qualitatively very good although the experimental spectrum as shown seems to have somewhat more intensity and is somewhat flatter in the tail region. Recently several people have been able to show that this agreement may be accounted for by including in the theory the effects of core orthogonalization [*13*]. Such effects (1 OPW) are in fact easy to include in the theory if their contribution to the total momentum distribution is small. For the details of this technique we refer the student to reference [*13*].

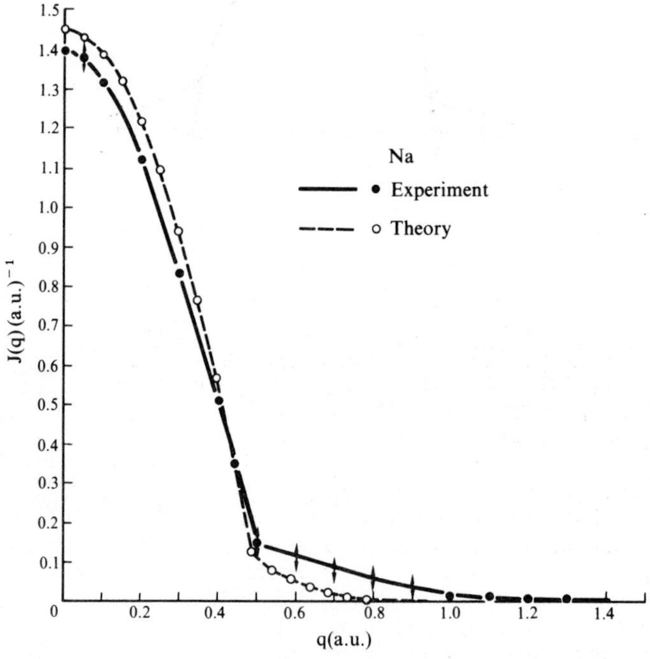

Figure 13 - Compton Profile for Na.

When both band structure and Coulomb effects may be important the problem of the momentum distribution becomes almost intractable [*10*,*11*]. In general

$$n_p = \frac{\partial E_p(\lambda)}{\partial \lambda} . \tag{14}$$

$E_p(\lambda)$ is the ground state energy of the inhomogeneous system characterized by the Hamiltonian $\mathcal{H}_\lambda{}^p$

$$\mathcal{H}_\lambda{}^p = \mathcal{H}_0 + \lambda a_{\vec{p}}^{\dagger} a_{\vec{p}}, \tag{15}$$

with

$$\mathcal{H}_0 = \sum_i \left[\frac{p_i^{\,2}}{2m} + \sum_G V_G e^{i\vec{G}\cdot\vec{r}_i} + \frac{1}{2}\sum_j \frac{e^2}{|\vec{r}_i - \vec{r}_j|}\right] . \tag{16}$$

The Hohenberg-Cohen formulation of the inhomogeneous electron gas provides us with a framework for calculating $E_p(\lambda)$. However, in practice one can really only evaluate equations (14-16) in the case where we treat the external potential as weak. In this case

$$N_p = n_p{}^0 + \Delta N_p,$$

with

$$\Delta N_p = \frac{1}{2}\sum_{\underset{\sim}{G}} \frac{1}{v_{\underset{\sim}{G}}}\left|\frac{V_{\underset{\sim}{G}}}{\varepsilon(\underset{\sim}{G},0)}\right|^2 \left(\frac{\partial \varepsilon_p(\underset{\sim}{G},0)}{\partial \lambda}\right)_{\lambda=0} , \tag{17}$$

where $(G,0)$ is the static dielectric constant of the homogeneous system, $v_G = 4\pi e^2/G^2$ and $\varepsilon_\lambda{}^p$ the dielectric constant of $\mathcal{H}_\lambda{}^p$ [*10*].

In Li band structure and Coulomb effects are both important and to provide ourselves with a qualitative understanding of them we will look at Compton spectra in this material with our weak external potential coupling RPA type of approximation which we have just discussed. The experimental results shown in figure 14 are rather interesting. They show a significant amount of momentum in the tail region and they are anisotropic. It turns out that one can understand the order of anisotropy, i.e. which of the curves in figure 14 are higher or lower at, for example, $q = 0$. This anisotropy comes only from the geometry of the reciprocal lattice and is quite independent of any details of the microscopic theory. The tail, on the other hand, cannot be accounted for by simply using a uniform field RPA result such as we employed in interpreting the Na data. However, it is possible to understand the size of this tail by using an

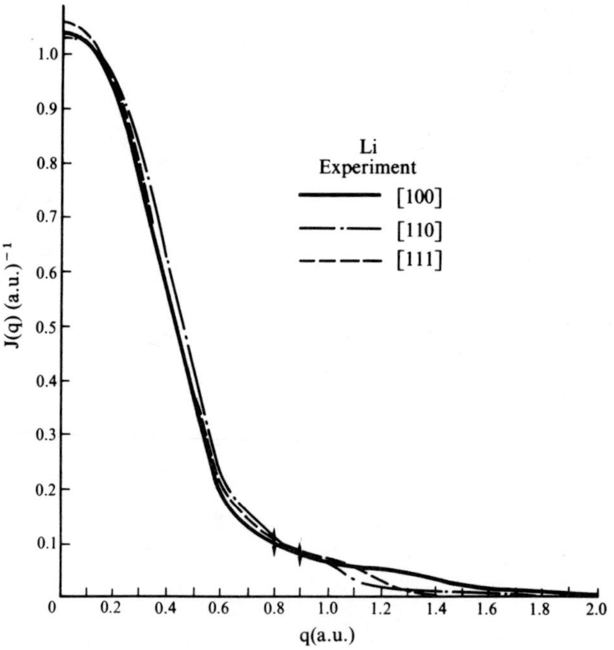

Figure 14 - Compton Profiles of single crystal Li.

equation of the form given on the preceding page. In figure 15 we show such a fit to the polycrystalline profile of Li. The various theoretical points on the curve represent different reasonable assumptions about the size of the various pseudo-potential coefficients in equation (17). It is clear from this figure that we can qualitatively understand the data but that quantitatively there is a significant amount of flexibility in the theory.

We now leave the high momentum regime and switch our attention to a low momentum regime where the characteristics of the spectrum are quite different and where the experimental problems are much more severe [*14*,*15*,*16*]. Qualitatively we expect that the spectrum will reflect the collective behavior of the system. At small k scattering amplitudes from different particles interfere with one another (see equation (2)) and a true representation involves many particles, i.e. the collective behavior of the system.

If we had an X-ray laser and a detector with infinite resolution we might expect to see a spectrum at long wavelengths, something like that which is shown schematically in figure 16. The sharp peak at zero energy is due to the coupling between density fluctuations in the electron gas and the density fluctuations in the lattice. The position of this peak occurs at the lattice

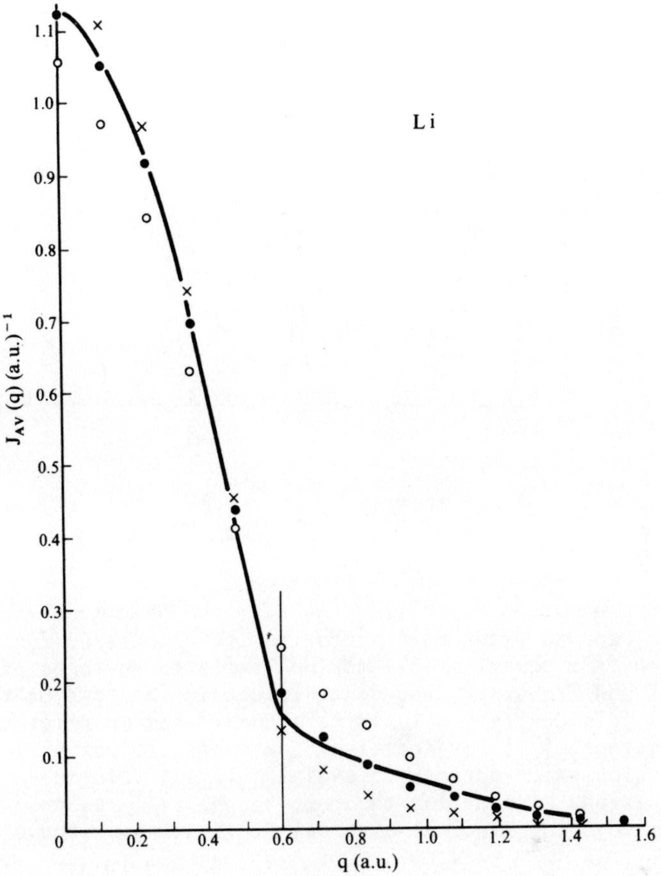

Figure 15 - Polycrystalline Compton Profile for Li with various theoretical (points).

phonon frequency (of the order of 0.01 eV). The peak at an energy of approximately 5 eV simply reflects the electron plasmon mode. The edge at 50 eV arises from the excitation of the core electrons to the conduction band continuum (see figure 6). It will never be possible to resolve with X-rays the low lying phonon portion of this spectrum. On the other hand, a careful investigation of the plasmon peak and the edge connected with the core electron is certainly feasible. In these talks I will concentrate exclusively on the behavior of the plasmon peak at long wavelengths and follow its evolution to somewhat shorter wavelengths where it takes on some of the characteristics of the single particle spectrum we have discussed earlier.

In order to give ourselves a theoretical framework in which to operate we would like to rewrite the expression for the cross

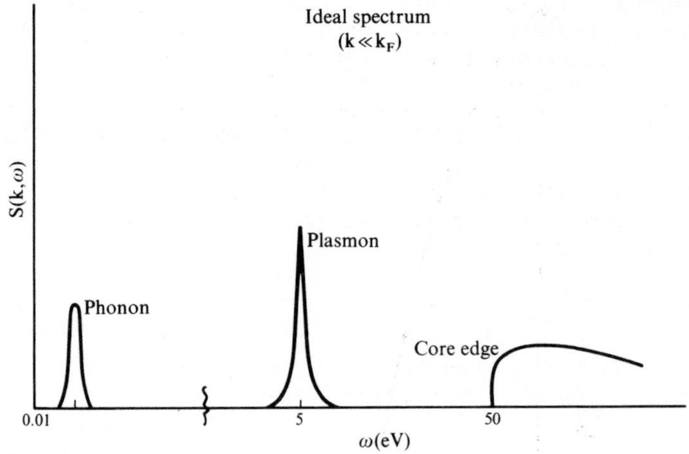

Figure 16 - Ideal scattering spectrum in the collective regime.

section, equation (2). It is quite rigorously true that this general second order Born approximation expression for the cross section (see equation (2)) may be rewritten in terms of the wave vector and frequency dependent dielectric constant of the system [*17*]. This derivation is quite standard and we refer the reader to reference [*17*] for details. The essential point is that the expression which appears in equation (2) is related, by the fluctuations dissipation theorem, to the real response of the system to a longitudinal external field at wave vector k so that

$$\frac{d\sigma}{d\Omega d\omega} = \left(\frac{e^2}{mc^2}\right)^2 \left(\frac{\omega_1}{\omega_2}\right) (\varepsilon_1 \cdot \varepsilon_2)^2 S(\vec{k},\omega), \tag{18}$$

where

$$S(\vec{k},\omega) = \frac{k^2}{4\pi e^2} \text{ Im } \left[\frac{1}{\varepsilon_M(\vec{k},\omega)}\right] \tag{19}$$

and

$$\varepsilon_M(\vec{k},\omega)^{-1} = \left[1 - \frac{4\pi e^2}{(\vec{k}+\vec{G})^2}\, \alpha(\vec{k}+\vec{G},\vec{k}+\vec{G}',\omega)\right]^{-1}. \tag{20}$$

The quantity α is the polarizability tensor, tensor in the reciprocal lattice vectors $\vec{G}$ of the system. It is simply proportional to the response of the system to a total field at wave vector $\vec{k}$. It is a tensor because of the fact that in an inhomogeneous periodic system a perturbation at wave vector $\vec{k}$ results in a response at all wave vectors $\vec{k} + \vec{G}$, where $\vec{G}$ is any reciprocal lattice vector.

Equation (20) is sufficiently complicated by both band structure and many body effects that for the purpose of these lectures we will choose to work completely within the so-called RPAB (RPA with band structure) approximation. This is equivalent to replacing α by the polarizability of a non-interacting set of Bloch electrons

$$\alpha^0(\vec{k} + \vec{G}, \vec{k} + \vec{G}, \omega)$$

$$= \sum_{\vec{p}, \ell\ell'} \frac{[n(\ell,\vec{p}) - n(\ell',\vec{p} + \vec{k})]}{[\omega + E(\ell,\vec{p}) - E(\ell',\vec{p} + \vec{k}) + i\delta]}$$

$$\times \langle \ell,\vec{p}| e^{-i(\vec{k}+\vec{G})\cdot\vec{r}} |\ell',\vec{p} + \vec{k}\rangle \langle \ell',\vec{p} + \vec{k}| e^{i(\vec{k}+\vec{G}')\cdot\vec{r}} |\ell,\vec{p}\rangle . \quad (21)$$

The wave functions which appear in equation (21) are the one electron Bloch states of the non-interacting system. The Coulomb equations are only included in so far as the response to the total field is taken to be non-interacting. Such an approximation while it gives us tractable expressions for the entire regime of energy and momentum space is not equivalent to the RPA expressions equation (13) and figure 13 for the momentum distribution. In fact at large momentum transfers equations (19-20) go over to equation 9 with a *non-interacting* (band structure included) expression for $n_{\vec{p}}$.

The evaluation of equation (21) depends on the details band structure. However, without going into any of the specifics concerning this band structure certain general characteristics of equation (21) are evident.

(1) In the absence of band structure α is diagonal. This implies that if we drive the system at wave vector k the response is only at k;

(2) All the non-direct interband transitions are present, i.e. the imaginary part of ε_M exists whenever

$$\omega = E(\ell,\vec{p}) - E(\ell',\vec{p} + \vec{k}); \quad (22)$$

(3) Provided the imaginary part of ε_M is small the zeros of this dielectric function show up as poles in the spectrum. In a system with band structure these singularities will be periodic in momentum space, i.e. the plasmon becomes a periodic disturbance with its own typical band structure character.

In order to clarify some of these points I would like to remind you of the well known results for the spectrum in the absence of band structure. Having done this we will describe, mostly qualitatively, how band structure modifies this simple picture. The essence of the uniform RPA results is contained in figure 17. The dashed curve in the figure is the plasmon, i.e.

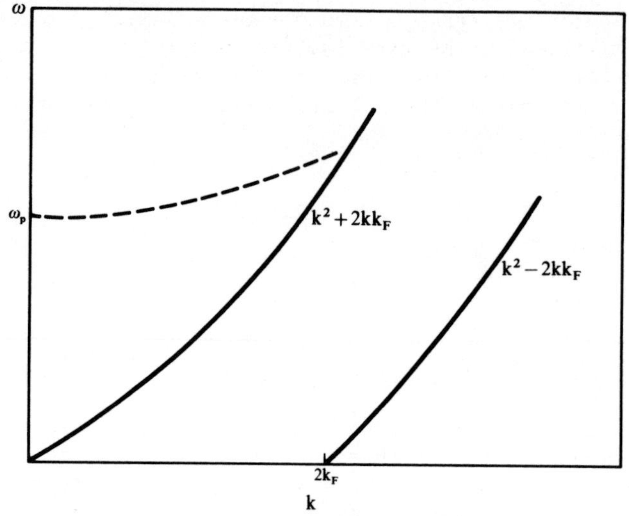

Figure 17 - RPA excitation spectrum.

the curve on which $\varepsilon_M(\vec{k},\omega) = 0$. The two parabolas in the figure define that region of energy and momentum space in which ε_M has an imaginary part. It is the region for which real *single pair* excitations with total momentum k and energy ω exist, i.e.

$$\omega(\vec{k},p) = \frac{k^2}{2m} + \frac{\vec{k}\cdot\vec{p}}{m} . \tag{23}$$

In this simple approximation the plasmon is non-damped until its dispersion curve intersects the continuum at a value of

$$k = k_c = \frac{\omega_p}{v_F} , \tag{24}$$

where upon it may decay into an electron hole pair. The dispersion relation of the plasmon at long wavelengths in RPA is given by

$$\omega_k = \omega_p \left[1 + \frac{3}{10} \frac{k^2 v_F^2}{\omega_p^2} + \ldots \right] . \tag{25}$$

It is worthwile to point out that at this time this RPA picture is in a very real sense a weak coupling (small deviation from ideal gas) approximation to the spectrum of fluctuations in metals. There is a good possibility that we are in fact faced with a strongly correlated liquid like problem and that this kind of approximation may in some region of momentum space break down qualitatively. None of the features of a real liquid such as the existence of short range spatial correlations have been included.

In figures 18 through 21 we show several theoretical uniform RPA spectra for a series of momentum transfers.

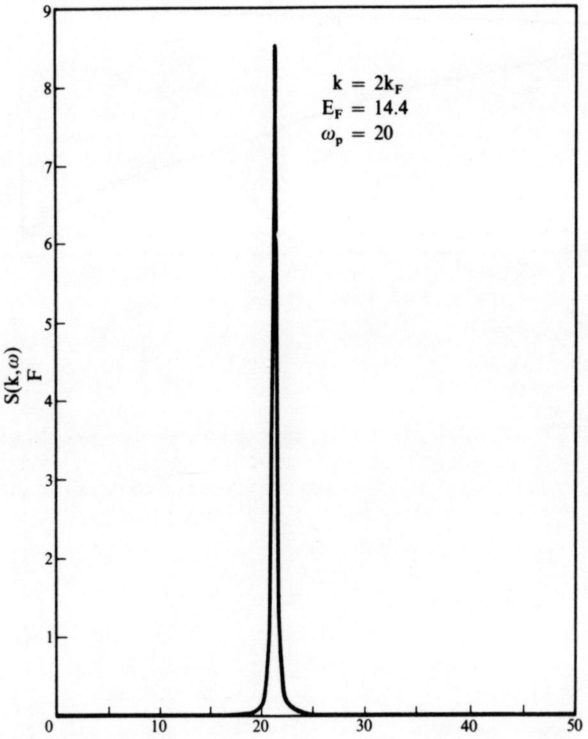

Figure 18 - RPA spectrum for Be (E_F = 14.3 eV, ω_p = 20 eV)

At low momentum transfer the spectrum is dominated by a sharp peak. At larger momentum transfer the spectrum has a mixed characteristic looking somewhat broad but still having a modest peak in the neighborhood of the plasmon like zero of the dielectric constant larger momentum transfer the spectrum looks very much like a free Fermi Compton profile, i.e. a parabola centered at the recoil energy $k^2/2m$.

Another function which is of interest is the so called

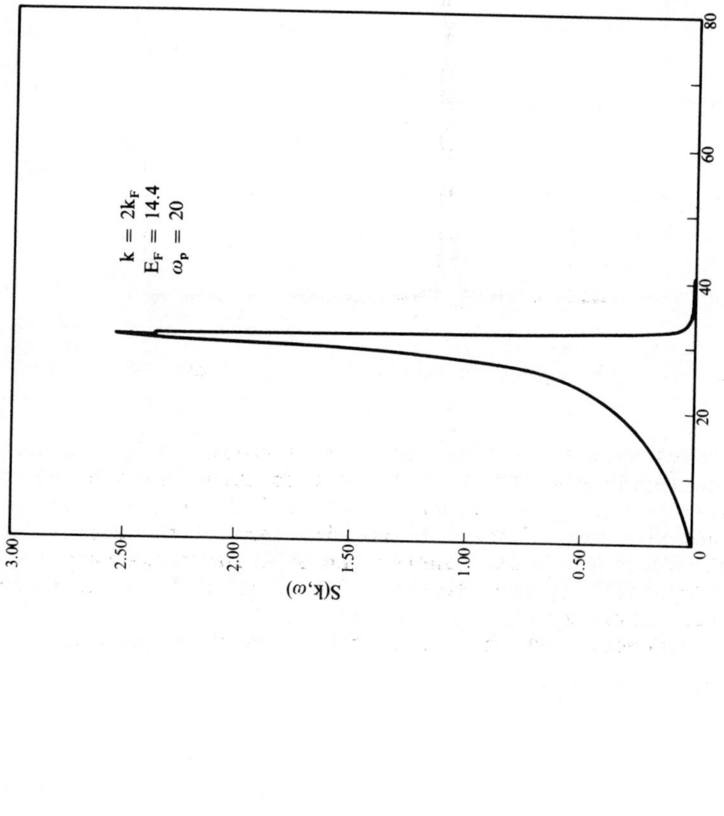

Figure 19

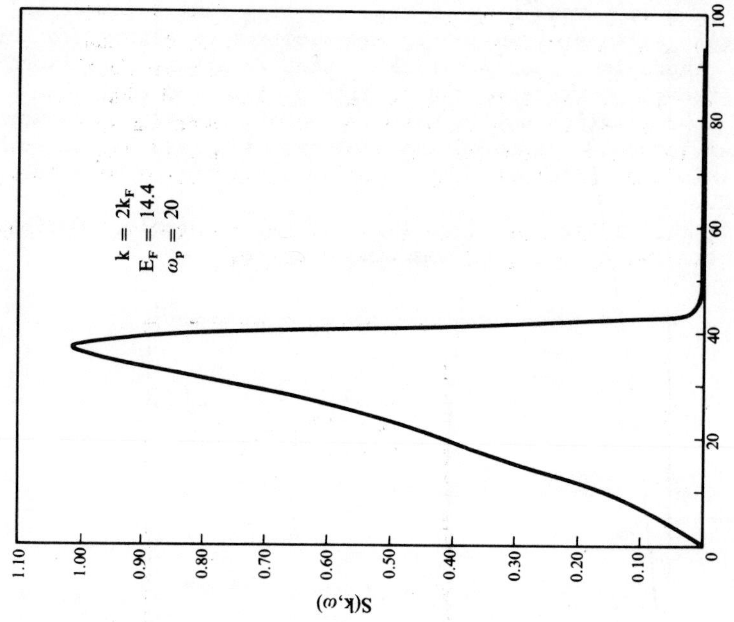

Figure 20

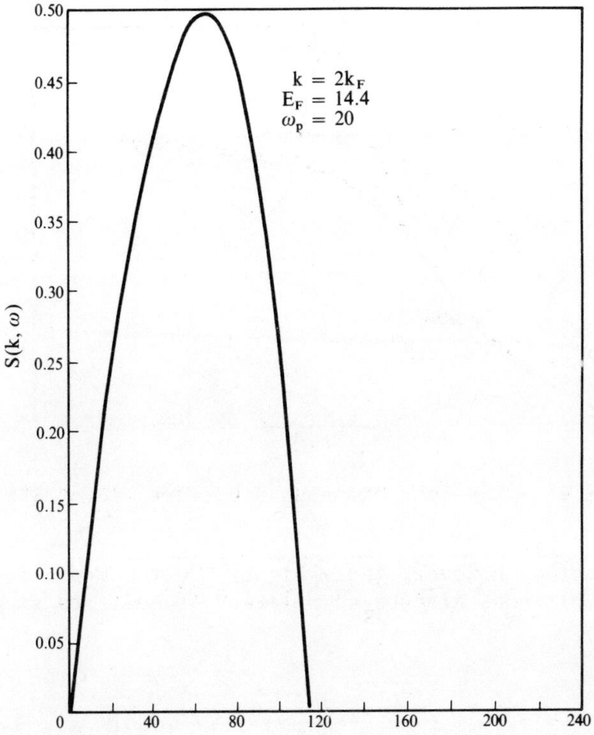

Figure 21

instantaneous structure factor $S(\vec{k})$ defined by,

$$S(\vec{k}) = \int S(\vec{k},\omega)d\omega, \tag{26}$$

where the quantity $S(\vec{k},\omega)$ is the dynamic structure factor defined in equation (18). In RPA it is given by the solid curve in figure 22 . The student should note that it is a smooth curve starting out quadratically at the origin and approaching one at large k. Since the structure factor is simply the Fourier transform of the density correlation function it is simply related to the probability of finding another particle at a distance r given a particle at the origin.

There have been several theoretical analysis which attempt to include corrections to this simple RPA picture [*18*,*19*]. In these treatments the Coulomb interaction is treated perturbatively and most of the corrections concern themselves with modifications of the plasmon behavior, at long wavelengths. As far as I know, there have been no computations which give us a quantitative description of the spectrum at intermediate momentum transfers ($k \sim k_F$). At

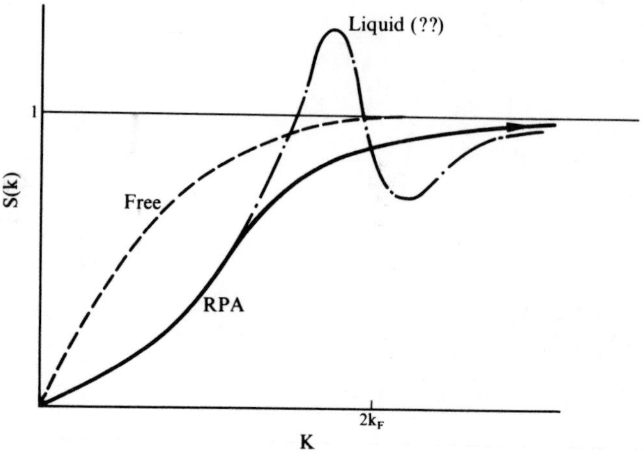

Figure 22 - The instantaneous structure factor $S(k)$.

long wavelengths, however, these higher order computations tell us that the plasmons dispersion relation is modified as given below

$$\omega = \omega_p\left[1 + \frac{3}{10}\,\frac{(1 - ar_s)k^2 v_f{}^2}{\omega_p{}^2} + \ldots\right] . \tag{27}$$

The parameter r_s is simply defined as the ratio of the inter-particle spacing $(4\pi/3)r_0{}^3 = n$ to the Bohr radius $a_0 = \hbar^2/me^2$. Physically it is simply the ratio of the mean potential energy of the particles to their mean kinetic energy, i.e.

$$r_s = \frac{\langle V\rangle}{\langle k\cdot E\rangle} . \tag{28}$$

The parameter a in equation (27) is typically of the order of 0.1. The r_s corrections on the plasmon dispersion arise from the exchange coupling between the electrons and they are of such a sign as to soften the plasmon dispersion. This is expected on physical grounds since the exchange keeps parallel spin electrons apart giving rise (in the presence of a uniform background) to an effective attractive exchange interaction. In addition to this change in the dispersion, a damping term proportional to k^2 arises from a process whereby the plasmon may decay into two electron hole pairs (see figure 23). After a lengthy computation Dubois finds that

$$\frac{\mathrm{Im}\omega_p}{\omega_p} = B\left(\frac{k}{k_F}\right)^2 , \tag{29}$$

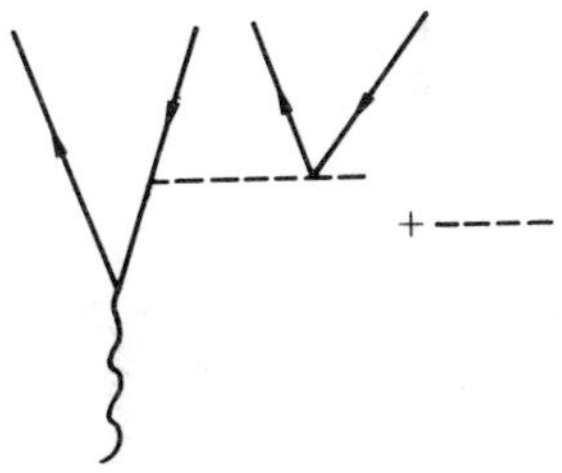

Figure 23 - Higher order corrections to plasmon damping.

where B is a function of r_s and is of order 0.1 for r_s at metallic densities.

Very little quantitative is known about the deviations of the structure factor from the solid RPA curve shown in figure 22. However, it is possible to make some interesting speculations concerning the behavior of this structure factor in the interacting system. It seems clear that the structure factor $S(k)$ must have a behavior, as a function of k, which is sketched qualitatively as the dotted curve in figure 22. The peak in this curve crudely corresponds to an incipient Bragg peak due to the presence of short range order in electron liquid. The electron liquid must, of course, be quite far from uniform since the rather strong Coulomb interactions will tend to keep the electrons somewhere near their mean separation. In fact, it is well known that at 'low density' the large Coulomb interactions dominate the behavior of the system and the gas must crystalize to a Wigner solid. After such crystallization $S(\vec{k})$ will consist of a series of δ functions, arising from elastic Bragg scattering, sitting on top of a smoothly varying thermal diffuse background. The peak in the liquid is a reflection of the incipient solid behavior which must be present.

Since we are speculating and some of these speculations may prove to be relevant to the data which I will show you, let us speculate a little further and see how the peaking of the structure factor might influence the behavior of the inelastic X-ray scattering spectrum, i.e. the dynamic structure factor. From the definition of $S(\vec{k},\omega)$ (see equations (26,19)). It is easy to show that it satisfies the well known sum rule,

$$\int d\omega S(\vec{k},\omega) = \frac{k^2}{2m} . \tag{30}$$

If in addition, we assume that the collective mode, i.e. pole in $S(k,\omega)$ exhausts this sum rule i.e.

$$S(\vec{k},\omega) = S(\vec{k})\delta(\omega - \omega(\vec{k})),$$

then it follows that

$$\omega(\vec{k}) = \frac{k^2}{2mS(\vec{k})} . \tag{31}$$

This is the famous Feynman expression for the excitation spectrum in liquid He. It is a good approximation at long wavelengths where the collective mode does indeed exist. The sum rule at shorter wavelengths is not so well justified but qualitatively correct. It certainly gives the right behavior at both long and short wavelengths. At long wavelengths $S(k) \sim k^2$ giving the plasmon and at short wavelengths (large k) it goes to unity giving us a pole at

$$\omega(k) = \frac{k^2}{2m} , \tag{32}$$

(not too bad an approximation). For RPA then the smooth behavior of $S(k)$ reflects the smooth monotonically increasing behavior of the quantity $\omega(k)$ the mean excitation frequency for the system. If $S(k)$ really peaks at a value of k such that $k \cong 2\pi/r_{int}$ where r_{int} is the mean interparticle spacing as shown in figure 22, then we would expect a decrease, in fact, a minimum in the dispersion curve at this value of k. This is the origin and the meaning of the roton minimum in liquid He^4.

Enough for speculation. We will now take a hard look at the effect of band structure within the RPAB on the spectrum. Apart from details band structure effects in the solid introduces two fundamental changes into our simple RPA picture. These two phenomena are schematically shown in figure 24.

(1) The presence of the periodic potential introduces Umklapp or interband transitions into the absorptive part of ε_M. This shows up in the ω- vs. k-plane as the folded parabolas shown dotted in figure 24.

(2) The tensor character of α forces the plasmon to become a periodic entity folding the peak at the Brillouin zone boundary and opening up gaps at this point in momentum space.

In order to evaluate equation (19) we need expressions for all of the electronic band energies and for the matrix element of the operator $e^{i\underset{\sim}{k}\cdot\underset{\sim}{r}}$ between the Bloch states. We want to avoid such a detailed calculation and for the purpose of these lectures we consider a weak pseudo-potential approach. This approximation has, as we shall see, both of the essential features described above and is even quantitatively correct for many simple materials. Defining

$$T_{GG'}(\vec{k},\omega) = \frac{4\pi e^2}{(k + G)^2} \alpha^0(\vec{k} + \vec{G},\vec{k} + \vec{G}',\omega), \tag{33}$$

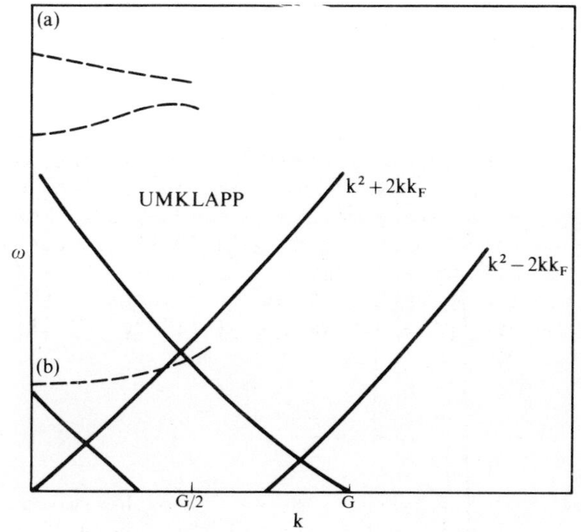

Figure 24 - Band structure modifications of RPA excitation spectrum

and keeping only terms quadratic in the weak pseudo-potential V_G. We find for k in the first Brillouin zone that

$$\varepsilon_M = 1 + T_{OO} + \sum_G \frac{T_{OG}T_{GO}}{1 + T_{GG}} . \tag{34}$$

Detailed estimates and some very crude arguments show that near $k = 0$ the off diagonal terms, i.e. the summation in equation (34) are unimportant near the zone boundary. However, they are the terms which determine the splitting of the plasmon its folding back into the first zone. We can see how this comes about by noting that

$$T_{OO}(-\tfrac{1}{2}G,\omega) = T_{GG}(-\tfrac{1}{2}G,\omega). \tag{35}$$

Let me remind the student that roughly speaking the zeros of ε_M determine the position of the plasmon pole. In the absence of band structure we look for the zeros of $1 + T_{OO}$. As we approach the zone boundary $1 + T_{OO}$ and $1 + T_{GG}$ have identical zeros by virtue of equation (35). The term $T_{OG}T_{GO}$ in equation (34) simply splits these two degenerate roots in the usual way, into two distinct branches. It is clear from this simple discussion that equation (34) while it is written down on the basis of a weak pseudo-potential arguments is equivalent to a two band model of the band structure. Such a model has been written down more exactly and results of calculations within the framework of these two band models show the general features which we have described. In figure 25 we give a typical spectrum

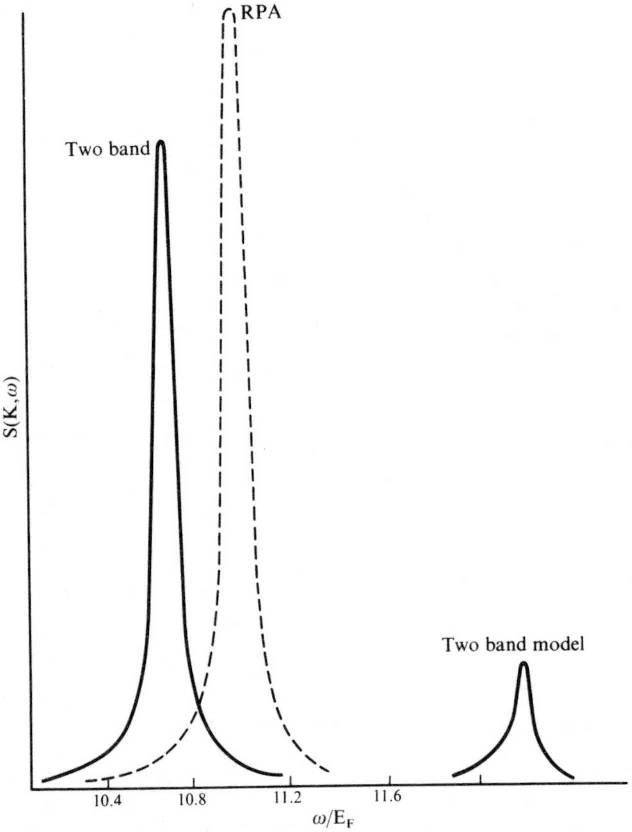

Figure 25 - Two band model (ω_p/E_F = 11 q/q_F = 0.4).

obtained within the framework of this model, for a particularly unreasonable set of parameters, (ω_p/ω = 11, E_F = 1 eV, G = 2.2 k_F, V_G = 0.2 eV) chosen to accentuate the banding and splitting of the single plasmon band. The two plasmon peaks are evident and the gap is simply proportional to the value of the pseudopotential we have arbitrarily chosen.

In order to go further and compare the theoretical predictions of RPAB with experimental data, particularly in Be, we are forced into making still another set of approximations in order to evaluate the components of the dielectric matrix given in equation (21). For materials like Be the plasmon frequency is known to be quite high compared to typical interband energies.

This suggests that we may expand the general expressions (equation (21)) for α^0 as a power series in $1/\omega^2$.

Using commutator techniques [17] to perform the intermediate state sums present in α^0 we find to order ω^{-4} in the diagonal terms and ω^{-2} in the off diagonal term, ($m = \hbar = 1$)

$$\alpha^0(\vec{k}+\vec{K},\vec{k}+\vec{K},\omega)$$

$$= \frac{q^2 N}{\omega^2} - \frac{1}{\omega^4}\sum_{\vec{G}} (\vec{q}\cdot\vec{G})^2 V_{\vec{G}}\rho_{\vec{G}}$$

$$- \frac{3q^2}{\omega^4}\sum_{\ell,\vec{p}} n(\ell,\vec{p})\langle \ell,\vec{p}|(\vec{q}\cdot\vec{\nabla})^2|\ell,\vec{p}\rangle + \frac{q^6 N}{4\omega^4},$$

$$\alpha^0(\vec{k}+\vec{K},\vec{k}+\vec{K}',\omega) = \frac{(\vec{k}+\vec{K})\cdot(\vec{k}+\vec{K}')}{\omega^2}\,\rho(\vec{K}'-\vec{K}) \tag{36}$$

where $\vec{q} = \vec{k} + \vec{K}$, N is the average electron density, and $V_{\vec{G}}$ and $\rho_{\vec{G}}$ are the G-th Fourier component of the lattice potential (including the structure factor) and the electron density.

(1) The leading term gives the free electron plasmon.

(2) The next term in ω^{-4} modifies the position of the zero wave vector plasmon. In a hexagonal crystal it is generally anisotropic while in a cubic material it is isotropic. It represents the effects of interband transition on the plasmon. Since the plasmon is assumed to be at high frequencies relative to the interband transitions the sign of this term is such as to push the plasmon *up* in frequency.

(3) The next term characterize the k^2 dispersion of the plasmon. In the absence of band structure effects it gives the usual dispersion $\omega_p^2(k) = \omega_p^2(0) + (3/5)k^2 v_f^2$. In general it will, give rise to an anisotropy in the plasmon dispersion.

(4) The off diagonal term comes in squared when one inverts the tensor given in equation (2). It is of such a sign as to decrease the frequency of the long wavelength plasmon i.e. the higher plasmon bands push down the lowest branch of the spectrum.

(5) This high frequency limit completely omits linewidth effects.

The experimental setup is similar to that used by many investigators except that in this study Ge(110) crystals (0.5 eV energy resolution) were used to analyze the ehergy of the radiation scattered at a scattering angle fixed by slits to $\pm\frac{1}{2}°$.

Copper X-rays, K_{α_1} and K_{α_2}, with about 2.5 eV, natural linewidths and separated by 19.95 eV were the source of radiation. At each scattering angle the energy of the radiation was analyzed by a double Bragg spectrometer in a 100 eV region around the K_{α_1} and K_{α_2} lines. In figure 26(a) we show the response of the system with no sample and only elastically (thermal diffuse) scattered K_{α_1} and K_{α_2} X-rays. In figure 26(b) is displayed the spectrum observed at the same scattering angle but with the Be sample in the beam. The spectrum in figure 26(a), was used in a trial and error technique to remove the elastically scattered X-rays from the spectrum in figure 26(b) with results shown in figure 26(c). Making use of the known K_{α_1} and K_{α_2} separation and linewidths we were able to calibrate the spectrometer (relate angle to energy).

The compiled experimental results are plotted in figure 27. The arrow on the scattering angle axis (momentum transfer squared) indicates the position of the RPA cutoff momentum $k_c = \omega_p/V_F = 1.24$ Å^{-1}. The dispersion data is consistent with an intercept at $k = 0$ for both C-and A-axis of 18.35 eV ± 0.35 eV and a dispersion which is about 30% different in the two directions. The linewidth data shows the plasmon to be considerably narrower along the C-axis with a width at $k = 0$ of about 3 eV along C and 5 eV along A. The long wavelength limit of the plasmons dispersion relation, for the high density homogeneous gas. (including exchange [*17*] is shown in figure 27(a). The intercept at $k = 0$ (18.2 eV) is within experimental error of the observed value at 18.35. The slope which has about a 10% exchange softening in it agrees with the experimental C-axis slope. The theoretical linewidth prediction [*9*] for the homogeneous electron gas originating from two pair final states, is not plotted since it gives a linewidth of 0.25 eV at $k/k_c = 1$ for Be. This width is a factor of twenty-five smaller than the observed value.

We have attempted to evaluate equation (19) to quantatively predict the observed features of the spectrum. The x at $k = 0$ in figure 27 corresponds to a computed value using equation (36) and a weak pseudo-potential [*10*] model for the band structure. A similar evaluation for the dispersion corrections due to band structure yields results which (for suitable choice of the pseudo-potential parameters) are anisotropic in qualitatively the correct way but almost an order of magnitude too small to account for the data. Using equation (36) we have also estimated the importance of core orthogonalization effects on this dispersion term and find these effects to be two orders of magnitude smaller than the observed anisotropy.

At the largest scattering angles studied, the spectrum has a rather peculiar and not understood shape. The two crystal directions for this 22° ($k = 1.55$ Å^{-1}) scattering angle are shown in figure 28 along with a plot of the homogeneous RPA

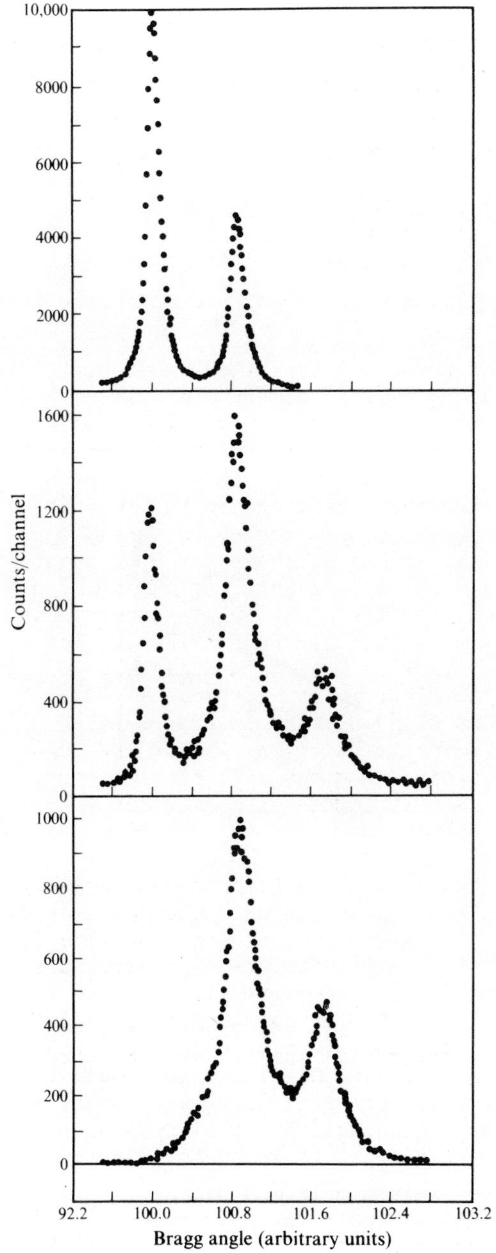

Figure 26 - Spectra obtained by sweeping double crystal spectrometer. (a) Elastically scattered Cu K_1, K_2 X-rays; (b) Be $k = 1$ Å^{-1}, $k \parallel C$; (c) Be spectrum after removal of elastic components. Plasmon loss lines from K_{α_1} and K_{α_2} input remain.

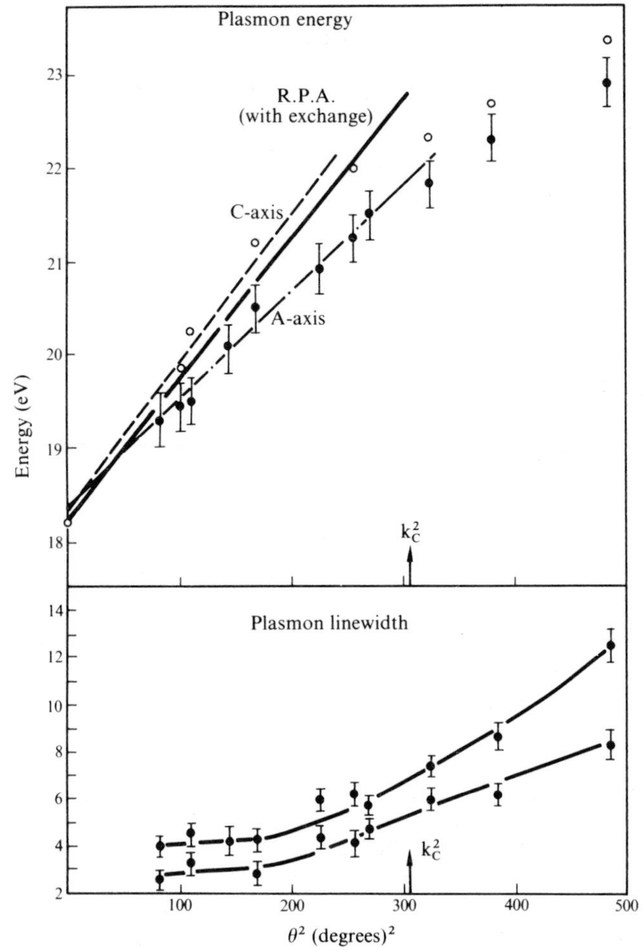

Figure 27 - Plasmon energy and linewidth for $k \| C$ - 0 and $k \| A$ - x as a function of the square of the scattering angle (θ).

result after it has been convoluted with the experimental resolution function. The two experimental spectra are centered at nearly the same energy but have different shapes. They both in turn look different from the homogeneous RPA result. While some of the features in this spectrum, no doubt depend on the details of the Be band structure others may very well be signaling a real breakdown of this simple RPA picture.

In Li the plasmon is at considerably lower energies and corresponds roughly to the curve marked (b) in figure 24. A typical experimental spectrum for Li is shown in figure 29. The compiled experimental results for the dispersion and damping of

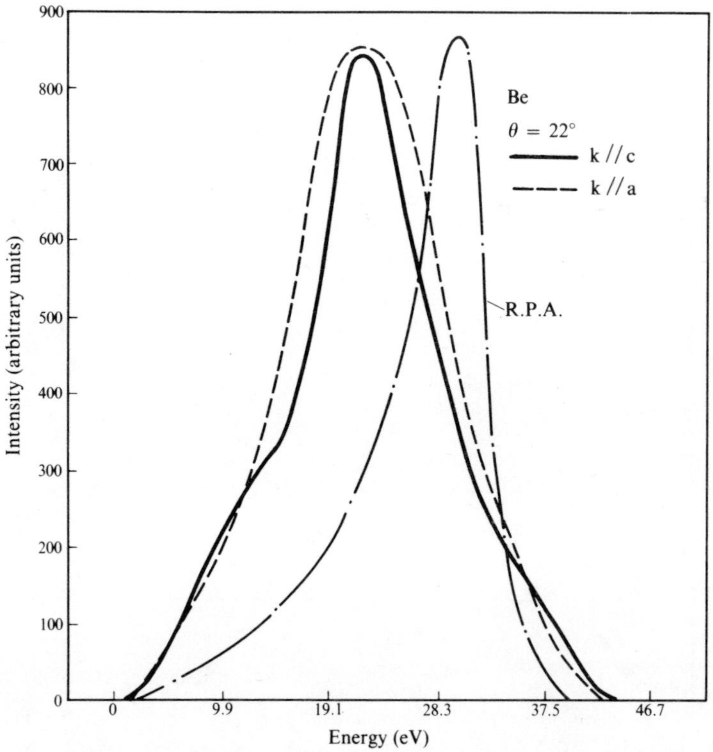

Figure 28 - K_1 spectra obtain for θ = 22° $k \parallel C$ and $k \parallel A$ after the K_2 contribution has been subtracted. They compared with the RPA reults which has been smeared with an instrumental resolution function.

the plasmon in Li is shown in figures 30-31. The plasmon frequency is depressed from its free (no band structure value) in quite reasonable agreement with the quantitative prediction in equation (36). The dispersion is considerably flatter than the predicted RPA curve shown plotted as the solid curve in figure 30. We see no experimental evidence for anisotropies. This is also in quite good agreement with equation (36) since Li is cubic.

At shorter wavelengths particularly in the range $k \sim k_f$ the spectrum as in Be lags behind the RPA results. The fact that it is similar to Be and because of the fact that the band structure is so different in this material we feel quite strongly that this lagging behind is a general feature of the spectrum in metals and is symptomatic of the breakdown of RPA as we have already discussed (see equations (31) and figure 22). In order to really assess the validity of this point of view we have begun a careful investigation of the second row of elements where band structure effects are much smaller put Coulomb effects

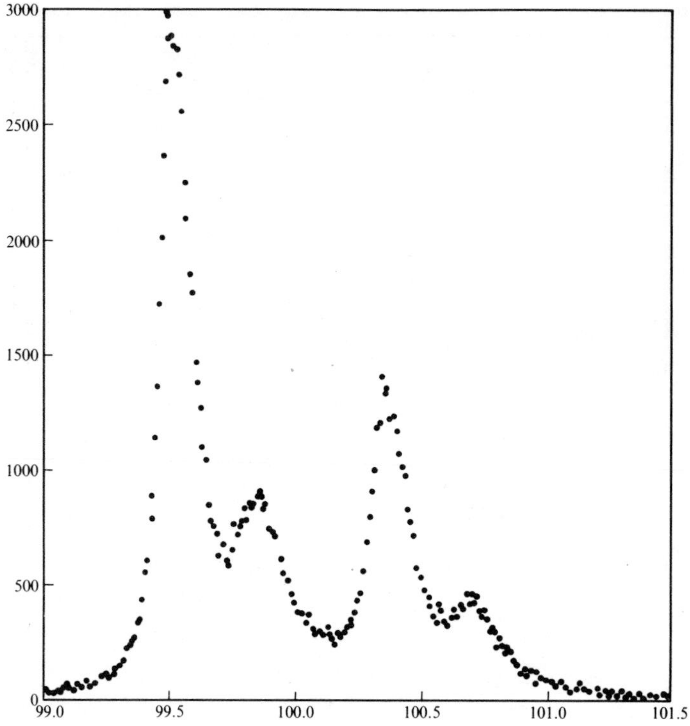

Figure 29 - Li spectrum obtained by sweeping double crystal spectrometer.

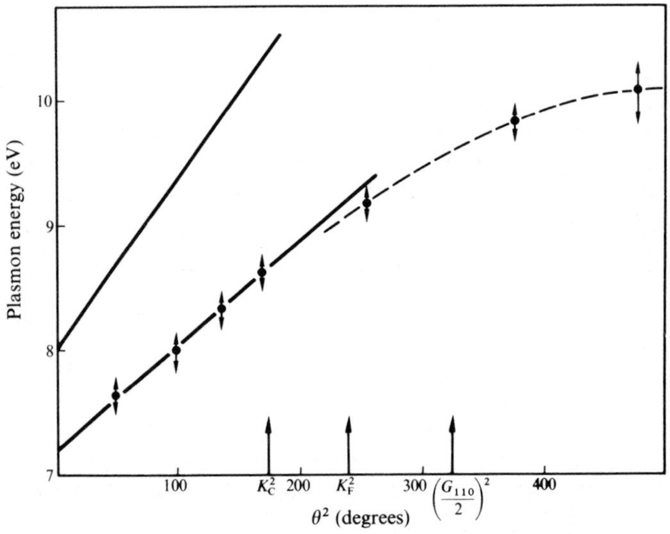

Figure 30 - Plasmon energy in Li.

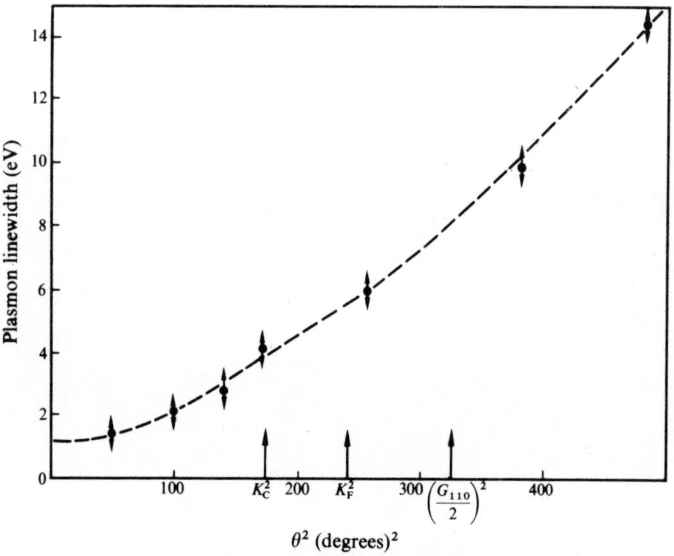

Figure 31 - Plasmon linewidth in Li.

comparable. The comparison between Li and Na and between Be and Mg should be illuminating.

REFERENCES

1. Egelstaff, P.A. (1965). *Thermal Neutron Scattering*, (ed. Egelstaff, P.A.), (Academic Press, New York).
2. Raether, H. (1965). *Solid State Excitations by Electrons, Ergebnisse der exakten Natur-Wissenchaften*, (Springer-Verlag, Berlin).
3. Wright, George B. (1969). *Light Scattering Spectra of Solids, Proceedings of the International Conference on Light Scattering Spectra of Solids*, (ed. Wright, George B.), (Springer-Verlag, New York).
4. Phillips, W.C. and Weiss, R.J. (1968). *Phys. Rev.*, **171**, 790,
5. Eisenberger, P. and Platzman, P.M. (1970). *Phys. Rev.*, **A2**, 415.
6. Eisenberger, P. (1970). *Phys. Rev.*, **A2**, 1678; (1972). *Phys. Rev.*, **5**, 628.
7. Eisenberger, P., Henneker, W.H. and Cade, P.E. (1972). *J. Chem. Phys.*, **56**, 1207.
8. Platzman, P.M. and Tzoar, N. (1965). *Phys. Rev.*, **139**, 410.
9. Luttinger, J.M. (1960). *Phys. Rev.*, **119**, 1153.
10. Eisenberger, P. Lam, L. Platzman, P.M. and Schmidt, P. (1972). *Phys. Rev.*, **B6**, 3671.
11. Lundqvist, B.I. and Lyden, C. (1971). *Phys. Rev.*, **B4**, 3360.

12. Daniel, E. and Vosko, S.H. (1960). *Phys. Rev.*, **120**, 2041.
13. Pandy, K.C. and Lam, L. (1973). *Phys. Lett.*, **A43**, 391.
14. Alexandropoulos, N.G. (1971). *J. Phys. Soc. Japan*, **31**, 1790.
15. Miliotiis, D.M. (1971). *Phys. Rev.*, **B3**, 701.
16. Eisenberger, P., Platzman, P.M. and Pandy, K.C., (to be published).
17. Platzman, P.M. and Wolff, P.A. (1973). *Waves and interactions in Solid State Plasmas*, (Academic Press, New York).
18. Kanuzawa, H., Misaww, S. and Fuirta, E. (1960). *Prog. Theor. Phys.*, **23**.
19. Dubois, D.F. and Kivelson, M.F. (1969). *Phys. Rev.*, **186**, 409.

INTRODUCTION TO COLLECTIVE EXCITATIONS IN SOLIDS

A.A. LUCAS

Institut de Physique, Université de Liège,
Liège, Belgium

and

Facultés des Sciences, Université de Namur,
Namur, Belgium

1. COLLECTIVE EXCITATIONS IN LOCAL APPROXIMATION

A GENERAL PROBLEM

An important class of collective excitations of solids are the so-called plasma oscillations i.e. collective motions that the electrons and ions as a whole can undergo either under the influence of external probes or as thermal or quantum mechanical zero-point fluctuations.

From the point of view of classical physics, plasma oscillations can be introduced as the following eigenmode problem. One starts with Maxwell's equations

$$\text{div}\,\vec{D} = 0, \tag{1}$$

$$\text{div}\,\vec{H} = 0, \tag{2}$$

$$\text{rot}\,\vec{E} + \frac{1}{c}\dot{\vec{H}} = 0, \tag{3}$$

$$\text{rot}\,\vec{H} - \frac{1}{c}\dot{\vec{D}} = 0, \tag{4}$$

governing the electromagnetic fields associated with the charge and current distributions in the absence of external sources.

These equations are complemented by a constitutive equation

$$\vec{D} = \varepsilon\vec{E}, \tag{5}$$

where in general the dielectric function ε, is an integral operator connecting the E and D fields in a nonlocal manner. In what follows, we consider mainly collective modes in the so-called *local approximation* in which $\varepsilon(\omega)$ is independent of wave-vector and in (5) reduces to a proportionality constant. The conditions of the validity of this approximation have been discussed recently by Economu [1]. Aspects of non-locality will be taken up by other speakers in this school [2].

The ω-dependence of ε is determined by the particular solid state plasma under consideration. We shall study the following two standard cases.

$$\varepsilon(\omega) = \varepsilon_\infty \frac{\omega^2 - \omega_{LO}{}^2}{\omega^2 - \omega_{TO}{}^2}, \tag{6}$$

$$\varepsilon(\omega) = 1 - \frac{\omega_p{}^2}{\omega^2}. \tag{7}$$

The form (6) applies to the lattice vibrations of a diatomic ionic crystal, where ω_{LO} and ω_{TO} are the longitudinal optical and transverse optical frequencies respectively and where ε_∞ is the high-frequency ($\omega \gg \omega_{LO}$) dielectric constant arising from the electronic polarizability. Expression (7) where $\omega_p = \surd(4\pi ne^2/m)$ is the electron plasma frequency, is appropriate to a free elctron-like metal. These dielectric functions have been written in a form which shows explicitly that in the local approximation and neglecting damping the bulk longitudinal collective modes are the zeros of the dielectric function whereas the bulk transverse modes are the poles of this function. For metals, the poles of $\varepsilon(\omega)$ represent the zero frequency shear motions corresponding to the vanishing shear modulus of the electron gas [4]. In a bulk metal, such shear collective motions are damped by multiple particle-hole excitations but we shall see that for finite samples, they give rise to well defined surface modes of non-zero frequency (see part C of section 3). The eigenmode problem is completed by the usual boundary conditions of electrodynamics to be satisfied by the fields at the surfaces and interfaces of the solid.

The complete solutions of this general problem has been discussed by Fuchs and Kliewer [5] for the case of an ionic stab, by Englman and Ruppin [6] for ionic solids of cylindrical and spherical shapes and by Economu [1] for metal-insulator interfaces in a variety of sandwich geometry. We refer the reader to these references for a more detailed study. Here, we shall examine a few of the simplest systems in order to convey the prominent aspects of the methods and solutions. A more complete exposition can also be found in Professor Mahan's lectures in this school [6'].

B SIMPLE EXAMPLES

1 One Metal-Vacuum Interface in Electrostatic Approximation

The metal occupies the semi-infinite, negative z-space as indicated in figure 1. In the electrostatic approximation ($c \to \infty$), Maxwell's equations reduce to

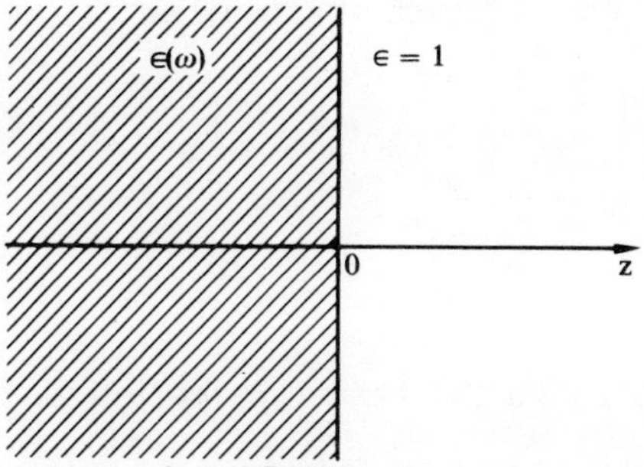

Figure 1

$$\text{rot}\ \vec{E} = 0, \tag{8}$$

$$\text{div}\ \varepsilon\vec{E} = 0. \tag{9}$$

Equation (8) implies that $\vec{E}$ derives from scalar potential V:

$$\vec{E} = -\ \text{grad}\ V. \tag{10}$$

Equation (9) can be satisfied in two ways:

$$\text{either}\ \varepsilon(\omega) = 0, \qquad \text{i.e.}\ \ \omega = \omega_p, \tag{11}$$

$$\text{or}\qquad \varepsilon(\omega) \neq 0, \qquad \text{div}\ \vec{E} = 0. \tag{12}$$

The solution $\omega = \omega_p$ is the bulk longitudinal plasmon mode and the field inside the metal can be taken as a sinewave in z to match the vanishing field outside. The condition (12) along with the equation (10) imposes to the potential V to satisfy the Laplace equation

$$\Delta V = 0, \tag{13}$$

and the corresponding solutions constitute *the surface* modes.

Translational invariance for continuous displacements parallel to the surface can be expressed by casting the potential Fourier transform

$$V(\vec{r}) = \int d\vec{k} V_k(z) e^{i\vec{k}\cdot\vec{\rho}}, \tag{14}$$

where $\vec{\rho}$ is the coordinate components of $\vec{r}$ along the surface and $\vec{k}$ is a two-dimensional real wave vector also parallel to the surface. Substitution of (14) into (13) yields

$$d^2V_k/dz^2 - k^2V_k = 0, \tag{15}$$

whose solutions are

$$V_k = Ae^{kz} + Be^{-kz}. \tag{16}$$

The arbitrary constants A, B are determined by the conditions that V should be finite everywhere and by the continuity conditions at $z = 0$:

$$\left.\frac{\partial V}{\partial \vec{\rho}}\right|_{z=-0} = \left.\frac{\partial V}{\partial \vec{\rho}}\right|_{z=+0} \quad \Rightarrow \quad A = B, \tag{17}$$

$$\varepsilon \left.\frac{\partial V}{\partial z}\right|_{z=-0} = \left.\frac{\partial V}{\partial z}\right|_{z=+0} \quad \Rightarrow \quad \varepsilon(\omega) + 1 = 0. \tag{18}$$

Introducing the form (7) into (18), one finds the solution

$$\omega = \omega_s = \frac{\omega_p}{\sqrt{2}} . \tag{19}$$

This is the frequency of the nonretarded surface plasmon mode. In the local approximation adopted here and with a step like boundary between metal and vacuum the instantaneous charge distribution associated with the plasma fluctuation is entirely concentrated at the surface $z = 0$ in the form of a sharp δ-function:

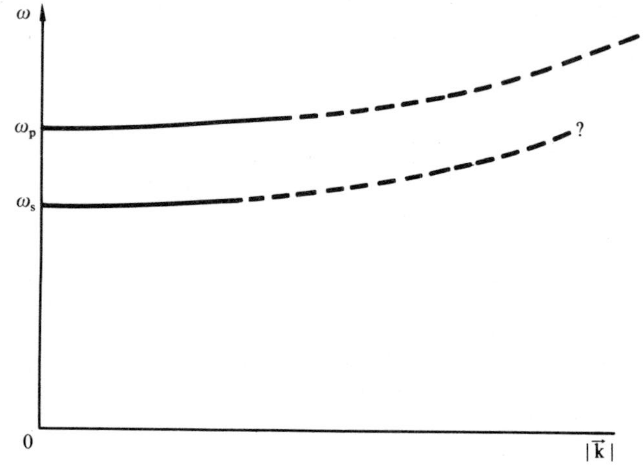

Figure 2

$$\rho_S(\vec{r}) = -\frac{1}{4\pi}\Delta V(\vec{r}) = -\frac{A}{4\pi}\delta(z)\int d\vec{k}\; k e^{i\vec{k}\cdot\vec{\rho}}. \qquad (20)$$

Each Fourier component of the potential and hence the field induced by the surface charge density fluctuation extends exponentially on both sides of the interface. The collective excitation spectrum of this simple system is illustrated in figure 2. As indicated in the figure, the dispersion (wavelength dependence) of the plasma frequencies, not included in the present simple local approximation, become important at higher values of the wave-vector, i.e. when the charge density fluctuation becomes sensitive to the atomic structure of the underlying lattice. For a study of these dispersion effects, we refer the reader to Professor March's lectures [2].

2 Optical Modes for Slabs

This case has been studied in great details in reference [5] for slabs of macroscopic sizes and more recently, de Wette and collaborators have studied the problem of very thin ionic slabs from the point of view of lattice dynamics [7].

In the local approximation, the method is similar to the one above and the excitation spectrum is implicitly contained in the equations

$$\text{Bulk modes} \qquad \varepsilon(\omega) = 0, \qquad \varepsilon(\omega) = \infty, \qquad (21)$$

$$\text{Surface modes} \qquad \frac{\varepsilon(\omega) - 1}{\varepsilon(\omega) + 1} = \pm\, e^{kd}, \qquad (22)$$

where d is the slab thickness and where one has to use form (6). There are two surface modes for each $\vec{k}$ parallel to the slab, one symmetrical (upper branch in figure 3) and one antisymmetrical (lower branch) with respect to the middle plane of the slab. The corresponding problem for a metal slab is solved by taking

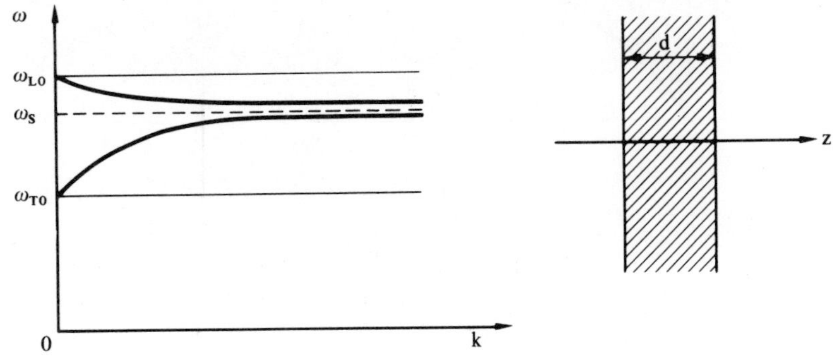

Figure 3

$\omega_{TO} = 0$ (zero-shear condition) and by identifying ω_{LO} with the bulk plasma frequency ω_p [4]. For a very thick slab, the two surface modes degenerate into a single surface frequency

$$\omega_S = \left(\frac{\varepsilon_\infty \omega_{LO}^2 + \omega_{TO}^2}{\varepsilon_\infty + 1}\right)^{\frac{1}{2}}. \quad (23)$$

When the number of boundaries increases, the number of surface mode branches increases accordingly [1] (n surface branches for n boundaries). An interesting simplification occurs when one considers a periodic array of slabs of width d separated by gaps of width l [1]. Here, a Block-Floquet theorem applies to the collective motions in the whole array. One obtains Block waves for the collective motions in the direction normal to the slab array and the boundary conditions satisfied at one interface are insured at all others by periodicity. The collective excitation spectrum for a metal-vacuum array is shown schematically in figure 4 and, as seen from this figure for each $\vec{k}$, one has a 'band structure' of surface plasmons characterized by a new wave-vector $\vec{q}$-normal to the array structure and whose band depends on the parameters d and ℓ.

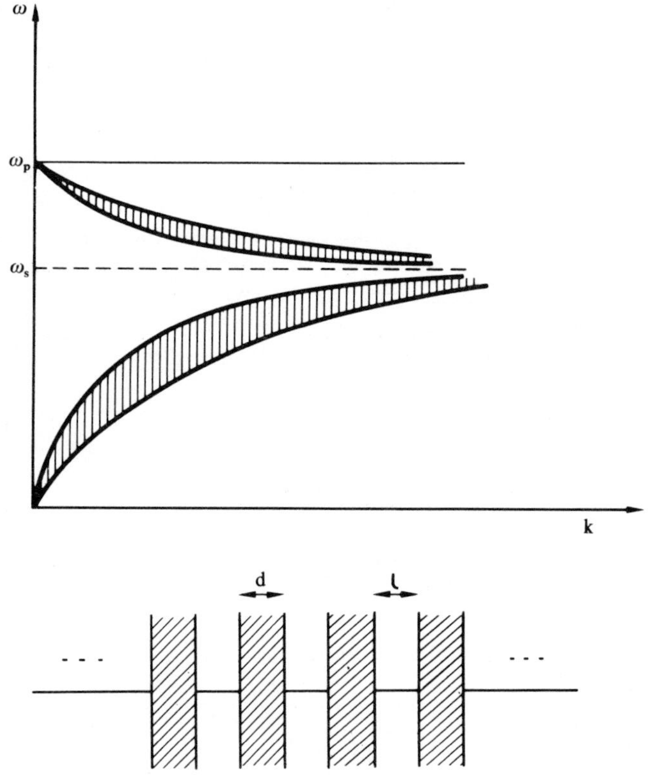

Figure 4

3. Spheres in Metals

Besides the usual bulk plasmons at ω_p, the surface plasmons in the electrostatic approximation will again result from the spherical boundary conditions applied to the solutions of Laplace equation $\Delta V = 0$ in spherical coordinates. The 'good quantum numbers' are here the angular momentum ℓ of the mode and its associated 'magnetic index' m, according to the separability [8] of Laplace's equation in spherical coordinates $\vec{r} = (r,\theta,\phi)$

$$V(\vec{r}) = \sum_{\ell,m} Y_{\ell m}(\theta,\phi)F(r), \tag{24}$$

where $Y_{\ell m}$ is a spherical harmonic. The radial function $F(r)$ satisfies the radial Laplace equation

$$\frac{1}{r^2}\frac{d}{dr}\left(r^2\frac{dF}{dr}\right) - \frac{\ell(\ell+1)}{r^2}F(r) = 0, \tag{25}$$

whose general solution is

$$F(r) = Ar^{\ell} + Br^{-(\ell+1)}. \tag{26}$$

Regularity requirement at $r = 0$ and $r = \infty$ and matching conditions at the sphere surface $r = R$ lead to the following surface plasmon spectra

$$\omega_\ell = \omega_p\left(\frac{1}{2\ell+1}\right)^{\frac{1}{2}} \qquad \text{(solid sphere in vacuum)}, \tag{27}$$

$$\omega_\ell = \omega_p\left(\frac{\ell+1}{2\ell+1}\right)^{\frac{1}{2}} \qquad \text{(empty sphere in metal)}. \tag{28}$$

These are illustrated in figure 5 along with a display of the eigenmodes of collective oscillations for the first few multipole modes.

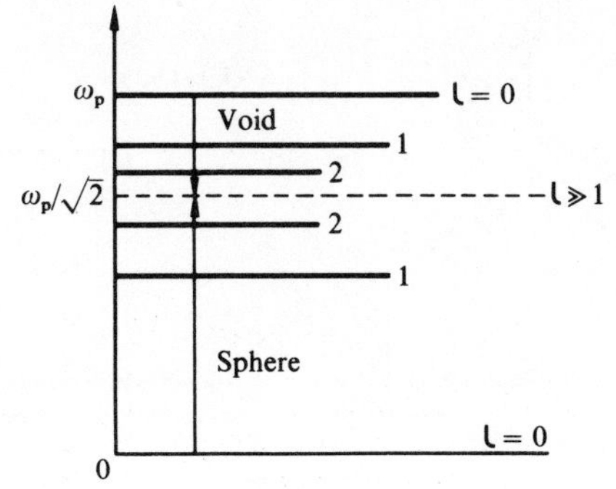

Figure 5 - (In part).

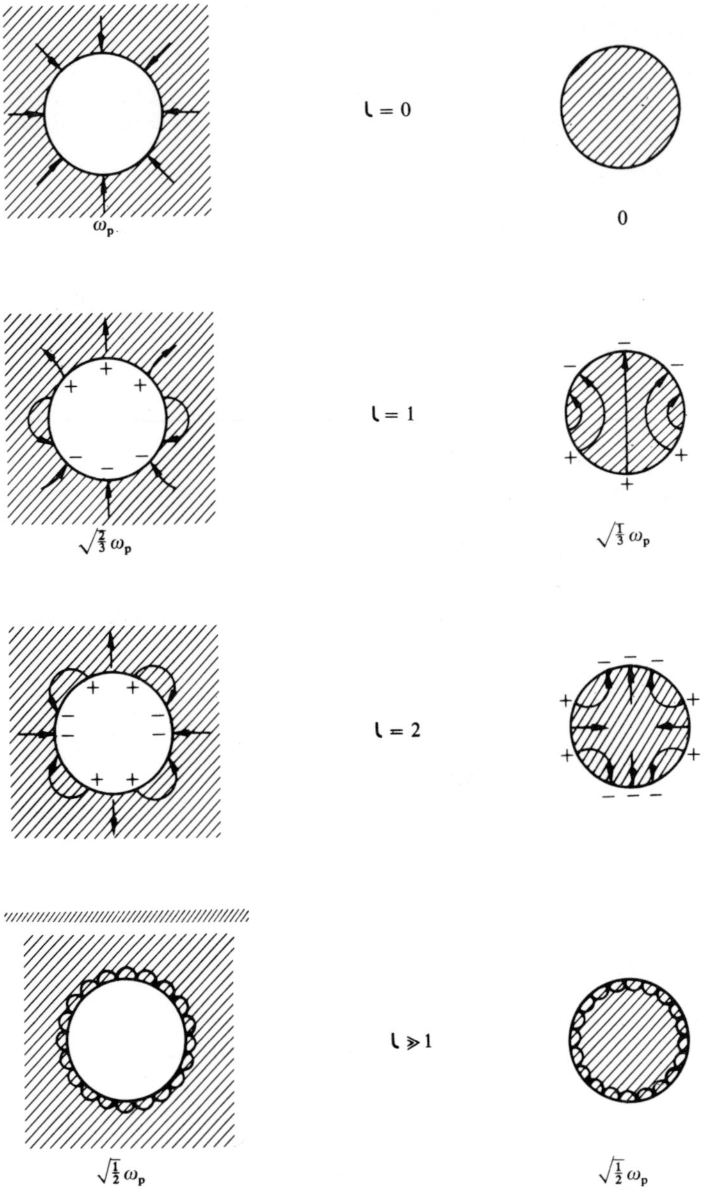

Figure 5 - (In part).

Each mode is $(2\ell + 1)$-degenerate.

For large ℓ, both sphere and void spectra converge to the frequency of the planar surface plasmon $\omega_p/\sqrt{2}$. This is expected since the high angular momentum modes become insensitive to the curvature of the surface. The existence of the dipole mode at

$\omega_p/\sqrt{3}$ for a small solid metallic particle was already implicit in Mie's calculation [*9*] of the optical absorption of small metal spheres and Fröhlich [*10*] predicted a corresponding effect in small ionic spheres. The dipole mode at $\omega_p\sqrt{(2/3)}$ characteristic of a spherical void has also been observed in electron microscope observations [*11*] of small bubbles in metals. The detailed features of these modes relevant for optical properties of solids have been extensively studied in reference [*6*].

2. COUPLING WITH ENERGETIC CHARGES

A GENERAL FORM OF INTERACTION HAMILTONIAN

In experiments using charged particles as probe of the elementary excitations of the solid state, one has a source of particles producing a beam with known intensity, energy and angular distributions. The most detailed information obtainable from an intensity measurement is then the function $I(\Omega,E,\Omega_0,E_0)$ giving the beam intensity scattered in the direction Ω and with energy E, Ω_0 and E_0 being the incident parameters.

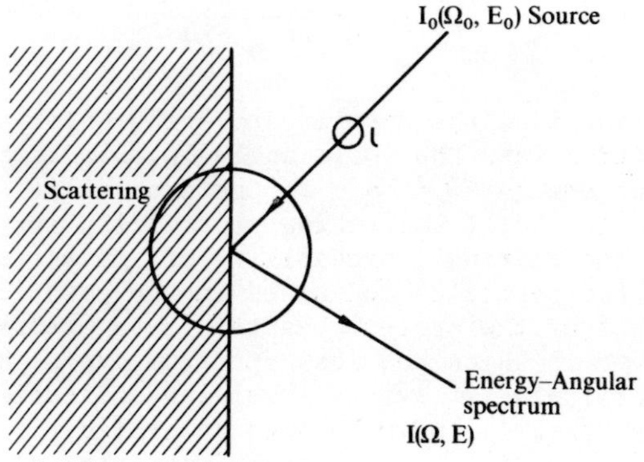

Figure 6

From this fundamental quantity, one can obtain a whole hierarchy of functions depending on the kind of detector used. Thus, one may measure the *Energy Profile* of scattered particles in a given angular domain D_Ω around the Ω direction (a diffraction spot, for instance)

$$I(E,D_\Omega) = \int_{D_\Omega} d\Omega I(\Omega,E,\Omega_0,E_0). \qquad (29)$$

A less refined measurement would provide an *Intensity Profile* in the same angular domain [*12*]:

$$I(D_\Omega,E_0) = \int_0^\infty dE I(E,D_\Omega). \qquad (30)$$

One may be satisfied with a *surface reflectivity* measurement

$$R(\Omega_0,E_0) = \int_{E_0-\Delta E}^{\infty} \mathrm{d}E_{\substack{\text{back-}\\ \text{scatt}}} \mathrm{d}\Omega I(\Omega,E,\Omega_0,E_0), \tag{31}$$

where ΔE is not too large so as to exclude the true secondary electron current.
The total *Secondary Yield* would be the total backscattered intensity

$$Y(\Omega_0,E_0) = \int_0^{\infty} \mathrm{d}E \int \mathrm{d}\Omega I(\Omega,E,\Omega_0,E_0). \tag{32}$$

In this chapter, we shall be talking about the *Loss Spectrum* which is defined as the ratio between energy profile and intensity profile:

$$P(E) = \frac{I(E,D_\Omega)}{I(D_\Omega,E_0)} . \tag{33}$$

This function gives the experimental probability for a particle to be scattered into the solid angle D_Ω, simultaneously losing an amount of energy $\hbar\omega = E_0 - E$, the probability being measured relative to the total scattering probability in this domain D_Ω. $P(E)$ gives the intrinsic probability of inelastic events (loss processes) for particles contained in the path (Ω_0,D_Ω).

The object of the present chapter is to describe an approximation scheme in which the loss spectrum can be calculated theoretically for inelastic events involving collective excitations.

The first task is to write down an explicit form for the Hamiltonian of the interfacing system (particles plus collective modes):

$$H = H_i + H_p + V_{ip}. \tag{34}$$

H_i is the 'free' particle (nonrelativistic) Hamiltonian

$$H_i = \frac{p^2}{2m} + V(\vec{r}) \tag{35}$$

where $V(\vec{r})$ may include the potential or pseudo-potential of the ions in the crystal, some external potential (an electric field for instance), etc. ... .

H_p is the Hamiltonian of the collective modes (phonons or plasmons)

$$H_p = \sum_\mu A \int \mathrm{d}\vec{k}\,\omega_{\vec{k}\mu}(a_{\vec{k}\mu}^{\dagger} a_{\vec{k}\mu} + \tfrac{1}{2}), \tag{36}$$

written in second-quantized form. A is the normalisation area of the modes and $\vec{k}$, as before, is a two-dimensional wave vector parallel to the surface. μ indicates the set of indices, other than $\vec{k}$, which are necessary for a complete specification of the mode. For volume excitations, μ may include the third component (normal to the surface) of the wave vector, a branch index (optical, acoustical branches ...), etc. ... . The only assumption entering the form (36) is that the collective modes can be described in the harmonic approximation with little or no damping. This appears to be justified for many physical situations of interest.

The interaction term V_{ip} has necessarily the form

$$V_{\text{ip}} = \sum_{\mu} A\int d\vec{k}C_{\mu}(\vec{k},z)(e^{i\vec{k}\cdot\vec{\rho}}a_{\vec{k}\mu}^{\dagger} + e^{-i\vec{k}\cdot\vec{\rho}}a_{\vec{k}\mu}), \tag{37}$$

where $\vec{r} = (\vec{\rho},z)$ are the particle coordinates parallel and normal to the surface, respectively and where $C_{\mu}(k,z)$ represents the 'coupling constant' between the charged probe and the mode $(\mu,\vec{k})$. The z-dependence of C_{μ} results from having performed only a two-dimensional Fourier transform of the interaction. The form (37) is dictated by the three assumptions that (i) V_{ip} should be Hermitian; (ii) V_{ip} is taken to be linear with respect to the oscillation amplitudes $(a,a^{\dagger})$ of the normal modes and (iii) V_{ip} is invariant for translations parallel to the scattering surface (factor $e^{i\vec{k}\cdot\vec{\rho}}$). This last condition implies that we are restricting the description to the local approximation where only small enough $\vec{k}$'s are involved. The smallest relevant wavelength depends both on the coupling function $C_{\mu}(\vec{k},z)$ and on the geometrical arrangement of the receiving spectrometer which can often be set in such a way as to collect only those particles which have been inelastically scattered at sufficiently small angles with respect to the elastic direction.

The total Hamiltonian (34) is analgous to the Fröhlich Hamiltonian discussed by other speakers in this conference. The explicit form of $C_{\mu}(k,z)$ has been worked out elsewhere [*13*]. For the surface plasmon of a semi-infinite metal, it takes the form

$$C_{\mu}(k,z) = \left(\frac{e^2\hbar\omega_s}{4\pi Ak}\right)^{\frac{1}{2}} e^{-k|z|}, \tag{38}$$

where the exponential dependence on z results from our previous considerations on the amplitude of the surface modes (see equation (16)).

B HIGH ENERGY APPROXIMATION

In writing the total Hamiltonian, we have adopted from the start the second-quantized form of the collective mode Hamiltonian whereas we have kept the configuration space for the free particle. This is because we want to treat the inelastic scattering problem by considering the collective modes as a quantized system and the charged probe as a semi-classical object. This point of view, as we shall see, naturally leads to a simple interpretation of multiple inelastic scattering processes, avoiding the somewhat artificial notion of mean free path between successive first order scattering processes (Born approximation). By semi-classical treatment of the probe motion, we mean the use of a high-energy approximation defined as follows. Assume that the unperturbed energy E_0 of the particle be much larger than its average interaction energy with the collective modes:

$$E_0 \gg \langle V_{ip} \rangle . \tag{39}$$

Moreover, assume that the de Broglie wavelength of the particle be much smaller than the wavelength of the modes relevant for the diffusion process:

$$k_0 \gg k . \tag{40}$$

Under these two conditions, it is clear that from the point of view of the collective modes, the particle provides a quasi-infinite source of energy and momentum. Then such a way that the particle can be regarded as a source of perturbation which is spatially well localized with respect to the spatial extension of the modes and whose temporal dependence is given by the elastic trajectory

$$\vec{r} = [\vec{\rho}(t), z(t)] . \tag{41}$$

This trajectory equation for the center of the wave packet representing the particle is determined by the elastic scattering phenomena (single of multiple) under the crystal or external potentials $V(\vec{r})$.
The two conditions (39,40) are realised in the two important extreme cases where the particle is either moving very rapidly (high kinetic energy $p^2/2m$, small potential energy V) or is forced to remain around a fixed position in space (small kinetic energy, high potential energy). The first case corresponds to situations such as encountered in fast electron spectroscopy (transmission or reflection) or in X-ray Photoelectron emission, while the second case finds its application e.g. in the quantum mechanical study of the image charge phenomena. These situations will be discussed here or in Professor Mahan's lectures [14].

Once the temperal dependence of $\vec{r}(t)$ is determined by the elastic events, we introduce this classical 'C-function' in the Hamiltonian for the collective modes thus obtaining

$$H = \sum_{\mu} A \int d\vec{k} \{ \hbar\omega_{\vec{k}\mu} (a_{\vec{k}\mu}^{\dagger} a_{\vec{k}\mu} + \tfrac{1}{2}) + C_{\mu}[k, z(t)] [e^{i\vec{k}\cdot\vec{\rho}(t)} a_{\vec{k}\mu}^{\dagger} + e^{-i\vec{k}\cdot\vec{\rho}(t)} a_{\vec{k}\mu}] \} + E_0 . \tag{42}$$

To simplify the notions, we set (k,μ) = i and, dropping constant terms, (42) writes

$$H = \sum_i [\hbar\omega_i a_i^\dagger a_i + f_i(t)a_i^\dagger + f_i^*(t)a_i], \tag{43}$$

where $f_i(t)$ is a known *function* of time (not an operator). H is the Hamiltonian of collection of independent harmonic oscillators, each driven by a linear time dependent force. This is a standard problem [*15*] in elementary quantum mechanics. The solution is expressed most clearly through the concept of coherent states introduced by Glauber [*16*] in the context of Quantum Optics. Before proceeding with the study of our scattering problem, we shall first give an introductory account of the Glauber states.

C GLAUBER STATES

Let us first remind us of the essential results of the quantum theory of the harmonic oscillator expressed in the representation of second quantization.

The canonically conjugated position and momentum operator (q,p) are replaced by operators $(a^\dagger,a)$ for the creation and destruction of an energy quantum $\hbar\omega_0$, by means of the linear substitution:

$$q = \left(\frac{\hbar}{2m\omega_0}\right)^{\frac{1}{2}}(a^\dagger + a), \tag{44}$$

$$p = i\left(\frac{\hbar m\omega_0}{2}\right)^{\frac{1}{2}}(a^\dagger - a). \tag{45}$$

The quantization condition is written

$$[q,p] = i\hbar, \tag{46}$$

or

$$[a,a^\dagger] = 1. \tag{47}$$

The oscillator Hamiltonian

$$H = \frac{p^2}{2m} + \tfrac{1}{2}m\omega_0^2 q^2 \tag{48}$$

$$= \hbar\omega_0(a^\dagger a + \tfrac{1}{2}) \tag{49}$$

has the eigenvalue equation

$$H|n\rangle = \hbar\omega_0(n + \tfrac{1}{2})|n\rangle, \tag{50}$$

where the non-negative integer n specifies the number of quanta of energy $\hbar\omega_0$ present in the excited eigenstate $|n\rangle$. The latter can be constructed from the ground state, defined as

$$a|0\rangle = 0, \tag{51}$$

by successive creation of n excitation quanta

$$|n\rangle = \frac{(a^\dagger)^n}{n!}|0\rangle, \tag{52}$$

where $\sqrt{n!}$ is the normalisation factor such that $\langle n|n\rangle = 1$. The operators $a, a^\dagger$ are not diagonal in this energy representation. Their matrix elements result from the properties

$$a|n\rangle = \sqrt{n}|n-1\rangle \tag{53}$$

and

$$a^\dagger|n\rangle = \sqrt{n+1}|n+1\rangle. \tag{54}$$

The set of eigenvectors $\{|n\rangle\}$ constitute a complete orthonormal set. Such is the energy representation in second quantization.

We are now in a position to introduce the coherent states. A general state $|\alpha\rangle$ of the oscillator can always be expressed as a linear combination of the basis states $|n\rangle$:

$$|\alpha\rangle = \sum_{n=0} C_n(\alpha)|n\rangle. \tag{55}$$

By definition, the coherent states are the linear combinations (55) satisfying the relation

$$a|\alpha\rangle = \alpha|\alpha\rangle \tag{56}$$

where α is any complex number. In other words, the coherent states are the eigenvectors of the 'half-amplitude' α. The complex number α is called coherent amplitude or coherent displacement. One easily finds the coefficients $C_n(\alpha)$ of (55) to be:

$$|\alpha\rangle = e^{-\frac{1}{2}|\alpha|^2}\sum_{n=0}^{\infty}\frac{\alpha^n}{\sqrt{n!}}|n\rangle, \tag{57}$$

where the norm, left arbitrary by the definition (56), has been chosen so that the coherent states be normalized:

$$\langle\alpha|\alpha\rangle = 1. \tag{58}$$

Since a does not commute with H, a coherent state does not have a well defined energy. The probability to find the excited state

$|n\rangle$, when an energy measurement is performed on the oscillator, is given by the number

$$P_n(\alpha) = |\langle n|\alpha\rangle|^2 = \frac{|\alpha|^{2n}}{n!} e^{-|\alpha|^2}. \tag{59}$$

If such a measurement is repeated a large number of times on an oscillator always prepared in the coherent state $|\alpha\rangle$, one will find a *Poisson distribution* (59) of the only possible results which are the quantized energy levels $(n + \frac{1}{2})\hbar\omega_0$. With the relation (58), this distribution is normalized

$$\sum_{n=0}^{\infty} P_n(\alpha) = 1. \tag{60}$$

The average energy of an oscillator prepared in the coherent state $|\alpha\rangle$ is given by the first order moment of the distribution, namely

$$\langle\alpha|H|\alpha\rangle = \sum_n \hbar\omega_0(n + \tfrac{1}{2})P_n(\alpha) = \hbar\omega_0(|\alpha|^2 + \tfrac{1}{2}). \tag{61}$$

One will notice that the coherent state $|\alpha\rangle$ is not an eigenstate of the creation operator $a^\dagger$ (otherwise an $a^\dagger$ would commute and q and p could be measured simultaneously). However, one has the dual relation

$$\langle\alpha|a^\dagger = \alpha^*\langle\alpha|. \tag{62}$$

When $\alpha = 0$ in the relation (56) one recovers the definition (51) of the oscillator ground state which is the only coherent state with a well defined energy.

The set of the coherent states $\{|\alpha\rangle\}$ can be shown to have the property of completeness or closure but it is not an orthogonal basis set. We refer to Glauber's paper [*16*] for more details on these questions.

A fundamental property of the coherent states, which could also serve to define them, is that they can be generated by applying to the ground state a unitary 'displacement' operator $D(\alpha)$:

$$|\alpha\rangle = D(\alpha)|0\rangle, \tag{63}$$

where

$$D(\alpha) = e^{\alpha a^\dagger - \alpha^* a}. \tag{64}$$

The identity of (63) and (57) can be demonstrated by using Baker Haussdorf formula

$$e^{A+B} = e^{A}e^{B}e^{-\frac{1}{2}[A,B]}, \tag{65}$$

(applicable when $[A,B]$ is a c-number) and expanding the exponential operators in power series.

That $D(\alpha)$ is a displacement operator is clearly seen when it is expressed in terms of q and p:

$$D(\alpha) = e^{-i(q_0p-p_0q)/\hbar} = e^{iq_0p/\hbar}e^{-ip_0q/\hbar}e^{-iq_0p_0/2\hbar} \tag{65}$$

where

$$q_0 = \left(\frac{2\hbar}{m\omega_0}\right)^{\frac{1}{2}}\mathrm{Re}\alpha, \tag{66}$$

$$p_0 = (2m\hbar\omega_0)^{\frac{1}{2}}\mathrm{Im}\alpha. \tag{67}$$

Expression (65) is indeed the product (up to a phase factor) of the unitary operators for displacements of amount q_0 and p_0 in the q and p spaces respectively. Hence, one has

$$D^{\dagger}(\alpha)qD(\alpha) = q + q_0, \tag{68}$$

$$D^{\dagger}(\alpha)pD(\alpha) = p + p_0, \tag{69}$$

and, from (44,45)

$$D^{\dagger}(\alpha)aD(\alpha) = a + \alpha, \tag{70}$$

$$D^{\dagger}(\alpha)a^{\dagger}D(\alpha) = a^{\dagger} + \alpha^{*}. \tag{71}$$

When α is real, $|\alpha\rangle$ represents the quantum state corresponding to an oscillator at *rest* but displaced from its equilibrium position by an amount q_0. This is clearly seen when one examines the wavefunction in the $\{q\}$ representation. Writing equation (63) in this representation, one has

$$\Psi(q,\alpha) \equiv \langle q|\alpha\rangle = \langle q + q_0|0\rangle = \left(\frac{m\omega_0}{\pi\hbar}\right)^{\frac{1}{4}}e^{-m\omega_0(q+q_0)^2/2\hbar} \tag{72}$$

i.e. the ground state Gaussian eigenfunction displaced by the quantity q_0.

Finally, we also mention the additivity property of successive coherent displacements. This property, common to quantum coherent amplitudes and classical amplitudes, is expressed by the theorem

$$D(\beta)D(\alpha) = D(\alpha + \beta)e^{\frac{1}{2}(\beta\alpha^{*}-\beta^{*}\alpha)}, \tag{73}$$

which can be proved by using the definition (64) and the formula (65). The phase factor in (73) has no other meaning than to set the phase of the state with superposed coherent amplitudes

$$|\alpha + \beta\rangle = D(\alpha + \beta)|0\rangle, \tag{74}$$

with respect to the phase of the state obtained by successive coherent displacements

$$|\alpha + \beta\rangle' = D(\beta)D(\alpha)|0\rangle. \tag{75}$$

Since a coherent state results from a displacement performed on the ground state or any other coherent state, the most general way to prepare an oscillator in such a state is to act on it with a linear perturbation in the a, $a^\dagger$:

$$H = \hbar\omega_0(a^\dagger a + \tfrac{1}{2}) + f(t)a^\dagger + f^*(t)a, \tag{76}$$

where $f(t)$ is a function depending on time t.

If $f(t)$ is time independent, one verifies immediately that the coherent displacement

$$a' = a + \frac{f}{\hbar\omega_0}, \tag{77}$$

transforms H to the diagonal form

$$H' = \hbar\omega_0(a'^\dagger a' + \tfrac{1}{2}) - \frac{|f|^2}{\hbar\omega_0}, \tag{78}$$

whose ground state

$$|0\rangle' = D(f)|0\rangle = |f\rangle, \tag{79}$$

is precisely the coherent state of amplitude $f/\hbar\omega_0$.
If $f(t)$ explicitly depends on time, the problem (76) can still be solved exactly. In the interaction representation of the unperturbed Hamiltonian H_0, the Schrödinger equation is written

$$i\hbar \frac{d}{dt}|\Psi\rangle = e^{iH_0t/\hbar}(fa^\dagger + f^*a)e^{-iH_0t/\hbar}|\Psi\rangle, \tag{80}$$

or

$$i\hbar \frac{d}{dt}|\Psi\rangle = (fe^{i\omega_0t}a^\dagger + f^*e^{-i\omega_0t}a)|\Psi\rangle. \tag{81}$$

As solution, one finds (some care must be exercised in the integration since the factor acting on $|\Psi\rangle$ is an operator)

$$|\Psi(t)\rangle = e^{-i[I(t)a^\dagger + I^*(t)a]/\hbar}|\Psi(t_0)\rangle, \tag{82}$$

where $|\Psi(t_0)\rangle$ is the initial state of the oscillator when the perturbation was turned on, and where

$$I(t) = \int_{t_0}^{t} f(t')e^{i\omega_0 t'}dt'. \tag{83}$$

One sees from (82) that if the initial state is coherent (the ground state for instance), it will remain coherent for all time since the evolution operator is in the present case a displacement operator. The coherent amplitude $I(t)$ changes with time, while the perturbation is acting, and eventually reaches an asymptotic value

$$I = \int_{-\infty}^{+\infty} f(t')e^{i\omega_0 t'}dt', \tag{84}$$

i.e. the Fourier transform of the perturbing function at the oscillator frequency.

Recall that in case an emergency measurement is performed on the oscillator at time $t = +\infty$, one will find the result $(n+\frac{1}{2})\hbar\omega_0$ with the Poisson probability distribution

$$P_n = \frac{1}{n!}\frac{|I|^{2n}}{\hbar^{2n}}e^{-|I|^2/\hbar^2}, \tag{85}$$

and the average excitation energy obtained over a large number of identical measurements will be, according to equation (61),

$$W = \frac{|I|^2}{\hbar^2}\hbar\omega_0. \tag{86}$$

It is interesting to compare these results with the prediction of the purely classical theory of the same problem. Assume to simplify that $f(t)$ be real. In this case the classical Hamiltonian is

$$H_d = \frac{p^2}{2m} + \frac{1}{2}m\omega_0^2q^2 + F(t)q, \tag{87}$$

where

$$F(t) = \frac{2m\omega_0}{\hbar}f(t). \tag{88}$$

The measurable quantity is the work done by the perturbation on the oscillator, or the total energy transfer. One finds

$$W_{c\ell} = \frac{|F(\omega_0)|^2}{2m} ,$$

where

$$F(\omega_0) = \int_{-\infty}^{+\infty} F(t')e^{i\omega_0 t'}dt', \tag{89}$$

results identical to (86) and (84) respectively.
One sees that the classical theory allows to compute correctly the average energy transfered to the oscillator from the perturbation source, i.e. the first order moment of the Poisson distribution of possible energies. It is however important to remark that the very existence of such a distribution and more particularly its discrete character find their origin in the quantization of the energy levels and possess no classical analogue. In other words, whereas classically an arbitrary and continuous amount of energy $W_{c\ell}$ can be transfered to the oscillator, quantum mechanically, the transfer can only take place by discrete quanta $\hbar\omega_0$.

D ENERGY LOSS SPECTRA

Returning now to equation (42) or (43) and comparing with (76), we see that in the limit of long times, each oscillator mode will find itself prepared in a coherent state by the passage of the external charged particle. The final step of our high energy scattering method consists in a straightforward application of energy conservation.
If there were only one oscillator mode, its energy distribution (i.e. the probability to find it in a given energy state) would furnish at the same time the loss spectrum of the scattered particles (i.e., the probability $P(\hbar\omega)$ defined in equation (33) for the particle to lose the energy $\hbar\omega$) since any excitation energy provided to the oscillator must be subtracted from the total energy of the particle. The generalization of this result to the present case where the energy lost by the particle can be distributed arbitrarily to the entire collection of independent modes $(\vec{k},\mu)$, does not pose special difficulties. Since energy conservation is a global property for the whole system, it is convenient to express it by a single Fourier transform of the loss spectrum $P(\omega)$:

$$P(\omega) = \frac{1}{2\pi}\int_{-\infty}^{+\infty} d\tau e^{i\omega\tau}P(\tau), \tag{90}$$

$$P(\tau) = P_0 \exp\left(\sum_{\mu} A\int d\vec{k} \, \frac{|I_\mu(k)|^2}{\hbar^2} \, e^{-i\omega_{k\mu}\tau}\right) , \tag{91}$$

$$P_0 = \exp\left[-\sum_\mu A\int d\vec{k}\,\frac{|I_\mu(k)|^2}{\hbar^2}\right] , \qquad (92)$$

$$I_\mu(k) = \int_{-\infty}^{+\infty} dt' C_\mu[k,z(t')]e^{i\vec{k}\cdot\vec{\rho}(t')}e^{i\omega_{k\mu}t'} . \qquad (93)$$

Expanding the exponential in (91) in powers of the coherent amplitudes $I_\mu(k)$ and the integrating (90) term by term, one recovers a distribution similar to (59). One verifies that the spectrum (90) is normalized

$$\int_{-\infty}^{+\infty} d\omega P(\omega) = 1, \qquad (94)$$

and that the average energy lost by a large number of particles following the trajectory (41) is given by

$$W = \int_{-\infty}^{+\infty} d\omega\hbar\omega P(\omega) = \sum_\mu A\int d\vec{k}\hbar\omega_{\mu k}\,\frac{|I_\mu(k)|^2}{\hbar^2} , \qquad (95)$$

in agreement with the result (86).

P_0 represents the probability for the particle following the trajectory (41) to escape any loss of energy. This is not to be confused with the probability that the particle effectively follows the trajectory (41), a quantity which depends explicitly on the elastic phenomena.

Two cases of Poisson distribution observed experimentally are sketched in figures 7,8 and have been discussed in details elsewhere [*17*]. Some considerations to these experimental date are also given in Professor Mahan's lectures [*14*].

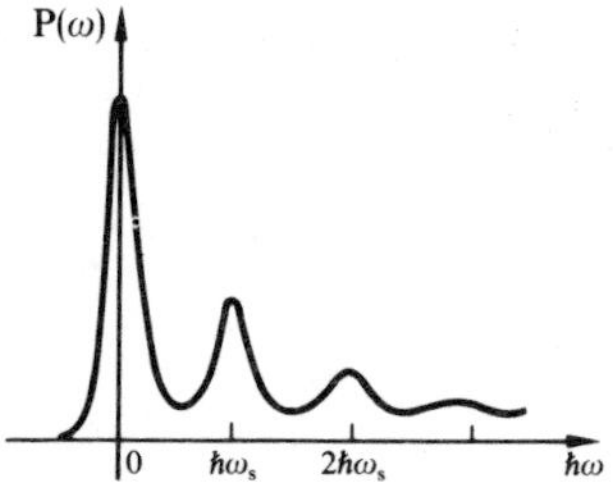

Figure 7

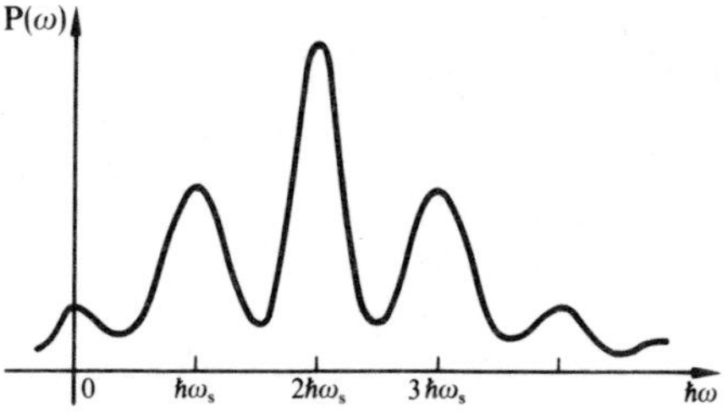

Figure 8

3. APPLICATIONS TO MACROSCOPIC SURFACE PROPERTIES

A Image Charge Theory

As pointed out by Mahan [*14*], several groups of workers have recently studied the problem of screening of external charges by metal surfaces from the point of view of virtual excitations of surface plasmons. We refer the reader to Professor Mahan's lectures for a detailed account of the theory.

B ADHESION THEORY

The long range van der Waals attraction between two neutral atoms is well understood since the work of London in 1930 [*19*]. Similar forces also exist between submacroscopic or macroscopic solid particles. They can be computed from the purely macroscopic point of view provided by the set of Maxwell's equations. The interaction takes place via the quantum fluctuations of the overlapping electromagnetic field of the two bodies at short separation. The theory was given by Lifshitz [*20*] in 1955 and the resulting adhesion energy is expressed as a functional of the frequency dependent dielectric function of the material.

In 1968, Van Kampen et al. [*21*] published an important paper in which they showed that the Lifshitz formula for adhesion forces between two semi-infinite dielectric media can be derived from the consideration of surface collective excitations alone.

The reasoning goes as follows. Consider the case of the two semi-infinite metals separated by a gap of width d. If d is much smaller than the plasma wavelength $\lambda_p = 2\pi c/\omega_p$, one can neglect the retardation in the propagation of the fields of plasma fluctuations from one medium to the other. From Maxwell's equations, the eigenfrequencies of surface mode oscillations turn out to be given by

$$\omega_\pm(k,d) = \frac{\omega_p}{\sqrt{2}}\left(1 \pm e^{-kd}\right)^{\frac{1}{2}}, \tag{95}$$

which are the explicit solutions of (22) when using the dielectric functions (7). The adhesion energy is then simply given by the shift in zero-point energies of the surface plasmons on going from d finite to $d = \infty$:

$$W(d) = \frac{1}{(2\pi)^2}\int d^2k\tfrac{1}{2}\hbar[\omega_+(d) + \omega_-(d) - 2\omega_s]. \tag{96}$$

Some algebraic manipulations then lead from this formula to the Lifshitz' result [*21*].

The explicit dependence on d is immediately obtained from (96) by the change of integration variable $y = kd$

$$W(d) = \frac{1}{d^2}\frac{\omega_p}{4\pi\sqrt{2}}\int_0^\infty y\,dy((1 + e^{-y})^{\frac{1}{2}} + (1 - e^{-y})^{\frac{1}{2}} - 2), \tag{97}$$

where the y-integral is now a number. This is the celebrated d^{-2} powerlaw, valid when $d \ll \lambda_p$. The bulk modes do not contribute to the adhesion property due to the fact that the fields of their quantum fluctuations do not leak out of the surfaces [6′] into the gap. Notice that in the y-integral of (97), there is a natural cutoff of the integration range at about $y \simeq 1$ due to the exponential convergence to zero of the integrand. This means that for the adhesion problem, the only relevant surface modes are those whose wave-vector k is no greater than $1/d$. When d is much larger than atomic distances, this justifies the use of the local approximation to describe the collective plasma motions.

We shall further illustrate these fundamental ideas by considering the adhesion problem in spherical geometry. Elaborate theories along the lines of the Lifshitz method have been developed [22] for the interacting of two spherical particles and recently, Langbein [23] has given a general perturbation expansion as a function of the ratio R/D of the sphere radius R to their distance D. The case of two identical metal spheres in vacuum such that $R,D \ll \lambda_p$ (no retardation) is particularly straightforward. The treatment given here is in fact entirely similar to the original derivation of London [19] for point harmonic oscillator models of molecules. Each sphere, when isolated, is capable of substaining surface plasma oscillations which can be classified as multipole modes (see equation (27) and figure 5). According to equation (26) the multipole mode of order ℓ induces potential fluctuations in vacuum which decay like $r^{-(\ell+1)}$ with distance from the sphere center. If the ratio R/D is small enough, one can limit ourselves to the consideration of the lowest order modes which for solid spheres are of dipolar nature ($\ell = 1$).
Each sphere 'sees' the other as a fluctuating dipole and their fluctuations dipole moment $\vec{q}_1$, $\vec{q}_2$ coupled through the ordinary dipole-dipole interaction. The Hamiltonian for the coupled modes thus has the simple form

$$H = \frac{1}{2R^3\omega_1{}^2}\left(\dot{\vec{q}}_1{}^2 + \omega_1{}^2\vec{q}_1{}^2\right) + \frac{1}{2R^3\omega_1{}^2}\left(\dot{\vec{q}}_2{}^2 + \omega_1{}^2\vec{q}_2{}^2\right)$$

$$+ \tfrac{1}{2}\vec{q}_1\cdot\frac{E - 3\vec{D}^0\vec{D}^0}{D^3}\cdot\vec{q}_2, \qquad (98)$$

where R^3 is the static polarizability of each sphere [24] and $\omega_1 = \omega_p/\sqrt{3}$ is their dipole plasma frequency (equation (27) with $\ell = 1$).

Equation (98) is a normal mode problem. The 6 eigenmodes are listed in figure 9. Each new coupled mode has a bonding or anti-bonding character for the two spheres according to whether its frequency is lower or higher than the unperturbed value ω_1. The symmetric modes Ω_1, Ω_3, Ω_5 are antibonding, whereas the antisymmetric ones Ω_2, Ω_4, Ω_6 are bonding. The frequencies are found to be

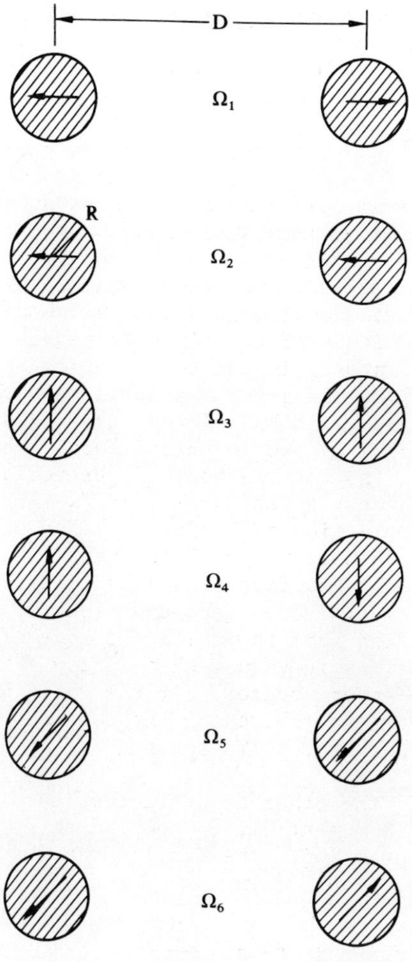

Figure 9

$$\Omega_{\substack{1\\2}} = \omega_1\left(1 \pm 2\,\frac{R^3}{D^3}\right) , \tag{99}$$

$$\Omega_{\substack{3\\4}} = \Omega_{\substack{5\\6}} = \omega_1\left(1 \pm \frac{R^3}{D^3}\right) . \tag{100}$$

The adhesion energy is again given by the shift in zero-point energies of the modes when D goes from a finite value to infinity:

$$W(D) = \tfrac{1}{2}\hbar \sum_{i=1}^{6} [\Omega_i(D) - \Omega_i(\infty)]. \tag{101}$$

Expanding the square roots in powers of R^3/D^3, the first non-vanishing term is of order R^6/D^6:

$$W(D) \simeq -\tfrac{3}{4}\,\hbar\omega_1 \frac{R^6}{D^6}\,, \qquad (102)$$

i.e. London's formula [*19*] in which the atomic polarizability is replaced by the metal sphere polarizability R^3. There is no point in retaining higher order terms in R^3/D^3 since other interactions of similar order such as dipole-quadrupole etc. ... have not been included in the initial Hamiltonian. The full series is given by Langbein [*23*].

C PLASMON COHESION OF METAL SURFACES

In a recent publication [*25*], the authors have shown that a major part of the surface energy of metals resides in the zero point energy of surface plasmons characteristic of the interface. That is, the energy required to create a unit area of new macroscopic surface is spent mostly in the creation of surface plasmons and the destruction of an equivalent number of bulk plasmons. The precise redistribution of collective modes into surface and bulk modes when the crystal is split, is governed by what has been called the 'Begrenzung (or confinement) sum rule', a rule first encountered in the not unrelated context of electron spectroscopy of collective excitations in thin films [*26*].

In this discussion, we shall first elucidate the physical content of the Begrenzung sum rule in terms appropriate to the present surface energy problem. Second, the prediction of the theory will be compared with experiment for more than fifty metallic elements on a single diagram clearly showing the universality of the $r_s^{-5/2}$ dependence of surface energies on the density parameter r_s. Third, we shall briefly discuss the connection between the plasmon surface cohesion and the long range adhesion forces between macroscopic bodies.

As a gas, or better as a liquid continuum, the valence electrons of a metal as a whole are susceptible to experience longitudinal or transverse collective motions compatible with their mutual Coulomb repulsion and their attraction by the ion positive background. For each wave vector $\vec{k}$, there is one longitudinal pressure or density wave of frequency ω_p (the zero of the frequency-dependent dielectric function $\varepsilon(\omega) = 1 - \omega_p^2/\omega^2$) and also two degenerate, transverse shear waves of zero frequency (the double pole of ε). These modes are illustrated in figure 10a and figures 11a,b, respectively. The waves corresponding to all possible orientations and values of $\vec{k}$ form a complete set with which a general collective motion of bulk plasma continuum can be represented in a Fourier analysis.

A similar situation exists for several other physical systems such as an elastic continuum [*27*] (one longitudinal, compressional and two transverse shear sound waves) or a diatomic ionic crystal [*28*] (one longitudinal optical phonon and two, degenerate transverse optical phonons) with, of course, different dispersion relations. However, in a real, nonsuperfluid liquid, a pure shear plane wave would rapidly be spoiled into vortices which in turn would quickly damp out by single-particle excitations but, for our problem, damping is unconsequential because we shall not consider real excitations but only ground state properties.

If a bulk sample is to be cut into two along a plane, the only relevant modes are clearly those with $\vec{k}$ parallel to the future cut (see figures 10,11). Indeed one can always assume, as a first approximation, that the modes with a non-vanishing component k_z of $\vec{k}$ along the normal to the cut will simply undergo specular reflection on the new surfaces and, apart from an eventual space quantization of k_z, they will remain true, undisturbed bulk modes.

If a gap of finite width d is introduced by a cut, one immediately sees from figure 10a that the initial bulk mode at ω_p, if it were to subsist undisturbed, would tend to induce surface charges symmetrically opposed to each other on each face of the gap. Therefore, this mode helps to separate further the two half-spaces. Equivalently, one can say that the restoring forces acting on the plasma electrons of the bulk are reduced by the fields of the new surface charge distribution superposed to the initial field of the density wave. This results in a smaller plasma frequency $\omega_+ < \omega_p$ and hence a negative zero-point energy expanditure.

Considering figure 11c on the other hand, it is seen that the p-polarized shear plasmon, whose flow lines are perpendicular to the cut, if it were to subsist, would tend to accumulate an antisymmetrical distribution of surface charges, resulting in a tendency to oppose the separation of the two solids. Said otherwise the electron plasma initially flowing freely in zero field (the only transverse solution of div $\varepsilon\vec{E}$ = rot $\vec{E} = 0$), is now subjected to the restoring field of accumulated surface charges which leads to a nonzero plasma frequency $\omega_- > \omega_T = 0$ and hence a positive zero-point energy expanditure. Finally, figure 11d shows that the S-polarized shear plasmon whose flow lines are parallel to the cut will not generate any surface charges and hence no energy is required or released from this mode which remains a pure (heavily damped) bulk mode of zero frequency.

In summary, starting from three bulk plasmons $(\omega_p,\omega_T,\omega_T)$ for each $\vec{k}$ parallel to the cut, one finds three 'surface' plasmons $(\omega_+,\omega_-,\omega_T)$ which are, respectively, antibonding, bonding and neutral for the two half-solids. This exhausts the Begrenzung sum rule.

The values of the frequencies $\omega_\pm(d) = \omega_p(1 \pm e^{-kd})^{\frac{1}{2}}/\sqrt{2}$ of the surface modes are easily obtained from electrostatics [*29*]. The actual pattern of flow lines of these modes is shown in figure 12. When d is finite, the energy to produce the gap is a creation energy while the residual work required to separate the two half-solids is an adhesion energy. The surface energy is the total shift

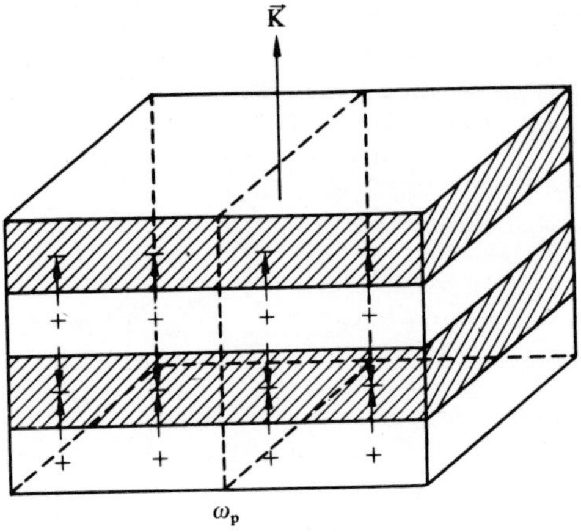

(a)

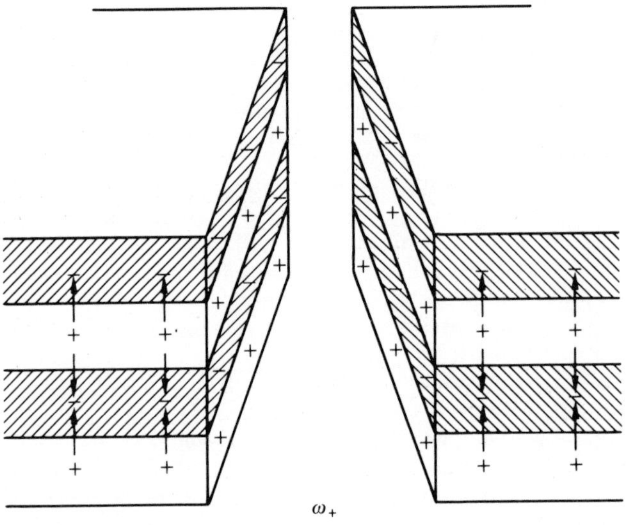

(b)

Figure 10 - (a) Representation of the charge distribution during a longitudinal bulk plasmon oscillation of wave vector $\vec{k}$. The crystal is supposed to be infinite in all directions and the external boundaries in this and the following figures are only meant to help the representation of the mode pattern. The broken line gives the location of the future planar cut. (b) The same mode, after the cut. Charges accumulate symmetrically, as shown on the faces of the gap. These fluctuating surface charges lower the plasma frequency ($\omega_+ < \omega_p$) and repel each other. The resulting surface mode ω_+ is antibonding.

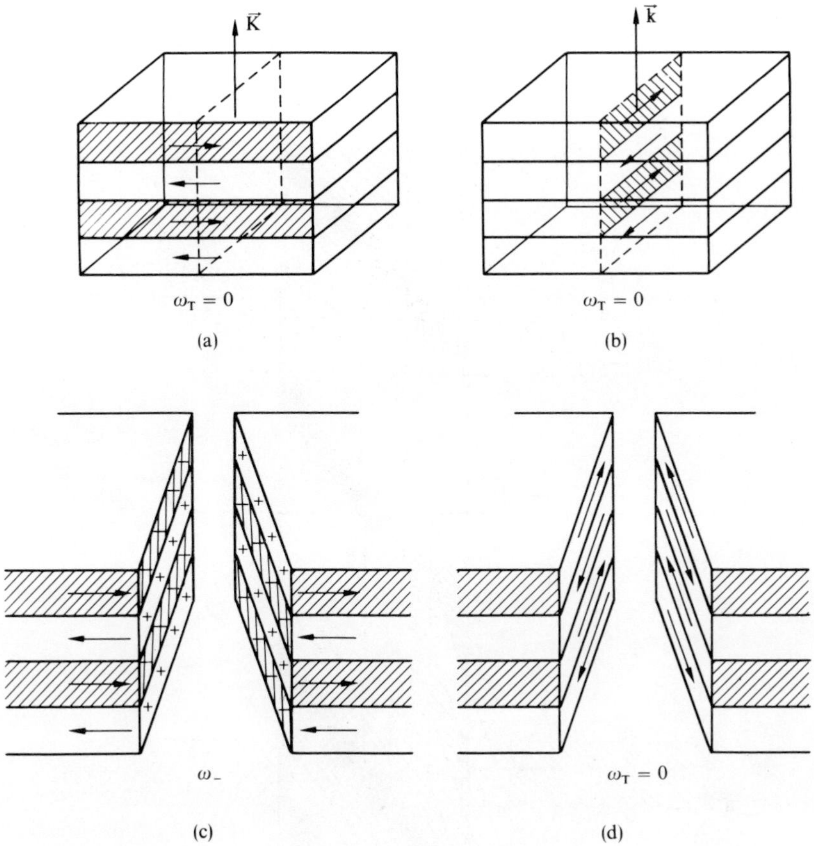

Figure 11 - (a) Representation of the flow pattern of the p-polarized shear mode (the polarization refers here to the plane of the future cut). As a liquid of flow viscosity, the electron plasma can be shared without energy ($\omega_T = 0$). (b) Same as in figure 11a for the S-polarized shear mode. (c) Antisymmetrical accumulation of charges on the faces of the gap induced by the p-polarized shear plasmon. The resulting surface mode ω_- is bonding. (d) The S-polarized shear plasmon leaves the gap faces neutral and therefore does not contribute to the binding or surface energy.

$$\sigma_{LR} = \frac{1}{2}\frac{1}{(2\pi)^2}\int d\vec{k}\tfrac{1}{2}\hbar(\omega_+ + \omega_- - \omega_p) \qquad \text{(creation)}$$

$$+ \frac{1}{2}\frac{1}{(2\pi)^2}\int d\vec{k}\tfrac{1}{2}\hbar[2\omega_s - (\omega_+ + \omega_-)] \qquad \text{(adhesion)} \qquad (103)$$

$$= \frac{1}{2}\frac{1}{(2\pi)^2}\int d\vec{k}\tfrac{1}{2}\hbar(2\omega_s - \omega_p) = Cr_s^{-5/2}, \qquad (104)$$

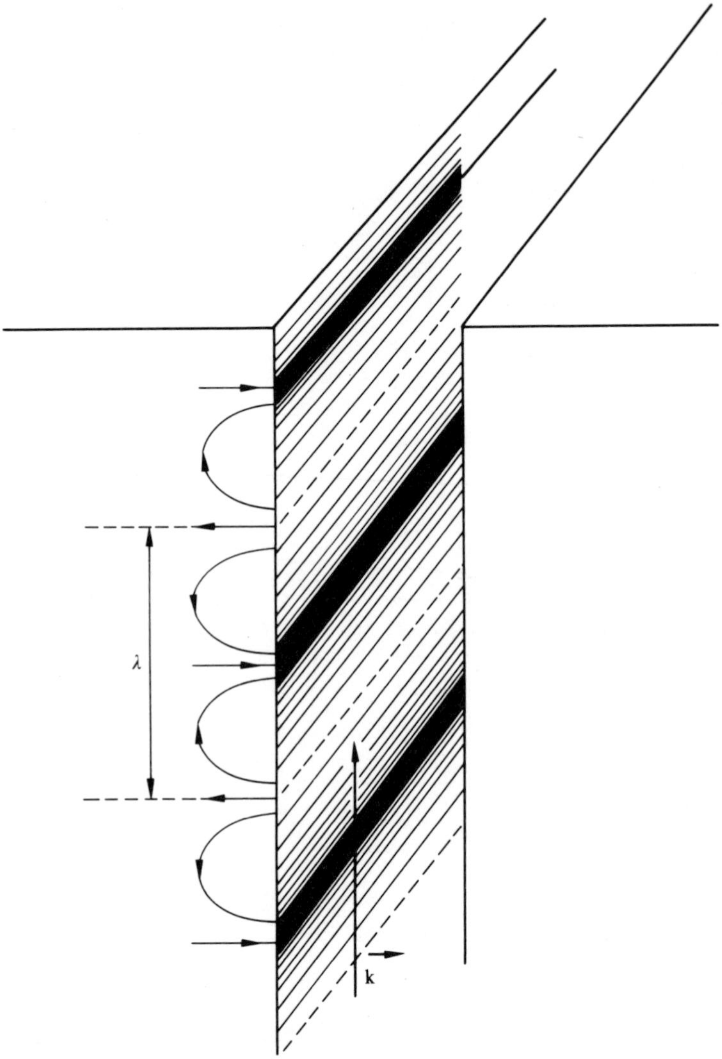

Figure 12 - Flow lines of the actual surface plasmons of wave vector $\vec{k}$. Fluctuations of density occur only at the surface by electron excursion into bulk. The instantaneous electric field decays exponentially as e^{-kz} on both sides of the surface (when the gap width d is much larger than k^{-1}).

where C is a universal constant, r_s is the usual electron density parameter and $\omega_s = \omega_p/\sqrt{2}$.

In figure 13, this $r_s^{-5/2}$ law is compared with experimental surface energies of more than 50 metals [*30*] (the latter should be affected by a vertical error bar of ±25%. One notices that the alkali metals (Li, Na, K, Rb, Cs, Fr) and the alkaline earths

(Be, Mg, Ca, Sr, Ba, Ra) already provide a group of 12 'simple' metals which fit rather closely the predicted power law over a wide range of r_s values ($1.7 < r_s < 6$). This fact provides sufficient confidence in the proposed mechanism of surface cohesion to attempt an extension of the model to transition and other metals, as discussed in detail in reference [*25*].

The correctness of the numerical proportionality factor in (104) has been checked by comparison with well known adhesion formulas [*31*,*32*]. This is illustrated in figure 14 for Al. One finds that the inverse square law d^{-2} of adhesion holds true down to quite small distances ($d \gtrsim 2.5$ Å). However, for smaller d, the interaction remains finite due to the finite value of the cutoff wave vector k_c, a feature also found in the image charge problem [*33*,*34*]. One notices the stiffness of the ω_- plasmon bonding for very small separation: the gap tends 'to snap' with infinite force as $d \to 0$ since $W(d) \sim d^{\frac{1}{2}}$, d referring here to a gap in the continuous positive background (see figure 14). This of course results from the neglect of the probability of tunnelling and electron exchange through the gap (or electron tailing into vacuum) by having implicitly assumed an infinite potential barrier at the surface of the jellium. One should emphasize however that the calculated plasmon cohesion merely utilizes a conservation property of degrees of freedom (the Begrenzung sum rule) and does not depend on the detailed forces involved in the cutting process (see the cancellation of intermediate, d-dependent terms in (1)).

In conclusion, the doubly degenerate surface plasmon branch characteristic of a large planar empty gap between two semi-infinite metals (or, equivalently, of a thick metal slab in vacuum) can be thought of as the vestige of two formerly bulk plasmon branches which have been destroyed by splitting the crystal (or extracting the slab). The zero-point energy shift provides a surprisingly good estimate of the observed surface energy of most metals.

REFERENCES

1. Economu, E.N. (1969). *Phys. Rev.*, **182**, 539.
2. March, N. (1974). (This conference).
3. Born, M. and Huang, K. (1954). *Dynamical Theory of Crystal Lattices*, (Clarendon Press, Oxford), p. 83.
4. Kittel, C. (1967). *Introduction to Solid State Physics*, (Wiley, New York).
5. Fuchs, R. and Kliewer, K.L. (1965). *Phys. Rev.*, **140**, 2076.
6. Englman, R. and Ruppin, R. (1970). *Rep. Prog. Phys.*, **33**, 149.
6'. Mahan, G.D. (1974). (This conference).
7. De Wette, F. (1974). (This conference); see also Chen, T.S., Alldredge, G.P., de Wette, F.W. and Allen, R.E. (1971). *Phys. Rev. Lett.*, **26**, 1543. Important progress in the study of phonon modes in micro-crystals have been made by several groups, see, e.g., Martin, T.P. (1973). *Phys. Rev.*, **7**, 3906, and references therein.

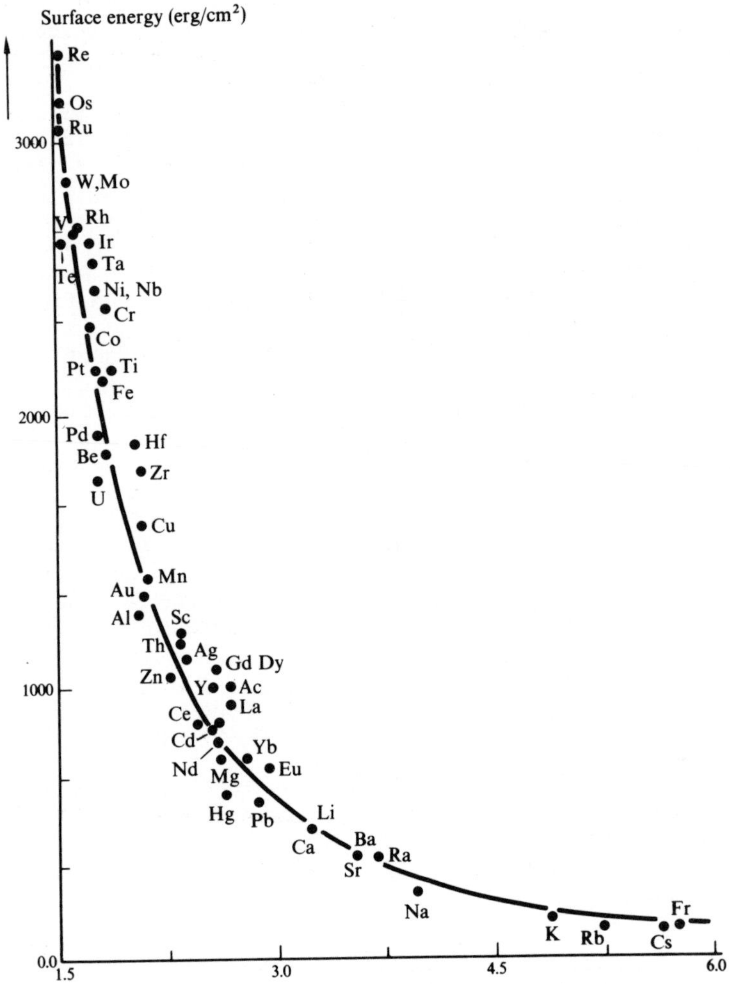

Figure 13 - Plot of the universal power law (104) giving the surface energy as a function of density parameter r_s. The experimental points have been obtained by extrapolation to zero temperature of the surface tension of the liquid metals around their melting points, using available temperature coefficients.

8. Morse, P. and Feshbach, H. (1953). *Methods of Theoretical Physics, Vol. II*, (Mc Graw-Hill, New York), p. 1264.
9. Mie, G. (1908). *Ann. Phys.*, **25**, 377.
10. Fröhlich, H. (1948). *Theory of Dielectrics*, (Oxford V.P.).
11. Henoc, P. and Henry, L. (1970). *J. de Phys.*, **31**, C1-55.

$$\frac{W(d)}{|W(0)|} = \frac{2}{2-\sqrt{2}} \cdot \frac{1}{k_c^2} \int_0^{k_c} [\sqrt{1+e^{-kd}} + \sqrt{1-e^{-kd}} - 2]\, k\, dk$$

where $k_c(\text{Å}^{-1}) = 0.334[\hbar\omega_p (\text{eV})]^{1/3}$
and $\hbar\omega_p = 15$ (eV) (Al)

$$\frac{W'(d)}{|W(0)|} = \frac{K\frac{\hbar\omega_p}{d^2}}{|W(0)|}$$

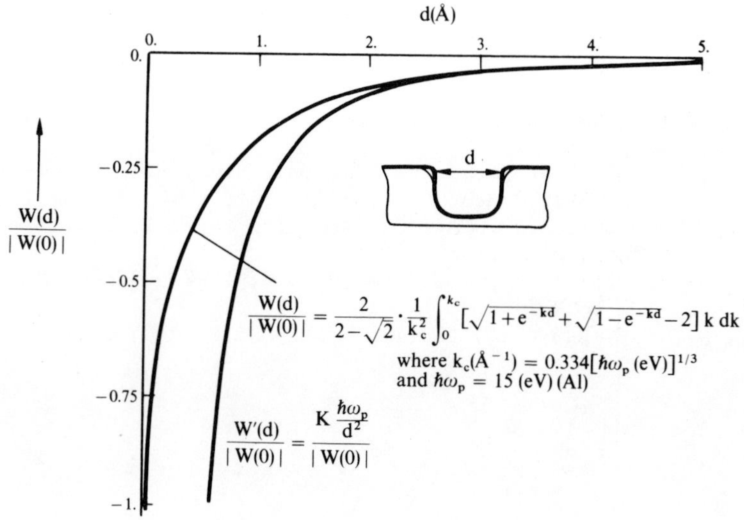

Figure 14 - Comparison between the adhesion law $W'(d) \sim d^{-2}$ and the actual plasmon binding energy $W(d)$ of two semi-infinite metals (Al) as a function of separation d. If $d \gg k_c^{-1}$, the power law d^{-2} holds true. For $d < k_c^{-1}$ the plasmon binding behaves like $d^{\frac{1}{2}}$. Short range repulsion from electron tunnelling and exchange across the gap will smooth out the small-d behavior.

12. In recent LEED literature, intensity profile may refer to the intensity of quasi-elastically scattered beam as a function of incident energy E_0.

13. Lucas, A.A., Kartheuser, E. and Badro, R. (1970). *Phys. Rev.*, **B2**, 2488; Sunjić, M. and Lucas, A.A. (1971). *Phys. Rev.*, **B3**, 719.

14. Mahan, G.D. (1974). (This conference).

15. Goldman, I.I. et al. (1960). *Problems in Quantum Mechanics*, (Infosearch, London), p.103.

16. Glauber, R.J. (1969). In *Quantum Optics, Proceedings International School 'Enrico Fermi', Varenna, Italy*, (Academic Press, New York), p. 15.

17. Lucas, A.A. and Sunjić, M. (1972). *Prog. Surf. Sci.*, **2**, 75.

18. Lucas, A.A. and Sunjić, M. (1972). *Surf. Sci.*, **32**, 439.

19. London, F. (1930). *Z. Phys. Chem.*, **11**, 222.

20. Lifshitz, E.M. (1956). *Sov. Phys. JETP*, **2**, 73.

21. Van Kampen, N.G., Nijboer, B.R.A. and Schram, K. (1968).*Phys. Lett.*, **A26**, 307.

22. Imura, H. and Okans, K. (1973). *J. Chem. Phys.*, **58**, 2763.

23. Langbein, D. (1971). *J. Phys. Chem. Solids*, **32**, 1657, and references therein.

24. Jackson, J.D. (1962). *Classical Electrodynamics*, (Wiley, New York), p. 115.

25. Schmit, J. and Lucas, A.A. (1972). *Solid State Commun.*, **11**, 415, 419.
26. See Lucas, A.A. et al. [*17*].
27. See Born, M. and Huang, K. *loc. cit.*.
28. See Born, M. and Huang, K. *Loc. cit.*.
29. See Economu, E.N. [*1*].
30. Allen, B.C. (1972). *Liquid Metals*, (ed. Beer, S.Z.), (Marcel Dekker, New York); Jones, H. (1972). *Metal Sci.*, *J.*, **5**, 15.
31. Krupp, H. (1967). *Adv. Colloid Interface Sci.*, **1**, 111.
32. See Van Kampen, N.G. et al. [*21*].
33. Sunjić, M., Toulouse, G. and Lucas, A.A. (1972). *Solid State Commun.*, **11**, 1629.
34. Ray, R. and Mahan, G.D. (1972). *Phys. Lett.*, **A42**, 301.

PLASMONS IN INHOMOGENEOUS ELECTRON GASES

N.H. MARCH

Department of Physics, Imperial College of Science and Technology, South Kensington, London, England

1. INTRODUCTION

In these lectures, a discussion will be given of collective modes in an inhomogeneous electron gas.

Particular attention will be directed to the propagation of plasmons in periodic metal lattices which has recently been discussed by March and Tosi (1972). As an application of the theory, the energy losses of fast electrons fired into a nearly-free-electron metal will be considered, and the possibility of experimental tests of the theory briefly examined.

The full quantum-mechanical discussion will be motivated by starting out from the semi-classical theory of collective oscillations in an electron gas, due to Bloch. Though Bloch's theory, as we shall see, has one rather serious limitation, nevertheless the major features of the theory are regained in the full quantum-mechanical treatment.

Some comments will also be made on the application of these theories to inhomogeneous electron gases whose density is not periodic; in particular to the electron gas near a metal surface.

2. HOMOGENEOUS ELECTRON GAS

It will be useful to begin the discussion of inhomogeneous gases by referring to the plasma oscillations which can occur in a uniform electron gas, due to the long-range Coulomb interactions. Though by now these are quite well understood, the many-electron problem posed by this Sommerfeld or jellium model is not exactly soluble. Nevertheless, in the high density limit, in which the mean interelectronic spacing tends to zero, the problem is susceptible to accurate solution, using the so-called random phase approximation.

Evidently, since approximations are necessary in jellium, the inhomogeneous problem cannot be solved exactly, and throughout the lectures we shall be returning to the homogeneous gas, to illustrate the methods used in the spatially varying electronic cloud.

In jellium, let us suppose that we make some displacement of the charge density of the electrons, which diminishes the electron density in a given region. The excess positive charge thereby created locally will cause electrons to rush into this region, to screen the positive charge. These electrons will overshoot, be drawn back again, and will proceed to oscillate with a frequency given by

$$\omega_p^2 = \frac{4\pi\rho_0 e^2}{m}, \tag{2.1}$$

ρ_0 being the uniform electron density. This plasma frequency is a high frequency in a metal, typically 10^{16} vibrations per second.

Let us express the above description of plasma oscillations quantitatively by writing down an equation of motion for the density change

$$\rho_1(\mathrm{R},t) = \rho(\mathrm{R},t) - \rho_0, \tag{2.2}$$

$\rho(\mathbf{R},t)$ being evidently the total electron density at position R at time t. Clearly we must have

$$\frac{\partial^2\rho_1(\mathrm{R},t)}{\partial t^2} = -\,\omega_p^2\rho_1(\mathrm{R},t). \tag{2.3}$$

Such a theory is correct, in jellium, in the long wavelength limit, that is at zero wave vector **k**. At long wavelengths, that is small **k**, the theory still can be expressed by a simple modification of equation (2.3), which is worth writing down, namely

$$\frac{\partial^2\rho_1(\mathrm{R},t)}{\partial t^2} + \omega_p^2\rho_1(\mathrm{R},t) = \alpha\nabla^2\rho_1(\mathrm{R},t). \tag{2.4}$$

If we now express $\rho_1(\mathrm{R},t)$ in the form

$$\rho_1(\mathrm{R},t) = n_{\mathbf{k}}\exp(i\mathbf{k}\cdot\mathrm{R})\exp(i\omega t), \tag{2.5}$$

then we see immediately that the relationship between ω and k, or the dispersion relation of the plasmons, is

$$\omega^2 = \omega_p{}^2 + \alpha k^2. \tag{2.6}$$

In the random phase approximation, it is worth recording the result that

$$\alpha = \frac{3p}{m\rho} = \frac{2\langle T\rangle}{m}, \tag{2.7}$$

where $\langle T\rangle$ is the mean kinetic energy per particle and p is the pressure. Incidentally, the same result holds for a classical perfect gas, namely

$$\alpha = \frac{3}{m\beta} \qquad (p = \rho/\beta;\ \beta = 1/k_BT).$$

Having introduced the problem by referring to the homogeneous gas, let us now turn to the approach pioneered by Bloch (1933).

3. BLOCH'S HYDRODYNAMIC THEORY

Bloch, in his pioneering work, considered the possible modes of oscillation of an inhomogeneous electron gas. He obtained the equation of motion of the gas from an action principle, by writing

$$\delta\int_{t_1}^{t_2} L dt = 0, \tag{3.1}$$

with

$$L = m\int \rho \frac{\partial w}{\partial t}\, d\mathbf{r} - H. \tag{3.2}$$

As before, ρ is the number of electrons per unit volume, while w is the velocity potential. This is related to the velocity $\mathbf{v}$ by

$$\mathbf{v} = -\ \mathrm{grad} w. \tag{3.3}$$

H is the energy of the electron gas in motion, and if we work in the semi-classical Thomas-Fermi-Dirac approximation, we may write

$$H = \tfrac{1}{2}m\int \rho(\mathrm{grad} w)^2 d\mathbf{r} - \int (V_N + \tfrac{1}{2}V_{el} + V_{ext})e\rho d\mathbf{r}$$
$$+ C_k\int \rho^{5/3} d\mathbf{r} - C_e\int \rho^{4/3} d\mathbf{r}, \tag{3.4}$$

C_k and C_e being the usual constant representing respectively the Fermi kinetic energy and the exchange energy. V_N is the potential of any nuclei or background positive charge, V_{el} is the electrostatic potential created by the electronic cloud while V_{ext} denotes the external time-dependent potential. Using the variational principle we find (cf. Lamb (1907))

$$m\frac{\partial w}{\partial t} = \tfrac{1}{2}m(\mathrm{grad}w)^2 - (V_{total} + V_{ext})e$$
$$+\frac{5}{3}C_k\rho^{2/3} - \frac{4}{3}C_e\rho^{2/3}, \qquad (3.5)$$

and

$$\frac{\partial \rho}{\partial t} = \mathrm{div}(\rho\,\mathrm{grad}w). \qquad (3.6)$$

If we put into equation (3.5) the expression for the pressure of the electron gas as a function of density, which in the Thomas-Fermi-Dirac approximation has the free electron form

$$p = \frac{2}{3}C_k\rho^{5/3} - \frac{1}{3}C_e\rho^{4/3}, \qquad (3.7)$$

then the equation of motion (3.5) can be expressed as

$$m\frac{\partial w}{\partial t} = \tfrac{1}{2}m(\mathrm{grad}w)^2 - (V_N + V_{el} + V_{ext})e + \int\frac{dp}{\rho}. \qquad (3.8)$$

Equation (3.6) is, of course, simply the continuity equation

$$\mathrm{div}\vec{j} = -\frac{\partial \rho}{\partial t}. \qquad (3.9)$$

Equations (3.8) and (3.6) represent the basis of Bloch's hydrodynamic theory.

We can now proceed to obtain an equation of motion for the 'displaced charge' $\rho_1(R,t)$, from equation (2.2), representing oscillations around the ground-state density $\rho_0(R)$, which is now, of course, a spatially varying density. The basic assumption to be made is that $|\rho_1| \ll \rho_0$, enabling the theory to be linearized for small density oscillations about the ground state density ρ_0.

It then turns out that, by differentiating equation (3.6) with respect to t, to obtain an equation for $\partial^2\rho_1/\partial t^2$, we can eliminate $\partial w/\partial t$ using equation (3.5) and obtain an equation of motion which generalizes equation (2.3) in two essential respects:

(i) $\omega_p^2 \to 4\pi e^2\rho_0(R)/m$, i.e. a 'local' plasma frequency appears;

(ii) Terms depending on grad$\rho_0(R)$ and on gradients of the potentials now appear.

We shall not write these terms out explicitly as it will become very clear what they are, from the quantal treatment which we discuss below.

Bloch's theory of collective oscillations is very appealing because it can be posed, in essence, in terms of the ground-state charge density $\rho_0(R)$.

4. QUANTUM-MECHANICAL THEORY

Thus, by putting together the equation of continuity and the semi-classical equation of motion of the electron gas, we can obtain an equation for $\partial^2\rho_1(R,t)/\partial t^2$.

But we want to turn now to discuss at a more fundamental level the way to treat plasma oscillations in a spatially varying electronic cloud, such as we have in a periodic metal crystal. One trouble with Bloch's theory, it turns out, is that it fails to give a quantitative result for α in equation (2.4) when applied to the homogeneous gas. His theory, while correct in order of magnitude, gets α wrong by a factor of about 2 and it is necessary to understand where this quantitative discrepancy arises.

Generalizing the work of Singwi et al. (1970) for a homogeneous gas, we can write down explicitly an equation of motion for the electron density $\rho(R,t)$. It will be convenient again to consider the gas in an external potential $V_{ext}(R,t) \equiv V_e(R,t)$ and then the equation of motion takes the form

$$\frac{\partial^2\rho(R,t)}{\partial t^2} - \frac{1}{m}\nabla_\alpha\nabla_\beta\Pi_{\alpha\beta}(R,t)$$

$$-\frac{1}{m}\int d\mathbf{x}\nabla_\alpha^{(R)}\{\rho(R,t)\nabla_\alpha^{(R)}V_{xc}(R,x,t)\}$$

$$= \frac{1}{m}\nabla_\alpha\{\rho(R,t)\nabla_\alpha V_H(R,t)\}. \qquad (4.1)$$

In equation (4.1), $\Pi_{\alpha\beta}(R,t)$ is the momentum flux tensor, which can be best expressed in terms of the Wigner distribution function or mixed density matrix as

$$\Pi_{\alpha\beta}(R,t) = \sum_\sigma \int \frac{d\mathbf{p}}{(2\pi)^3}\frac{p_\alpha p_\beta}{m} f_\sigma(\mathbf{p},R,t)$$

$$= -\frac{\hbar^2}{m}\nabla_\alpha^{(\mathbf{r})}\nabla_\beta^{(\mathbf{r})}\rho(R+\tfrac{1}{2}\mathbf{r},R-\tfrac{1}{2}\mathbf{r},t)\Big|_{\mathbf{r}=0}, \qquad (4.2)$$

the second line of this equation simply rewriting $\Pi_{\alpha\beta}$ in terms of the single-particle density matrix $\rho(\mathrm{R} + \frac{1}{2}\mathbf{r},\mathrm{R} - \frac{1}{2}\mathbf{r},t)$.

$V_H(\mathbf{R},t)$ in equation (4.1) is the Hartree potential

$$V_H(\mathrm{R},t) = V_e(\mathrm{R},t) + V_N(\mathrm{R}) + \int d\mathbf{x}\phi(\mathrm{R} - \mathbf{x})\rho(\mathbf{x},t), \qquad (4.3)$$

where $V_N(\mathrm{R})$ is the potential of the ions as before while $\phi(\mathrm{R} - \mathbf{x})$ is the Coulomb potential. Finally, $V_{xc}(\mathrm{R},\mathbf{x},t)$ is the exchange and correlation potential defined by

$$\nabla_\alpha^{(\mathrm{R})} V_{xc}(\mathrm{R},\mathbf{x},t) = \nabla_\alpha \phi(\mathrm{R} - \mathbf{x}) \frac{\langle \rho(\mathrm{R},t)\rho(\mathrm{R},t)\rangle_c}{\rho(\mathrm{R},t)}, \qquad (4.4)$$

where we have written

$$\langle \rho(\mathrm{R},t)\rho(\mathbf{x},t)\rangle_c = \langle \rho(\mathrm{R},t)\rho(\mathbf{x},t)\rangle - \rho(\mathrm{R},t)\rho(\mathbf{x},t). \qquad (4.5)$$

If again we write

$$\rho(\mathrm{R},t) = \rho_0(\mathrm{R}) + \rho_1(\mathrm{R},t), \qquad (4.6)$$

where $\rho_0(\mathrm{R})$ is the ground-state periodic density and $\rho_1(\mathrm{R},t)$ is its distortion produced by the external field, then the plasmon eigenmodes are obtained by linearizing the above equation with respect to ρ_1 and seeking the non-vanishing solutions when the external potential is put to zero.

As we shall see below, the linearization of the Hartree term in equation (4.1) can be carried out exactly. However, the momentum flux tensor term and the exchange and correlation potential term can only be treated at present by approximate procedures which we shall describe below.

4.1 LINEARIZATION OF HARTREE TERM

Considering the terms linear in ρ_1 and writing

$$V_i(\mathrm{R}) = V_N(\mathrm{R}) + \int d\mathbf{x}\phi(\mathrm{R} - \mathbf{x})\rho_0(\mathbf{x}), \qquad (4.7)$$

and

$$\begin{aligned} V_1(\mathrm{R},t) &= \int d\mathbf{x}\phi(\mathrm{R} - \mathbf{x})\rho_1(\mathbf{x},t), \\ \nabla^2 V_1(\mathrm{R},t) &= -4\pi e^2 \rho_1(\mathrm{R},t); \end{aligned} \qquad (4.8)$$

we find for the Hartree term

$$\frac{1}{m}\sum_\alpha \nabla_\alpha\{\rho(\mathrm{R},t)\nabla_\alpha V_H(\mathrm{R},t)\} \rightarrow \qquad \text{(Contd)}$$

$$\rightarrow \frac{1}{m} \sum_{\alpha} \nabla_\alpha \{\rho_1(R,t) \nabla_\alpha V_i(R)\}$$

$$+ \frac{1}{m} \sum_{\alpha} \nabla_\alpha \{\rho_0(R) \nabla_\alpha [V_e(R,t) + V_1(R,t)]\}$$

$$= \frac{1}{m} \rho_1(R,t) \nabla^2 V_i(R) + \frac{1}{m} \nabla \rho_1(R,t) \cdot \nabla V_i(R)$$

$$+ \frac{1}{m} \rho_0(R) \nabla^2 [V_e(R,t) + V_1(R,t)]$$

$$+ \frac{1}{m} \nabla \rho_0(R) \cdot \nabla [V_e(R,t) + V_1(R,t)]. \qquad (4.9)$$

Using the relation between V_1 and ρ_1, and the corresponding relation between V_e and ρ_e, it follows that

$$\frac{1}{m} \sum_{\alpha} \nabla_\alpha \{\rho(R,t) \nabla_\alpha V_H(R,t)\}$$

$$\rightarrow - \frac{4\pi e^2}{m} \rho_0(R)[\rho_e(R,t) + \rho_1(R,t)]$$

$$+ \frac{1}{m} \nabla \rho_0(R) \cdot \nabla [V_e(R,t) + V_1(R,t)]$$

$$+ \frac{1}{m} \rho_1(R,t) \nabla^2 V_i(R) + \frac{1}{m} \nabla \rho_1(R,t) \cdot \nabla V_i(R). \qquad (4.10)$$

At this point, we utilize the Euler equation determining the ground-state density $\rho_0(R)$. If the single-particle kinetic energy density is denoted by $t_R[\rho]$ and the exchange correlation energy density is similarly written as $\varepsilon_R^{xc}[\rho]$ then we have

$$\frac{\delta t_R[\rho]}{\delta \rho} + V_i(R) + \frac{\delta \varepsilon_R^{xc}[\rho]}{\delta \rho} = \mu, \qquad (4.11)$$

where μ is the chemical potential.

As an example of the above, Bloch's theory, based on the Thomas-Fermi-Dirac approximation would write

$$t_R[\rho] = C_k \rho^{5/3}, \qquad (4.12)$$

and

$$\varepsilon_R^{xc}[\rho] = - C_e \rho^{4/3}. \qquad (4.13)$$

The Thomas-Fermi three-halves relation between density ρ and potential V_i is then regained if we neglect the exchange energy. Including this latter correction, we find the Euler equation of the Thomas-Fermi-Dirac method.

Thus, we find for the Hartree term the result

$$\frac{1}{m}\sum_{\alpha}\nabla_{\alpha}\{\rho(R,t)\nabla_{\alpha}V_{H}(R,t)\}$$

$$\to -\frac{4\pi e^{2}}{m}\rho_{0}(R)[\rho_{e}(R,t)+\rho_{1}(R,t)]$$

$$+\frac{1}{m}\nabla\rho_{0}(R)\cdot\nabla[V_{e}(R,t)+V_{1}(R,t)]$$

$$-\frac{1}{m}\rho_{1}(R,t)\nabla^{2}\left[\frac{\delta t_{R}[\rho]}{\delta\rho}+\frac{\delta\varepsilon_{R}^{xc}[\rho]}{\delta\rho}\right]$$

$$-\frac{1}{m}\nabla\rho_{1}(R,t)\cdot\left\{\nabla\left[\frac{\delta t_{R}[\rho]}{\delta\rho}\right]+\nabla\left[\frac{\delta\varepsilon_{R}^{xc}[\rho]}{\delta\rho}\right]\right\}. \qquad (4.14)$$

We have retained the functional derivative of the exchange and correlation energy density in this equation, since our aim here is to base an approximate treatment of $\rho(R,t)$ on the exact ground-state density $\rho_0(R)$.

The above Hartree term simplifies very greatly, of course, for the uniform electron gas. Thus the last three terms in the above equation vanish because of homogeneity, while the first term on the right-hand side is simply

$$-\omega_{p}^{2}[\rho_{e}(R,t)+\rho_{1}(R,t)].$$

It is clear that to study the eigenmodes of the system, the external charge density is made zero, and we regain the elementary equation of motion, without dispersion, for the plasmons.

Evidently, in the general case of an inhomogeneous gas, the local plasma frequency $\omega_p(\mathbf{R}) = (4\pi e^2/m)^{\frac{1}{2}}\{\rho_0(\mathbf{R})\}^{\frac{1}{2}}$, determined by the ground-state density $\rho_0(\mathbf{R})$ now enters the problem. We turn next to the kinetic term, involving the momentum flux tensor $\Pi_{\alpha\beta}(\mathbf{R},t)$.

4.2 Kinetic Term

Following the method outlined by Kadanoff and Baym (1962), it can be proved that the functional derivative

$$\frac{\delta\langle\psi_{\sigma}^{\dagger}(R+\frac{1}{2}\mathbf{r},t)\psi_{\sigma}(R-\frac{1}{2}\mathbf{r},t)\rangle}{\delta V_{e}(\mathbf{x}',t')} = \qquad \text{(Contd)}$$

$$= -\frac{i}{\hbar}\{\langle T[\psi_\sigma^\dagger(\mathrm{R}+\tfrac{1}{2}\mathbf{r},t)\psi_\sigma(\mathrm{R}-\tfrac{1}{2}\mathbf{r},t)\rho(\mathbf{x}',t')]\rangle$$

$$-\langle\psi_\sigma^\dagger(\mathrm{R}+\tfrac{1}{2}\mathbf{r},t)\psi_\sigma(\mathrm{R}-\tfrac{1}{2}\mathbf{r},t)\rangle\rho(\mathbf{x}',t')\}. \qquad (4.15)$$

Thus, the change in the momentum flux tensor Π due to a time-dependent external potential $V_e(\mathbf{x}',t')$ takes the linearized form

$$\Delta\Pi_{\alpha\beta}(\mathrm{R},t) = \int d\mathbf{x}'dt' \left.\frac{\delta\Pi_{\alpha\beta}(\mathrm{R},t)}{\delta V_e(\mathbf{x}',t')}\right|_{V_e=0} V_e(\mathbf{x}',t')$$

$$= \frac{i\hbar}{m}\sum_\sigma \int d\mathbf{x}'dt' V_e(\mathbf{x}',t')\nabla_\alpha^{(\mathbf{r})}\nabla_\beta^{(\mathbf{r})}$$

$$\times\{\langle T[\psi_\sigma^\dagger(\mathrm{R}+\tfrac{1}{2}\mathbf{r},t)\psi_\sigma(\mathrm{R}-\tfrac{1}{2}\mathbf{r},t)\rho(\mathbf{x}',t')]\rangle$$

$$-\langle\psi_\sigma^\dagger(\mathrm{R}+\tfrac{1}{2}\mathbf{r},t)\psi_\sigma(\mathrm{R}-\tfrac{1}{2}\mathbf{r},t)\rangle\rho(\mathbf{x}',t')\}\Big|_{\mathbf{r}=0}. \qquad (4.16)$$

At this stage, appeal is made to an often-used approximation in the uniform electron gas (Singwi et al., (1970)); namely to evaluate the above functional derivative for a gas of non-interacting particles and to use this value to write the change in Π as

$$\Delta\Pi_{\alpha\beta}(\mathrm{R},t) = \int d\mathbf{x}'dt' \left.\frac{\delta\Pi_{\alpha\beta}(\mathrm{R},t)}{\delta V_e(\mathbf{x}',t')}\right|_{\substack{\text{free}\\ \text{particles}}} V_{eff}(\mathbf{x}',t'). \qquad (4.17)$$

The functional derivative entering this equation is now to be evaluated for a gas of non-interacting Fermions. This implies that the kinetic term in the equation of motion for the electron density is approximated by assuming that the electrons move as independent particles in an effective potential $V_{eff}(\mathbf{x}',t')$ into which the effects of the electron-electron interactions are incorporated.

The effective potential is then chosen by using the equation

$$\nabla_\alpha V_{eff}(\mathbf{x}',t') = \nabla_\alpha V_e(\mathbf{x}',t') + \nabla_\alpha V_1(\mathbf{x}',t')$$

$$+ \int d\mathbf{x}''\nabla_\alpha^{(\mathbf{x}')} V_{xc1}(\mathbf{x}',\mathbf{x}'',t), \qquad (4.18)$$

(see Parrinello and Tosi, (1972)), where $\nabla_\alpha V_1(\mathbf{x}',t')$ and

$\nabla_\alpha V_{xc1}(\mathbf{x}',\mathbf{x}'',t)$ are the changes in the Hartree field and in the exchange and correlation field, produced by the external potential.

Explicitly we have, therefore,

$$V_1(\mathbf{x}',t') = \int d\mathbf{x}''\phi(\mathbf{x}' - \mathbf{x}'')\rho_1(\mathbf{x}'',t')$$

and

$$\nabla_\alpha^{(\mathbf{x}')} V_{xc1}(\mathbf{x}',\mathbf{x}'',t') = \nabla_\alpha\phi(\mathbf{x}' - \mathbf{x}'')\Delta\left\{\frac{\langle\rho(\mathbf{x}',t')\rho(\mathbf{x}'',t')\rangle_c}{\rho(\mathbf{x}',t')}\right\}. \quad (4.19)$$

Then the linearized form of the kinetic term is given by

$$-\frac{1}{m}\nabla_\alpha\nabla_\beta\Pi_{\alpha\beta}(\mathrm{R},t)$$

$$\rightarrow \frac{1}{m^2}\int d\mathbf{x}'dt' D(\mathrm{R},\mathbf{x}',t - t')$$

$$\times\left\{V_e(\mathbf{x}',t') + V_1(\mathbf{x}',t') + \int d\mathbf{x}'' V_{xc1}(\mathbf{x}',\mathbf{x}'',t')\right\} \quad (4.20)$$

where V_{xc1} is defined above, while the function $D(\mathrm{R},\mathbf{x}',t - t')$ is given by

$$D(\mathrm{R},\mathbf{x}',t - t') = \hbar^2\nabla_\alpha^{(\mathrm{R})}\nabla_\beta^{(\mathrm{R})}\nabla_\alpha^{(\mathbf{r})}\nabla_\beta^{(\mathbf{r})}$$

$$\times\left.\frac{\delta\rho(\mathrm{R} + \tfrac{1}{2}\mathbf{r},\mathrm{R} - \tfrac{1}{2}\mathbf{r},t)}{\delta V_e(\mathbf{x}',t')}\right|_{\substack{\mathbf{r}=0\\ \text{free}\\ \text{particles}}} . \quad (4.21)$$

4.3 Exchange and Correlation Term

In a similar manner, the exchange and correlation term, after linearization, takes the form

$$-\frac{1}{m}\int d\mathbf{x}\nabla_\alpha^{(\mathrm{R})}\{\rho(\mathrm{R},t)\nabla_\alpha^{(\mathrm{R})}V_{xc}(\mathrm{R},\mathbf{x},t)\}$$

$$\rightarrow -\frac{1}{m}\int d\mathbf{x}\nabla_\alpha^{(\mathrm{R})}\{\rho_1(\mathrm{R},t)\nabla_\alpha^{(\mathrm{R})}V_{xc}^0(\mathrm{R},\mathbf{x})\}$$

$$-\frac{1}{m}\int d\mathbf{x}\nabla_\alpha^{(\mathrm{R})}\{\rho_0(\mathrm{R})\nabla_\alpha^{(\mathrm{R})}V_{xc1}(\mathrm{R},\mathbf{x},t)\}$$

$$= -\frac{1}{m}\int d\mathbf{x}\nabla_\alpha^{(\mathrm{R})}\{\nabla_\alpha\phi(\mathrm{R} - \mathbf{x})\Delta\langle\rho(\mathrm{R},t)\rho(\mathbf{x},t)\rangle_c\}, \quad (4.22)$$

where

$$\nabla_\alpha^{(R)} V_{xc}(R,\mathbf{x}) = \nabla_\alpha \phi(R - \mathbf{x}) \frac{\langle \rho_0(R)\rho_0(x)\rangle_c}{\rho_0(R)} . \qquad (4.23)$$

We shall discuss the evaluation of $D(R,\mathbf{x}',t - t')$ for the inhomogeneous gas below. If we now make what is essentially the random-phase approximation, we neglect the exchange and correlation term above completely, and also that appearing in the kinetic term. Parrinello and Tosi (1972) have show how estimates can be made of these terms, and since they appear to make quantitative, rather than qualitative changes in the theory, we shall refer the reader to their paper for further details.

4.4 Form of Kinetic Term for Uniform Electron Gas

To see how to deal with the kinetic term explicitly, let us consider the homogeneous gas first (also putting $\rho_e = 0$ and $V_e = 0$) and write

$$\frac{1}{m} \sum_{\alpha\beta} \nabla_\alpha \nabla_\beta \Delta\Pi_{\alpha\beta}(R,t)$$

in Fourier transform as

$$-\frac{1}{m} \sum_{\alpha\beta} k_\alpha k_\beta \Delta\Pi_{\alpha\beta}(\mathbf{k},\omega)$$

But in the homogeneous gas we can write

$$\Delta\Pi_{\alpha\beta}(\mathbf{k},\omega) = -\frac{1}{m} C_{\alpha\beta}(\mathbf{k},\omega) V_1(\mathbf{k},\omega), \qquad (4.24)$$

where $C_{\alpha\beta}(\mathbf{k},\omega)$ is dependent solely on the unperturbed problem. Thus, the above Fourier transformed expression has the form

$$\frac{1}{m^2} \sum_{\alpha\beta} k_\alpha k_\beta C_{\alpha\beta}(\mathrm{k},\omega) V_1(\mathrm{k},\omega),$$

and returning to $\mathbf{r}$-space this can be formally expressed as

$$-\frac{1}{m^2}\int d\mathbf{x}' D(R - \mathbf{x}',\omega) V_1(\mathbf{x}',\omega).$$

Thus the equation of motion becomes

$$\frac{\partial^2 \rho_1(R,t)}{\partial t^2} = -\omega_p{}^2 \rho_1(R,t) - \frac{1}{m^2}\int d\mathbf{x}' dt' D(R - \mathbf{x}',t - t') V_1(\mathbf{x}',t'). \qquad (4.25)$$

But

$$V_1(\mathbf{x}',t') = e^2\int\frac{\rho_1(\mathbf{x}'',t')}{|\mathbf{x}'' - \mathbf{x}'|}\,d\mathbf{x}'', \tag{4.26}$$

and we seek a solution for ρ_1 of the form

$$\rho_1(\mathrm{R},t) = n_{\mathbf{k}}\exp\{i(\mathbf{k}\cdot\mathrm{R} - \omega t)\}. \tag{4.27}$$

Then we find

$$\omega^2 = \omega_p^2 + \frac{4\pi e^2}{m^2}\frac{D(\mathbf{k},\omega)}{k^2}\ . \tag{4.28}$$

Explicitly, $D(\mathbf{k},\omega)$ has the form

$$D(\mathbf{k},\omega) = -\sum_{\alpha\beta} k_\alpha k_\beta C_{\alpha\beta}(\mathbf{k},\omega). \tag{4.29}$$

As $k \to 0$ for finite ω, $D(\mathbf{k},\omega)$ can be shown to be proportional to k^4, the proportionality factor being $(2\rho_0/\omega^2)\langle T\rangle$. This then leads back to the dispersion relation of the random-phase approximation, when we put $\omega = \omega_p$ in the term of $O(k^2)$ in the above dispersion relation. Thus, it is clear that we have now a theory which avoids the error in the dispersion relation present in the original Bloch theory.

5. EQUATION OF MOTION FOR INHOMOGENEOUS GAS

In the general case, we can similarly write (with ρ_e and V_e again put to zero)

$$\frac{\partial^2\rho_1(\mathrm{R},t)}{\partial t^2} + \omega_p^2(\mathrm{R})\rho_1(\mathrm{R},t) - \frac{1}{m}\nabla\rho_0(\mathrm{R})\cdot\nabla V_1(\mathrm{R},t)$$

$$= \frac{1}{m}\rho_1(\mathrm{R},t)\nabla^2 V_i(\mathrm{R}) + \frac{1}{m}\nabla\rho_1(\mathrm{R},t)\cdot\nabla V_i(\mathrm{R})$$

$$- \frac{1}{m^2}\int d\mathrm{x}'dt' D(\mathrm{R},\mathrm{x}',t - t')V_1(\mathrm{x}',t'). \tag{5.1}$$

This, then, is the basic equation transcending Bloch's theory and involves the lattice through:

(i) The local plasma frequency $\omega_p(\mathrm{R})$;

(ii) The derivatives $\nabla\rho_0$, ∇V_i and $\nabla^2 V_i$. These terms in V_i, as we have seen, can, however, be re-expressed in terms of the functional derivatives of $t_{\mathbf{R}}[\rho]$ and $\varepsilon_{\mathbf{R}}^{xc}[\rho]$.

(iii) The kernel D,

In fact, in Bloch's hydrodynamic theory, the term involving the kernel D would take the local form

$$\sum_{\alpha} \nabla_{\alpha}[\rho_0(\mathbf{R})\nabla_{\alpha}\{Q(\rho_0)\rho_1\}],$$

where, as we saw earlier, $Q(\rho_0)$ was obtained by him using the Thomas-Fermi-Dirac approximation. In the present treatment, instead of the quantity $Q(\rho_0)$ being determined solely by the ground-state energy functionals, we need also information for $\omega \neq 0$.

If we were to make the approximation of calculating the kernel D at $\omega = 0$, then further progress is possible, since Stoddart, Beattie and March (1971) have then shown how the density matrix $\rho(R + \frac{1}{2}\mathbf{r},\mathbf{x}')$ can be expressed explicitly in terms of the electron density of the periodic lattice, using gradient expansions. Then the analogy with Bloch's formulation is very close. However, as we have remarked earlier, this formalism does not lead to the random-phase approximation to the plasmon dispersion relation in the homogeneous electron gas.

To ensure the correct dispersion relation for the uniform gas, D could be chosen at finite frequency to be the result for the homogeneous case, with k_f replaced by a local density approximation, $k_f \to \{3\pi^2\rho_0(R)\}^{1/3}$. Such a treatment evidently assumes that gradient corrections are small. Specifically in the long wavelength limit, we would write for the time Fourier transform of the term involving the kernel D the expression

$$\frac{4\pi e^2}{m^2\omega^2}\,\frac{3\hbar^2}{5m}\,(3\pi^2)^{2/3}\{\rho_0(R)\}^{5/3}\nabla^2\rho_1(R,\omega). \tag{5.2}$$

However, for one case, nearly-free electrons, we can already make progress without such an assumption. Therefore we shall turn to discuss the theory of the energy losses of fast electrons fired into metal foils, as an application of this theory. We shall be content with understanding how the Fourier components of a weak lattice potential enter the energy loss problem, and with some order of magnitude estimates.

6. UNIFORM ELECTRON GAS WITH EXTERNAL CHARGE

We now consider the electron gas perturbed by a fast electron ploughing through the metal. To establish the approach, let us neglect the inhomogeneity in the ground state density. Then, using the superscript zero to indicate that we are treating the uniform gas, we have for the equation of motion, including the external perturbation with density $\rho_e(R,t)$,

$$\frac{\partial^2 \rho_1{}^0(\mathrm{R},t)}{\partial t^2} = -\ \omega_p{}^2[\rho_1{}^0(\mathrm{R},t) + \rho_e(\mathrm{R},t)]$$

$$-\ \frac{1}{m^2}\int d\mathbf{x}'dt'D_0(\mathrm{R} - \mathbf{x}',t - t')$$

$$\times[V_1{}^0(\mathbf{x}',t') + V_e(\mathbf{x}',t')]. \qquad (6.1)$$

As usual, $V^0(\mathrm{R},t)$ is the electrostatic potential created by the electron density $\rho_1{}^0(\mathrm{R},t)$.

As we saw earlier, in the long wavelength limit and in the random-phase approximation

$$D_0(\mathbf{k},\omega) = \frac{2\rho_0\langle T\rangle k^4}{\omega^2}. \qquad (6.2)$$

For a fast electron, with velocity $\mathbf{v}$, injected into the metal, we have explicitly

$$\rho_e(\mathrm{R},t) = \delta(\mathrm{R} - \mathbf{v}t), \qquad (6.3)$$

and we can readily solve the above equation of motion to find

$$\rho_1{}^0(\mathbf{k},\omega) = 2\pi\delta(\mathbf{k}\cdot\mathbf{v} - \omega)\ \frac{\omega_p{}^2 + \dfrac{4\pi e^2}{m^2k^2}D_0(\mathbf{k},\omega)}{\omega^2 - \omega_p{}^2 - \dfrac{4\pi e^2}{m^2k^2}D_0(\mathbf{k},\omega)}. \qquad (6.4)$$

Before calculating the energy loss, let us sketch out the generalization of this argument to include a weak lattice potential.

7. INCLUSION OF WEAK LATTICE POTENTIAL

Let us write for the ground-state density

$$\rho_0(\mathrm{R}) = \rho_0 + \rho_L(\mathrm{R}), \qquad (7.1)$$

where ρ_0 is the uniform gas density, and similar expressions for V_i and ρ_1, namely

$$V_i(\mathrm{R}) = V_i{}^0 + V_i{}^L(\mathrm{R}), \qquad (7.2)$$

and

$$\rho_1(\mathrm{R},t) = \rho_1{}^0(\mathrm{R},t) + \rho_1{}^L(\mathrm{R},t), \qquad (7.3)$$

The weak lattice potential means that $\rho_L << \rho_0$. We can then develop the equation of motion to first-order in the lattice potential. If we use the explicit expression for $\rho_1{}^0(\mathbf{k},\omega)$ obtained above, we obtain after some calculation the result (March and Tosi, 1973)

$$\rho_1{}^L(R,\omega) = \frac{1}{m}\sum_{\mathbf{q}}\sum_{\mathbf{K}_n} \frac{\exp(i\{\mathbf{q} - \mathbf{K}_n\}\cdot R)}{[\omega^2 - \Gamma(\mathbf{q} - \mathbf{K}_n,\omega)][\omega^2 - \Gamma(\mathbf{q},\omega)]}$$

$$\times\Big\{(\mathbf{q}\cdot\mathbf{K}_n - \mathbf{K}_n{}^2)V_i{}^L(\mathbf{K}_n)\Gamma(\mathbf{q},\omega)\rho_e(\mathbf{q},\omega)$$

$$- 4\pi e^2\omega^2\rho_L(\mathbf{K}_n)\rho_e(\mathbf{q},\omega)$$

$$+ \frac{\omega^2}{m} D_L(\mathbf{q} - \mathbf{K}_n,\mathbf{q},\omega)V_e(\mathbf{q},\omega)\Big\} , \qquad (7.4)$$

with

$$\Gamma(\mathbf{k},\omega) = \omega_p{}^2 + \frac{4\pi e^2}{m^2k^2} D_0(\mathbf{k},\omega). \qquad (7.5)$$

The K_n's are, of course, reciprocal lattice vectors.

7.1 Energy Loss

The energy loss per unit path in the medium may be written as

$$-\frac{dW}{dx} = eE_x\Big|_{\mathbf{r}=\mathbf{v}t}$$

$$= \frac{e}{v}\,\mathbf{v}\cdot E\Big|_{\mathbf{r}=\mathbf{v}t}, \qquad (7.6)$$

where $E(\mathbf{r},t)$ is the electric field created by the density $\rho_1(\mathbf{r},t)$.

The probability of energy transfer $\hbar\omega$ at position r, say, $P(\mathrm{r},\omega)$ is now readily expressed as

$$P(\mathbf{r},\omega) = \frac{e^2}{v}\int dR\, \frac{\mathbf{v}\cdot(\mathbf{r} - R)}{|\mathbf{r} - R|^3}\,\rho_1(R,\omega), \qquad (7.7)$$

which, in Fourier transform, we can write as

$$P(\mathbf{r},\omega) = -\frac{ie^2}{v}\sum_{\mathbf{k}} \frac{\mathbf{v}\cdot\mathbf{k}}{k^2}\exp(i\mathbf{k}\cdot\mathbf{r})\rho_1(\mathbf{k},\omega). \qquad (7.8)$$

Inserting the above nearly-free electron theory of $\rho_1(\mathrm{k},\omega)$, we can write

$$P(\mathbf{r},\omega) = \sum_{\mathbf{q}} \exp(i\mathbf{k}\cdot\mathbf{r})P(\mathbf{r},\mathbf{q},\omega), \tag{7.9}$$

where $P(\mathbf{r},\mathbf{q},\omega)$ has the physical interpretation that at $\mathbf{r}$ it is the probability of energy transfer $\hbar\omega$ and momentum transfer $\hbar\mathbf{q}$ from the fast electron to the metal.

In terms of the deviation from the homogeneous electron gas loss function $P_0(\mathbf{q},\omega)$ we can now write

$$\begin{aligned} P(\mathbf{r},\mathbf{q},\omega) - P_0(\mathbf{q},\omega) \\ &\equiv \Delta P(\mathbf{r},\mathbf{q},\omega) \\ &= \frac{e^2}{mv} \operatorname{Im} \sum_{\mathbf{K}_n \neq 0} \exp(-i\mathbf{K}_n\cdot\mathbf{r}) f(\mathbf{q},\mathbf{K}_n,\omega)\rho_e(\mathbf{q},\omega), \end{aligned} \tag{7.10}$$

where the function $f(\mathbf{q},\mathbf{K}_n,\omega)$ involves $\Gamma(\mathbf{q},\omega)$, $V_i^L(\mathbf{K}_n)$, $D_L(\mathbf{q}-\mathbf{K}_n, \mathbf{q},\omega)$, etc..

Whereas in the homogeneous electron gas, the energy loss function P_0 is, of course, independent of $\mathbf{r}$, we must now integrate along the trajectory of the fast electron, in order to calculate the energy loss per unit path length, due to the first-order lattice effects.

Suppose first that the incident electron beam is at an arbitrary orientation to the axes of a metal single crystal. Then we would integrate the factor $\exp(-i\mathbf{K}_n\cdot\mathbf{r})$ over an essentially infinite path, yielding zero as the path length tends to infinity (we are neglecting surface effects in this discussion).

Thus, lattice effects in polycrystalline specimens are second order in the lattice potential, and, in nearly free electron metals, the usual uniform gas theory should be a very good approximation.

However, when the velocity $\mathbf{v}$ of the fast electron is along either a direct or a reciprocal lattice vector, then a *finite* periodicity enters the problem along the electron trajectory.

Thus the average $e^{-i\mathbf{K}_n\cdot\mathbf{r}}$ over this period vanishes unless $\mathbf{K}_n$ is perpendicular to the electron trajectory, and we find on integration over the trajectory the expression for the energy loss

$$\Delta P(\mathbf{q},\omega) = \frac{e^2}{mv} \sum_{\mathbf{K}_n \neq 0}{}' f''(\mathbf{q} - \mathbf{K}_n,\omega)\rho_e(\mathbf{q},\omega), \tag{7.11}$$

where f'' is the imaginary part of f and *only* reciprocal lattice vectors which are orthogonal to the trajectory enter the (primed) sum.

7.2 ANGULAR DISTRIBUTION

We wish now to discuss the probability of scattering through a definite angle.

For a fast electron with momentum $m\mathbf{v}$, this corresponds to a vector $\mathbf{q}$ with components $q_{||}$ along $\mathbf{v}$ and a vector $\mathbf{q}_{\perp}$ defining $\mathbf{q}$ in a lattice where cylindrical symmetry around $\mathbf{v}$ is not present, such that

$$|\mathbf{q}_{\perp}| = \frac{mv\theta}{\hbar} . \tag{7.12}$$

Now $\mathbf{q}$ and ω are related by energy conservation expressed through the delta function $\delta(\mathbf{q}\cdot\mathbf{v} - \omega)$ entering through $\rho_e(\mathbf{q},\omega)$. Using this relation between $\mathbf{q}$ and ω, we can write the probability of scattering through an angle θ, at azimuth ϕ say, as

$$\Delta P(\theta,\phi,\omega) = \frac{(mv/\hbar)^2}{(2\pi)^3}\int dq_{||}\,\Delta P(\mathbf{q},\omega) . \tag{7.13}$$

We can now integrate over all energy transfers to obtain

$$\Delta P(\theta,\phi) = \int_0^{\infty} d\omega \Delta P(\theta,\phi,\omega) . \tag{7.14}$$

We get then as a first approximation

$$\frac{\Delta P}{P_0} \sim \frac{m^2v^2\theta^2}{\hbar^2 K_1{}^2}\,\frac{\rho_L(|\mathbf{K}_1|)}{\rho_0} , \tag{7.15}$$

$\rho_L(\mathbf{K}_1)$ being the Fourier component of the ground state density for the shortest vector $\mathbf{K}_1$.

If we take the ratio ρ_L/ρ_0 as about 0.1 and θ typically as 10^{-2} radians, the lattic effects amount to some few percent of the free electron term. The angular distribution is of the form

$$\frac{1}{\theta^2 + (\frac{\hbar\omega_p}{mv^2})^2}\,[1 + \text{constant}\ \theta^2] , \tag{7.16}$$

the constant depending directly on the Fourier component $\rho_L(|\mathbf{K}_1|)$. Thus, for nearly free electron metals, there should be small quantitative effects introduced by the lattice for electrons fired along lattice vectors.

8. APPLICATION TO SURFACES

We shall conclude these lectures with some brief remarks on the application to metal surfaces.

To deal with the collective modes (surface plasmons) in this case, we ought to start from the ground-state electron density in the presence of the metal surface. Theories of this exist in a uniform positive background cut off abruptly to zero at the surface (Smith, 1969; Lang and Kohn, 1970). Though Lang and Kohn's work is fully quantum mechanical and includes the Friedel oscillations induced by the surface, we shall summarize Smith's results here, based on gradient expansions.

Thus, for the single-particle kinetic energy density $t_R[\rho]$, Smith corrects the simple Thomas-Fermi result by writing, following Kirzhnits (1957),

$$t_R[\rho] = C_k\rho^{5/3} + \frac{\hbar^2}{72m}\frac{(\nabla\rho)^2}{\rho} . \tag{8.1}$$

For the exchange term, we take the Dirac-Slater form

$$\varepsilon_x[\rho] = -C_e\rho^{5/3}, \tag{8.2}$$

while for the correlation result of Wigner (1938) is employed, namely

$$\varepsilon_c[\rho] = -\frac{0.056\rho^{4/3}}{0.079 + \rho^{1/3}} . \tag{8.3}$$

If $E[\rho]$ is the total energy thus obtained as a functional of the electron density $\rho(\mathrm{R})$, then the Euler equation for ρ in the inhomogeneous gas is obtained from the variational principle

$$\delta(E[\rho] - \mu N) = 0, \tag{8.4}$$

where μ is the chemical potential, while

$$N = \int\rho(\mathrm{R})d\mathrm{R}, \tag{8.5}$$

is the total number of electrons. If we take the z-axis perpendicular to the metal surface, then the Euler equation obtained by Smith (1969) is

$$\frac{d^2\rho}{dz^2} - \frac{1}{2\rho}\left(\frac{d\rho}{dz}\right)^2 = 36\left[\tfrac{1}{2}(3\pi^2)^{2/3}\rho^{5/3} + (\phi - \mu)\rho - \left(\frac{3}{\pi}\right)^{1/3}\rho^{4/3} - \frac{0.056\rho^{5/3} + 0.0059\rho^{4/3}}{(0.079 + \rho^{1/3})^2}\right], \tag{8.6}$$

where $\phi(\mathbf{r})$ is the electrostatic potential given by

$$\phi(\mathbf{r}) = v_{ext}(\mathbf{r}) + \int \frac{\rho(\mathbf{r}')d\mathbf{r}'}{|\mathbf{r} - \mathbf{r}'|}, \qquad (8.7)$$

$v_{ext}(\mathbf{r})$ being the 'external' potential of the positive background.

Smith has obtained an approximate solution to (8.6) using a variational trial function

$$\left.\begin{aligned} \rho(z) &= \rho_0 - \tfrac{1}{2}\rho_0 e^{\beta z}, & z < 0 \\ &= \tfrac{1}{2}\rho_0 e^{-\beta z}, & z > 0 \end{aligned}\right\}, \qquad (8.8)$$

with β as a variational parameter.

It is clear, in principle, that this ground-state density could be used in the quantum theory of collective oscillations outlined above. One would obtain in the extreme long wavelength limit the usual surface plasma frequency $\omega_p/\sqrt{2}$. However, the dispersion of the surface plasmon will depend on the density profile in (8.8). No detailed calculations using this method have yet been made, but would be of interest. Bloch's theory has been used, however, (see, for example, Ritchie, 1957), but we must refer to the original paper for details of this.

Finally, interesting ideas on the way the surface energy of metals depend on the zero-point energy of the plasma modes have been put forward recently by Lucas and Schmit (1972); see, however, Jonson and Sriwasan (1973). Though it is, no doubt, an extreme approximation to put all the emphasis on collective behaviour in calculating surface energy, non-local forms of the correlation energy which reflect the effect of the ground-state energy of the electron gas of the plasma modes will obviously have to be incorporated into the density functional approach as used by Smith and by Lang and Kohn, as a next step.

REFERENCES†

Bloch, F. (1933). *Z. Phys.*, **81**, 363; (1934). *Helv. Phys. Acta*, **7**, 385.

Janson, M. and Srinivasan, G. (1973). *Phys. Lett.*, **43A**, 427.

Hopfield, J.J. (1965). *Phys. Rev.*, **139A**, 419.

March, N.H. and Tosi, M.P. (1972). *Proc. Roy. Soc.*, **A330**, 373.

March, N.H. and Tosi, M.P. (1973). *Phil. Mag.*, (in press).

Parrinello, M. and Tosi, M.P. (1972). *Nuovo Cim.*, **B12**, 155.

Rethick, C.J. (1970). *Phys. Rev.*, **B2**, 1789.

Raether, H. (1965). Solid State Excitations by Electrons, in *Springer Tracts in Modern Physics, Vol. 38*, (Springer-Verlag, Berlin).

† Not all of these references are cited in the text.

Singwi, K.S., Sjölander, A., Tosi, M.P. and Land, R.H. (1970). *Phys. Rev.*, **B1**, 1044.

Smith, J.R. (1969). *Phys. Rev.*, **181**, 522.

Stoddart, J.C., Beattie, A.M. and March, N.H. (1971). *Int. J. Quantum Chem.*, **4**, 35.

Kadanoff, L.P. and Baym, G. (1962). *Quantum Statistical Mechanics*, (W.A. Benjamin, Inc., New York).

Kirzhnits, P.A. (1957). *Sov. Phys., JETP*, **5**, 64.

Lang, N. and Kohn, W. (1970). *Phys. Rev.*, **B1**, 4555.

Lucas, A. and Schmit, J. (1972). *Solid State Commun.*, **11**, 415.

HARTREE-FOCK MODEL CALCULATION FOR ATOMS AND CRYSTALS PLUS CORRECTIONS TO THE HARTREE-FOCK EXCITATION ENERGIES

T.C. COLLINS, R.N. EUWEMA, G.G. WEPFER, G.T. SURRATT, N.E. BRENER, J.L. IVEY and D.L. WILHITE

Aerospace Research Laboratories, Wright-Patterson Air Force Base, Ohio 45433

1. INTRODUCTION

One is interested in being able to describe many body systems of electrons and nuclei and their response to external perturbations . The systems are of course atoms, molecules, solids and gases and liquids. In doing so, one generally divides this problem into three parts: the description of the electronic system; the description of the nuclear motion; and the interaction of the two systems. The non-relativistic Hamiltonian of the total system is

$$\mathcal{H} = \sum_{b} \frac{P_b^{\,2}}{2M_b} + e^2 \sum_{b<c} \frac{Z_b Z_c}{|\vec{R}_b - \vec{R}_c|} + \sum_{i} \frac{p_i^{\,2}}{2m} + e^2 \sum_{i<j} \frac{1}{|\vec{r}_i - \vec{r}_j|} - e^2 \sum_{b,i} \frac{Z_b}{|\vec{r}_i - \vec{R}_b|} \,. \qquad (1.1)$$

The capital letters refer to the nuclei and the small letters refer to the electrons. To obtain the separation of the electronic and nuclear motion we will make the Born-Oppenheimer approximation [*1*]. As you will soon see, this is just the first approximation in a very long list of approximations before numerical results are generated. In the Born-Oppenheimer approximation, one assumes that the nuclei are stationary so that the first term on the right hand side of (1.1) is dropped and the second term is just a constant. The Hamiltonian for the electronic system then is

$$\mathcal{H}_e = \sum_{i=1}^{N} \frac{p_i^2}{2m} - e^2 \sum_{i,b} \frac{Z_b}{|\vec{r}_i - \vec{R}_b|} + e^2 \sum_{i<j} \frac{1}{|\vec{r}_i - \vec{r}_j|}, \quad (1.2)$$

and one would like to obtain the solution of

$$\mathcal{H}_e \Psi(\vec{k},\{\chi\}_N,\{\vec{R}\}_M) = E(\vec{k},\{\vec{R}\}_M)\Psi(\vec{k},\{\chi\}_N,\{\vec{R}\}_M), \quad (1.3)$$

where {} denotes the complete set of variables, and χ includes the coordinates and the spin variables of the electrons. The total wave function of the system is

$$\mathcal{X}_{\vec{q}}(\{\chi\},\{\vec{R}\}) = \sum_{\vec{k}} \Phi_{\vec{q}}(\vec{k},\{\vec{R}\}_M)\Psi(\vec{k},\{\chi\}_N,\{\vec{R}\}_M). \quad (1.4)$$

The solution for the Φ's will be the subject of other talks in the school [2]. I will drop this part and look at obtaining approximate solutions to equation (1.3) for the ground state and at obtaining excitation energies of the electronic system.

Most current solutions of (1.3) for real systems have been based upon the Hartree-Fock Model; that is, solutions to the model itself, approximations to the Hartree-Fock Model, and simple extensions of the Hartree-Fock Model. Thus it is important to review the Hartree-Fock Model. This is done in section 2. Some of the more common approximations and atomic and crystalline results using these approximations are also given. In section 3 a way to obtain a Hartree-Fock crystal is outlined and the results of the calculation are discussed. Some extensions of the Hartree-Fock Model are given in section 4 and the conclusions are presented in section 5.

2. REVIEW OF THE HARTREE-FOCK MODEL

Hartree-Fock Derivation

The Hartree-Fock Model [3] is well known and we will only give an outline of the method, but we will discuss some of the aspects of the Model in detail. It has been shown that any anti-symmetric wave function can be expanded in terms of anti-symmetric products of a complete set of one-electron functions taking N functions at a time [4]. An anti-symmetric product of the N one-electron functions, of course, forms a determinant; namely a Slater Determinant. Thus the many-body wave function is

$$\Psi(\{\chi\}_N) = \sum_{k} C_k D_k(\{\chi\}_N), \quad (2.1)$$

where

$$D_k(\{\chi\}_N) = \frac{1}{\sqrt{N!}} \det[\phi_i(\chi_j)] = \sqrt{N!}\,\hat{A}_N \prod_{i=1}^{N} \phi_i(\chi_i). \quad (2.2)$$

The ϕ_i's are one-electron functions and $\hat{A}_N = (N!)^{-1} \sum_P (-1)^P \hat{P}$. $\hat{P}$'s are the permutation operators. The Hartree-Fock Model [5] consists of finding the best single determinant which will minimize the total energy of the system. There are two main conditions which are placed on the trial wave function. The first is that the one-electron orbitals remain orthogonal, i.e., $\langle i|j\rangle = \delta_{ij}$. The second consists of requiring the spatial part of the orbitals for the same level be the same. The spin part will be different. This has acquired the name of the Restricted Hartree-Fock Model (RHF). We will discuss relaxing this last requirement later.

To find the single Slater determinant which will minimize the total energy of the system, one uses the variational principle with the constraint that the orbitals remain orthogonal. This is insured by using Lagrange undetermined multipliers. One thus varies with respect to the orbitals the following equation:

$$\delta\phi_k^*\{\langle D|\mathcal{H}_e|D\rangle - \sum_{i,j} \lambda_{ij}\langle\phi_i|\phi_j\rangle\}. \quad (2.3)$$

Now

$$\langle D|\mathcal{H}_e|D\rangle = N!\langle \prod_{i=1}^{N} \phi_i|\hat{A}_N\mathcal{H}_e\hat{A}_N|\prod_{j=1}^{N} \phi_j\rangle = 0. \quad (2.4)$$

and $[\hat{A}_N,\mathcal{H}_e] = 0$ with $\hat{A}_N^2 = \hat{A}_N$. Using these relations, one finds

$$\langle D|\mathcal{H}_e|D\rangle = \sum_P (-1)^P\langle \prod_{i=1}^{N} \phi_i|\mathcal{H}_e\hat{P}|\prod_{j=1}^{N} \phi_j\rangle$$

$$= -\frac{\hbar^2}{2m}\sum_{i=1}^{N}\int \phi_i^*(\chi)\nabla_r^2\phi_i(\chi)d\chi$$

$$- e^2\sum_{i=1}^{N}\int \phi_i^*(\chi)\sum_{b=1}^{n}\frac{Z_b}{|\vec{r} - \vec{R}_b|}\phi_i(\chi)d\chi$$

$$+ \tfrac{1}{2}e^2\sum_{i,j}{}'\int \phi_i^*(\chi)\phi_j^*(\chi')\frac{1}{|\vec{r} - \vec{r}\,'|}\phi_j(\chi')\phi_i(\chi)d\chi d\chi' -$$

(Contd)

$$\text{(Contd)}\ -\tfrac{1}{2}e^2 \sum_{i,j}{}' \int \phi_i^*(\chi)\phi_j^*(\chi') \frac{1}{|\vec{r} - \vec{r}\,'|} \phi_i(\chi')\phi_j(\chi)d\chi d\chi'. \quad (2.5)$$

An interesting point to be made is that one may remove the prime from the last two sums since the term $i = j$ exactly cancels.

Carrying out the variation denoted in equation (2.3) gives

$$\left[-\frac{\hbar^2}{2m}\nabla_{r_1}{}^2 - e^2 \sum_b \frac{Z_b}{|\vec{r}_1 - \vec{R}_b|}\right]\phi_i(\chi_1)'$$

$$+ \sum_j \int d\chi_2 \phi_j^*(\chi_2) \frac{1}{|\vec{r}_1 - \vec{r}_2|}[1 - \hat{P}_{12}]\phi_j(\chi_2)\phi_i(\chi_1)$$

$$= \sum_j \lambda_{ij}\phi_j(\chi_1). \quad (2.6)$$

In the case of double occupancy of the top degenerate energy levels one can make a unitary transformation on the ϕ_i's which diagonalizes the λ_{ij} matrix [6]. Thus one has

$$\left[-\frac{\hbar^2}{2m}\nabla_{r_1}{}^2 - e^2 \sum_b \frac{Z_b}{|\vec{r}_1 - \vec{R}_b|}\right]\psi_i(\chi_1)$$

$$+ \sum_j \int d\chi_2 \psi_j^*(\chi_2) \frac{1}{|\vec{r}_1 - \vec{r}_2|}[1 - \hat{P}_{12}]\psi_j(\chi_2)\psi_i(\chi_1)$$

$$= \varepsilon_i\psi_i(\chi_1). \quad (2.7)$$

If the wave function of the system is a single Slater determinant, the first-order density matrix may be written as

$$\rho(\chi_1,\chi_1') = \int D(\{\chi\}_N)D^\dagger(\{\chi\}_N)d\chi_2\ldots d\chi_N$$

$$= \sum_{i=1}^{N} \psi_i(\chi_1)\psi_i(\chi_1'). \quad (2.8)$$

This is called the Fock-Dirac density matrix. Substituting this definition in equation (2.7), one obtains the so-called Fock Operator:

$$F(\chi_1) = \left[-\frac{\hbar^2}{2m}\nabla_{r_1}{}^2 - e^2 \sum_b \frac{Z_b}{|\vec{r}_1 - \vec{R}_b|}\right] + \quad \text{(Contd)}$$

$$\text{(Contd)} \quad + \int d\chi_2 \frac{1}{|\vec{r}_1 - \vec{r}_2|}[1 - \hat{P}_{12}]_{\chi_2 = \chi_2'} \rho(\chi_2, \chi_2'), \tag{2.9}$$

where $\chi_2 = \chi_2'$ subscript in the last term means perform the operation then remove the prime and integrate. One thus has an effective one-body operator F and

$$F(\chi_1)\psi_i(\chi_1) = \varepsilon_i \psi_i(\chi_1). \tag{2.10}$$

One now can solve equation (2.10) by guessing the initial ψ_i's, generate the Fock operator, solve for new ψ_i's and continue until no appreciable change takes place. That is

$$\psi_i\text{'s} \rightarrow \rho \rightarrow F \rightarrow \psi_i\text{'s}$$

and one has self-consistent solutions.

Let us now turn our attention to the physical meaning of the ε_i's. Form the equation for the total energy of the N-particle system using a single Slater determinant made up of the

$$E_{\mathrm{N}}^{\mathrm{HF}} = \langle D|\mathcal{H}_e|D\rangle$$

$$= \sum_{i=1}^{N} \langle i|\left\{-\frac{\hbar^2}{2m}\nabla_1^2 - e^2 \sum_{b}^{n} \frac{Z_b}{|\vec{r}_1 - \vec{R}_b|}\right\}|i\rangle$$

$$+ \tfrac{1}{2}\sum_{i,j}^{N} \langle ij|\frac{1}{|\vec{r}_1 - \vec{r}_2|}[1 - \hat{P}_{12}]|ij\rangle. \tag{2.11}$$

Now remove the k-th orbital, without letting the system relax (frozen-orbital approximation), and form the total energy of the $(N - 1)$-particle system.

$$E_{\mathrm{N-1,k}}^{\mathrm{HF}} = \sum_{i \neq k}^{N} \langle i|\left\{-\frac{\hbar^2}{2m}\nabla_1^2 - e^2 \sum_{b}^{n} \frac{Z_b}{|\vec{r}_1 - \vec{R}_b|}\right\}|i\rangle$$

$$+ \tfrac{1}{2}\sum_{i,j \neq k}^{N} \langle ij|\frac{1}{|\vec{r}_1 - \vec{r}_2|}[1 - \hat{P}_{12}]|ij\rangle. \tag{2.12}$$

Take the difference of equations (2.11) and (2.12). Using equation (2.10) one finds

$$E_{\mathrm{N}}^{\mathrm{HF}} - E_{\mathrm{N-1,k}}^{\mathrm{HF}} = \langle k|F|k\rangle = \varepsilon_k. \tag{2.13}$$

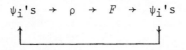

Thus the parameter ε_k is related to the energy to remove an electron from the N-particle system provided one doesn't allow the ψ_i's to change. This is known as Koopman's Theorem [7]. Note, nothing has been said about exciting the system other than removing the electron to infinity.

Let's see if one can get a handle on other excited states of the system. To do this, we will introduce the idea of expanding the ψ_i's in terms of basis sets. This will lead to the alphabet soup which is very common in the literature. For example, in calculating crystalline properties one uses orthogonalized plane waves (OPW) [8], linear combinations of atomic orbitals (LCAO) [9], and sometimes mixed basis sets of, say, plane waves and Gaussians (PWG) [10]. Of course the correct basis set is LCMBF, namely linear combination of my basis functions. The most important part of any basis set is that one can obtain convergence; that is, use a correct set so that the orbitals are properly represented. In doing this, one often has a basis larger than the number of particles in the system. One forms the matrix of the Fock operator with this basis; namely

$$\langle \chi_i | F | \chi_j \rangle,$$

and diagonalizes this matrix. One obtains the N lowest eigenvalues which are the Hartree-Fock values. These are called occupied values and the others are called virtual. But what are the other eigenvalues? Let's assume that we have diagonalized the Fock operator and look at the k-th matrix element where $k > N$

$$\varepsilon_k = \langle k | \left\{ - \frac{\hbar^2}{2m} \nabla_1^{\,2} - e^2 \sum_b \frac{Z_b}{|\vec{r}_1 - \vec{R}_b|} \right\} | k \rangle$$

$$+ e^2 \int d\chi_1 d\chi_2 \psi_k^*(\chi_1) \left\{ \sum_{j=1}^{N} \frac{\psi_j^*(\chi_2)\psi_j(\chi_2)}{|\vec{r}_1 - \vec{r}_2|} \right\} \psi_k(\chi_1)$$

$$- e^2 \int d\chi_1 d\chi_2 \psi_k^*(\chi_1) \left\{ \sum_{j=1}^{N} \frac{\psi_j^*(\chi_2)\psi_k(\chi_2)}{|\vec{r}_1 - \vec{r}_2|} \right\} \psi_j(\chi_1). \qquad (2.14)$$

Examining the last two terms one sees that there is no cancellation of any terms because $j \neq k$. In the case of occupied orbitals, there was cancellation of the self-energy term and the electron saw the potential due to all the other electrons, namely a V^{N-1} potential. However, the virtual state sees N electrons and thus a V^N potential. This points out that the virtual eigenvalues correspond to adding an electron to the system and do not represent excited states of the N-particle system. Another important point is that there is no relaxation included. There are several methods to correct this, one of which is to form a potential for the virtual states so that they see a V^{N-1} potential [11,12].

The above discussion causes the atomic and molecular calculators to worry about this problem. However, some people who calculate energy bands for crystals make the following statements. If one excites a Bloch electron, one takes $1/N$ electronic charge away from each atom and redistributes it in the unit cell; thus, $V^N \approx V^{N-1}$ and the virtual eigenvalue represents a reasonable excitation energy. This neglects any localization of the hole and electron in the system.

At this point one should also be reminded that eigenvalues of the Fock operator for a free electron gas give very bad results when compared to metallic systems. There is a zero density of states at the Fermi Surface and the conduction band width is 50% too wide. However, we will not let these facts stand in our way, but press on to make approximations to the Hartree-Fock Model.

APPROXIMATIONS TO THE HARTREE-FOCK MODEL

Approximations

Until the last few years one did not solve the Hartree-Fock Model because the exchange operator made the problem too difficult. Now, however, using methods developed by quantum chemists, the ARL theoretical group [13] has been able to do a Hartree-Fock Crystal.* This work will be outlined in the next section. Also Kunz [14] will describe his Hartree-Fock Model in this school. The major part of the literature reports calculations made with different approximations to this operator.

Most of the electron energy bands of crystals make the 'free-electron gas' approximation. Let's multiply and divide the exchange term of the Fock operator (equation (2.9)) by

$$\frac{[\psi_i^*(\chi_1')\psi_i(\chi_1')]}{[\psi_i^*(\chi_1')\psi_i(\chi_1')]_{\chi_1'=\chi_1}} .$$

This exchange operator times the wave function in the Hartree-Fock eigenvalue equation (2.10) becomes

$$\left\{ -e^2 \sum_j \frac{\int d\chi_2 \psi_j^*(\chi_2)\psi_i^*(\chi_1) \frac{1}{|\vec{r}_1 - \vec{r}_2|} \psi_i(\chi_2)\psi_j(\chi_1)}{\psi_i^*(\chi_1)\psi_i(\chi_1)} \right\} \psi_i(\chi_1)$$

$$= V_{ex}^i(\chi_1)\psi_i(\chi_1). \quad (2.15)$$

One can take the weighted average of this term with $\psi_i^*(\chi_1) \times \psi_i(\chi_1)/\rho$ being the weight and substitute plane waves for the ψ_i.

*Publisher's note: reference is made to the fact that various other groups have also solved the Hartree-Fock problem.

The result is the Slater exchange approximation [*15*]. That is

$$V_{\mathrm{ex}}^{\mathrm{S}}(\vec{r}) = -6\left\{\frac{3}{8\pi}\rho(\vec{r})\right\}^{1/3}. \tag{2.16}$$

Instead of making this approximation in the Fock operator, one can make the approximation in the total energy equation before one performs the variation operation. One obtains

$$E_{\mathrm{ex}} = -\tfrac{1}{2}C\int\rho^{4/3}d\tau, \tag{2.17}$$

where C are the constant terms in the $V_{\mathrm{ex}}^{\mathrm{S}}$ definition. Varying equation (2.17) with respect to the orbitals gives

$$-\frac{1}{2}C\left(\frac{4}{3}\rho^{1/3}\right)\psi_i(\chi_1) = \left(-\frac{2}{3}C\rho^{1/3}\right)\psi_i(\chi_1). \tag{2.18}$$

The value on the right-hand side is the operator obtained by Gaspar [*16*] and by Kohn and L.J. Sham [*17*]. This has led to a large number of guesses for the correct value in front of the exchange. What has been done is that a parameter, α, which can vary most of the time between 2/3 and 1 times Slater's exchange approximation is used. This is sometimes called the X_α Method [*18*].

$$V_{X_\alpha} = -\alpha C\rho^{1/3} \tag{2.19}$$

If the weighted average is not taken in equation (2.15) but the ψ_i's are again assumed to be plane waves one has the so-called Liberman exchange [*19*].

$$V_{\mathrm{ex}}^{\mathrm{L}}(\chi_1) = -8F(\eta_i)\left\{\frac{3}{8\pi}\rho(\vec{r})\right\}^{1/3}, \tag{2.20}$$

where

$$F(\eta_i) = \frac{1}{2} + \frac{1-\eta_i^2}{4\eta_i}\ln\left|\frac{1+\eta_i}{1-\eta_i}\right|; \qquad \eta_i = \frac{k_i}{k_F}.$$

One generally makes the choice that

$$k_i = [\varepsilon_i - V(\vec{r})]^{1/2} \quad \text{and} \quad k_F = \{3\pi^2\rho(\vec{r})\}^{1/3}.$$

The question now arises does one have something like a Koopman's theorem for these eigenvalues which are derived using the approximations for the exchange term. First taking the energy difference $E_N - E_{N-1,i}$ and then making the approximations one obtains the Hartree-Fock Slater Operator matrix element. Thus one has that the parameter, ε_i, derived using the Slater Approximation in the Fock operator does obey a Koopman-like theorem.

Next make the approximation and then take the energy difference. One has for the approximate exchange energy difference

$$E_{ex}^{N} - E_{ex}^{N-1,i} = -\tfrac{1}{2}C\int \rho_N^{4/3} d\tau + \tfrac{1}{2}C\int \rho_{N-1}^{4/3} d\tau. \quad (2.21)$$

Add and subtract the missing orbital from the last term gives

$$E_{ex}^{N} - E_{ex}^{N-1,i} = -\tfrac{1}{2}C\int \rho_N^{4/3} d\tau + \tfrac{1}{2}C\int (\rho_N - \phi_i^*\phi_i)^{4/3} d\tau. \quad (2.22)$$

Expanding the last term gives

$$E_{ex}^{N} - E_{ex}^{N-1,i} = -\tfrac{1}{2}C\int \rho_N^{4/3} d\tau + \tfrac{1}{2}C\int \rho_N^{4/3}\left[1 - \frac{4}{3}\frac{\phi_i^*\phi_i}{\rho_N} + \ldots\right] d\tau$$

$$= -\frac{2}{3}C\langle i|\rho^{1/3}|i\rangle + O\left(\frac{1}{N^{2/3}}\right). \quad (2.23)$$

Thus for large systems as $N \to \infty$, one also has a Koopman-like theorem for making the approximation in the total energy expression.

Since the systems are not homogeneous, some people have added gradient terms to the $\rho^{1/3}$ approximations [*19*]. We will now look at the results of calculations with these approximations.

Calculation Results

The first step is to calculate the total exchange energy in the above described approximations. A typical atomic example is Ar where one finds the values listed in Table I. In this

TABLE I

The exchange contribution to the total energy of atomic Ar. The self-energy contribution has been subtracted from the Liberman and Slater approximation. Units in Ry.

$\langle \frac{1}{2}V_{ex}^{HF}\rangle$	$\langle \frac{1}{2}V_{ex}^{Liberman}\rangle$	$\langle \frac{1}{2}V_{ex}^{Slater}\rangle$
13.86	14.69	11.91

Table the self-energy term has been subtracted out, and one finds that the Slater value underestimates exchange by ∿15% . The Liberman's approximation on the other hand overestimates the exchange energy by ∿5%. In figure 1, the total energy of atomic Kr is shown. The energy given is obtained in the Hartree-Fock

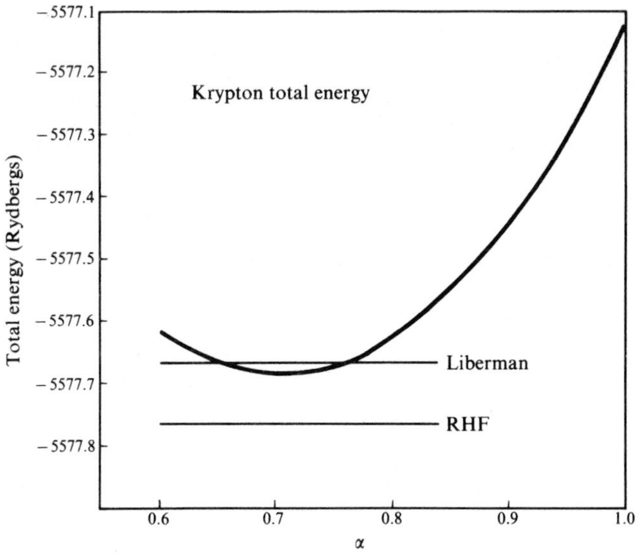

Figure 1 - Total Hartree-Fock Energy of atomic Kr derived using Hartree-Fock wave functions (RHF): Liberman wave function; and X_α wave functions.

approximation from wave functions using: (1) self-consistent restricted Hartree-Fock (RHF); (2) self-consistent Liberman approximation (liberman); and (3), α times the Slater approximation (α varies between 2/3 and 1). One sees that the difference between using the orbitals derived using Slater's approximation and the RHF value is three or four times the error made using either orbitals obtained from Liberman's approximation or Kohn-Sham-Gaspar approximations. One can, of course, make the substitution of α times the Slater approximation and obtain either the Hartree-Fock energy or the experimental total energy of the electronic system.

What about the eigenvalues of these different approximations? Table II gives the experimental ionization energies of Atomic Ar as well as the eigenvalues of the different approximations. The column labled I_{RHF} is obtained by doing self-consistent calculations on the N- and the $(N - 1)$-systems in the restricted Hartree-Fock approximation. This matches experiment the best. Of the approximations, the Slater eigenvalues come closer to matching experiment than the rest. However, as can be seen this

TABLE II

Eigenvalues for Various Exchange Approximations

The experimental values are from Siegbahn, M. (1967). *Nova Acta Regiee Socialates Scientiarium Upsatiensen Series IV, Vol. 20*. The last experimental values (†) are from More, C.E. (1952). *Atomic Energy Levels*, (National Bureau of Standards, Circular No. 467, Vol. II), (U.S. Government Printing Office, Washington, D.C.).

State	Exp (Ry)	I_{RHF} (Ry)	ε_{HF} (Ry)	ε_{Lib} (Ry)	ε_S (Ry)	ε_{KSG} (Ry)
$1S^{1/2}$	235.42	235.85	238.25	237.88	233.47	228.41
$2S^{1/2}$	23.52	24.04	24.82	24.58	22.96	21.62
$2P^{1/2}$	18.18	18.40	19.26	19.33	18.25	16.87
$2P^{3/2}$	18.02	18.18	19.09	19.19	18.08	16.71
$3S^{1/2}$	1.86	2.46	2.57	2.51	2.07	1.68
$3P^{1/2}$	1.17†	1.09	1.19	1.19	1.02	0.68
$3P^{3/2}$	1.16†	1.07	1.18	1.18	1.00	0.66

treatment generally underestimates the experiment. The Hartree-Fock and Liberman values match, but consistently overestimates experiment. The Kohn-Sham-Gaspar values underestimate experiment as much as Hartree-Fock overestimates. Thus when relaxation is taken into account in atomic systems as was done in column I_{RHF} the values are quite good. We shall see next that the same general trends hold for crystals that hold for atoms.

To test the ground state properties of the different approximations, one can compare the Fourier transforms of the charge density, ρ_k, with that measured from X-ray data. In particular the difference lines are compared, namely the lines which have the core contributions minimized. In zincblende the first difference line is $\vec{k} = (0,0,2)$, and the results are given in Table III.

TABLE III

ZnSe Form Factors

Experimental form factors are taken from P.M. Raccah, et al. [*20*]. Theoretical form factors are given based on relativistic Hartree-Fock, and SCOPW using different exchange approximations.

hkℓ	Exp	RHF	Liberman	Slater (Non Rel.)	Slater (Rel.)	KSG
002	14.86 ±0.5	11.22	11.44	11.72	12.16	11.15

One sees that all the values for ZnSe are outside experimental error. Also all the non-relativistic results are about the same. By using the Dirac Hamiltonian with Slater exchange approximation for the valence electrons and atomic Hartree-Fock core functions one reduces the error by 50%. But one is still outside the experiment in this case also.

In the comparison of the eigenvalues, one finds in crystals, what was found in the atomic case. In Table IV, one sees that the Liberman eigenvalues for Si are far apart. The Slater approximation eigenvalues are not too far off, however they underestimate experiment. The Kohn-Sham-Gaspar values are much too close together. It is interesting to note that when one uses the relativistic Dirac Hamiltonian, one finds that fairly large shifts occur between the *s*-like, *p*-like and *d*-like states. For example, in third-row elements the shift between *s*-like and *p*-like states is on the order of $\frac{1}{2}$ Ry. Thus one sees that the relativistic effects are just as large as the shifts caused by the different approximations.

TABLE IV

Comparison of the experimental indirect band gap and ε_2 peaks for silicon with SCOPW calculations with Slater, Kohn-Sham-Gaspar and Liberman exchange approximations. Units are in eV.

	Exp	Slater	KSG	Liberman
Indirect band gap	1.13	1.10	0.16	0.94
ε_2 peaks	4.3-4.5	4.1	3.3	6.1
	5.3	5.0	4.4	8.6

This concludes the brief outline of what has been done until the last few years. We shall now turn our attention to ways that are being employed to correct the shortcomings of these models. We shall next outline the method and give some results of Hartree-Fock crystal calculations in the next section. Then we shall show ways of obtaining more correct excitations energies in section IV.

3. HARTREE-FOCK CRYSTAL

General Crystalline Hartree-Fock Formalism

The fact that none of the approximations discussed in the last section lead to correct excitations in atoms or crystals, shows a great need to go beyond the approximate Hartree-Fock models. However, the Slater, Kohn-Sham-Gaspar, and the X_α method are more or less at a conceptual dead end. As pointed out, these approximations underestimate the true exchange. The Hartree-Fock model is correct through first order (this is called Brillouin's Theorem [*21*]). The second-order corrections would give a screened exchange plus a Coulomb hole correction [*22*]. Thus carrying these models further would mean one would be screening an exchange which is already too small.

On the other hand, if one is able to do an accurate Hartree-Fock calculation there are various well-investigated and well-understood techniques which have been developed and applied to atoms and molecules to improve upon the Hartree-Fock formalism. Examples are projected Hartree-Fock [*23*] and configuration interaction methods [*24*]. Also, one can calculate a dielectric function to be used in the second-order corrections to obtain good excitation energies. We shall return to this latter point in a latter section.

The method described here will use spatially localized basis functions centered at various locations in the crystal. One important point is, that in crystals, very long-range basis functions do not need to be employed because of the periodicity of the lattice. In particular a crystalline diamond computation will be described for which only *s*- and *p*-like functions are used. For *s*-symmetry basis functions, contracted sets [*25*] of primitive Gaussian functions are centered on the various atomic locations:

$$\phi(\vec{r}) = \sum_{\alpha} A_{\alpha} e^{-\alpha|\vec{r}|^2}. \tag{3.1}$$

For *p*-symmetry basis functions, Gaussian lobe functions, originally proposed by Pruess [*26*] and later developed independently by Whitten [*27*], are used. Such functions have been succesfully employed in a variety of *ab initio* Hartree-Fock calculations on molecules [*27*].

The lobe functions of *p*-symmetry are constructed as the differences of two contracted Gaussian lobes, centered distances $\pm \vec{R}_{\alpha}$ from the origin of the *p* function. The displacement is designed to reproduce the dipolar angular dependence of an atomic *p* function:

$$P_x(\vec{r}) = \sum_{\alpha} A_{\alpha}\{e^{-\alpha|\vec{r}-\vec{R}_{\alpha}|^2} - e^{-\alpha|\vec{r}+\vec{R}_{\alpha}|^2}\} \tag{3.2}$$

where

$$|\vec{R}_{\alpha}| = \frac{C}{\sqrt{\alpha}}; \qquad 0.005 \leqslant C \leqslant 0.01. \tag{3.3}$$

Atomic studies have established the lobe displacement range given in equation (3.3). A series of SCF calculations on the Ar atom were performed [*28*], allowing *C* to vary over the given range with the resulting total energies stable to 0.001 a.u. The value of 0.09 is used here.

Tables are available of optimized Gaussian fits to atomic Hartree-Fock wave functions, including contracted as well as fully uncontracted functions [*29*]. The short-range (large α) groups may, in general, be carried over for use in crystal calculations. The intermediate-range groups (those with appreciable overlap with neighboring sites) should be fully uncontracted to allow for maximal distortion from atomic character. The long-range groups, which in the atom are necessary to fit the exponential tail of the wave functions, may be discarded for the crystal because of the intermediate-range functions on neighboring sites.

By fully utilizing crystal symmetry, the Hartree-Fock problem is made much easier. Diamond, our example, has two atoms in the unit cell. The atom at (0,0,0) will be labelled type 1, and the atom at $(\frac{1}{4},\frac{1}{4},\frac{1}{4})$ will be labelled type 2. The coordinates with respect to the basis vectors are

$$\vec{t}_1 = \tfrac{1}{2}a(\hat{j} + \hat{k}); \qquad \vec{t}_2 = \tfrac{1}{2}a(\hat{i} + \hat{k}); \qquad \vec{t}_3 = \tfrac{1}{2}a(\hat{i} + \hat{j}). \tag{3.4}$$

There are twenty four rotation-reflection operators which transform each atom into an equivalent atom. The inversion about the point $\vec{r} = (\frac{1}{8},\frac{1}{8},\frac{1}{8})$ interchanges atoms of type 1 with atoms of type 2. In the remainder of this section, the set $\{X\}$ will include all the 48 rotation-reflection-inversion operators.

As is well known, only the first-order density matrix (the Dirac density matrix [*31*]),

$$\rho(\chi,\chi') = \sum_{i=1}^{N} \psi_i(\chi)\psi_i^*(\chi') \tag{3.5}$$

is needed in the Hartree-Fock formalism. Should one want to extend this procedure beyond the Hartree-Fock formalism, then the symmetry properties of the second-order density matrix should be considered.

The first-order density matrix can be expanded using our local basis set (assuming double occupancy of the Hartree-Fock orbitals)

$$\rho(\vec{r},\vec{r}\,') = \sum_{\alpha\beta ab} P_{\alpha\beta}^{ab}\phi_\alpha(\vec{r} - \vec{R}_a)\phi_\beta^*(\vec{r}\,' - \vec{R}_b). \tag{3.6}$$

For closed-shell ground states (semiconductors and insulators), the first-order density matrix has full crystal symmetry. Thus

$$\rho(\vec{r},\vec{r}\,') = \rho(\vec{r}\,',\vec{r}) = \rho(X\vec{r},X\vec{r}\,') = \rho(\vec{T} + \vec{r},\vec{T} + \vec{r}\,') \tag{3.7}$$

Now operating on any product of two functions

$$\phi_\alpha(\vec{r} - \vec{R}_a)\phi_\beta^*(\vec{r} - \vec{R}_b)$$

with all of the symmetry operators of the crystal will produce a set of product functions. By taking all possible values of α and β on each site and by taking different values for $|\vec{R}_a - \vec{R}_b|$, we can generate distinct product sets containing all possible product pairs for $|\vec{R}_a - \vec{R}_b|$ less than a given value. We only need to acquire information about one member of each product set since information about all other members may then be generated by using crystal symmetry operators. Also the symmetry requirement of equation (3.7) says that the $P_{\alpha\beta}^{ab}$ in equation (3.6) will be the same (except for sign) for all the pairs of functions of s and p character in one product set. This gives

$$\rho(\vec{r},\vec{r}\,') = \sum_I P_I \sum_{(\alpha\beta ab)} \epsilon_{\alpha\beta}^{ab}\phi_\alpha(\vec{r} - \vec{R}_a)\phi_\beta^*(\vec{r} - \vec{R}_b), \qquad (3.8)$$

where I runs over the different product sets which are labelled symmetry sets, $\epsilon = \pm 1$, and where $(\alpha\beta ab)$labels the different product pairs of a symmetry set. It should be noted that the P_I's do not have complete variational freedom since ρ must be idempotent and must satisfy the normalization condition

$$\int \rho(\vec{r},\vec{r})d\vec{r} = 2NZ. \qquad (3.9)$$

N is the number of unit cells, each with a total electronic charge of $2Z$.

Table V lists the members of the first six symmetry sets. These sets include all of the one-center and nearest-neighbor two-center contributions to the first-order density matrix. The symmetry analysis reduces 68 coefficients to 6. Table VI gives a representative member of each symmetry set for the first six shells [32] of atoms. The total number of symmetry-independent coefficients can be found by multiplying the number of symmetry-independent s-s local basis function products by $\frac{1}{2}N_s(N_s+1)$, the s-p products by N_sN_p, etc., where there are N_s (N_p) separate basis functions of s (p) character.

For example, good calculations can be done using six shells of atoms in diamond with four s-like basis functions and three with p-like character. One then has 270 symmetry-independent coefficients characterizing the first-order density matrix. It will be shown later that only 270 one-electron integrals of each type need be done, and that all of the two-electron integrals can be collected in a symmetry array $(270)^2$.

In order to determine the symmetry-independent coefficients of the first-order density matrix, the method of Roothan [33] is used. For each local basis function $\phi_\alpha(\vec{r} - \vec{R}_a)$, where ϕ_α is a contracted set of Gaussian lobe functions, construct a Bloch function

$$\Phi_\alpha^{\vec{k}}(\vec{r}) = \frac{1}{\sqrt{N}} \sum_a e^{i\vec{k}\cdot\vec{R}_a} \phi_\alpha(\vec{r} - \vec{R}_a), \qquad (3.10)$$

where $\vec{k}$ labels the Brillouin zone point and N is the number of unit cells in the crystal. Separate Bloch functions are needed for each of the two atoms in the unit cell. The one-electron wave functions $\Psi_n^{\vec{k}}$ are then expanded in terms of the Bloch functions

$$\Psi_n^{\vec{k}}(\vec{r}) = \sum_\alpha C_{n\alpha}^{\vec{k}}\Phi_\alpha^{\vec{k}}(\vec{r}), \qquad (3.11)$$

TABLE V

First Six Symmetry Sets for Diamond Symmetry

The s stands for an LBF of s-symmetry, while x, y, and z are p-symmetry LBF's. The second atom's coordinates are (0,0,0), $(\frac{1}{4},\frac{1}{4},\frac{1}{4})$, $(\frac{1}{4},-\frac{1}{4},-\frac{1}{4})$, $(-\frac{1}{4},\frac{1}{4},-\frac{1}{4})$, and $(-\frac{1}{4},-\frac{1}{4},\frac{1}{4})$ relative to the basis vectors given in equation (4). The first atom is at (0,0,0).

Set	Atom	[μν]
1	1	$xx + yy + zz$
2	1	ss
3	2	$xx + yy + zz$
	3	$xx + yy + zz$
	4	$xx + yy + zz$
	5	$xx + yy + zz$
4	2	$xy + yx + xz + zx + yz + zy$
	3	$- xy - yx - xz - zx + yz + zy$
	4	$- xy - yx + xz + zx - yz - zy$
	5	$xy + yx - zx - xz - yz - zy$
5	2	$xs - sx + ys - sy + zs - sz$
	3	$xs - sx - ys + sy - zs + sz$
	4	$- xs + sx + ys - sy - zs + sz$
	5	$- xs + sx - ys + sy + zs - sz$
6	2	ss
	3	ss
	4	ss
	5	ss

TABLE VI

Representative members of the diamond symmetry sets for the first six shells of atoms. The *s* stands for an *s*-symmetry LBF, while *x*, *y*, and *z* are *p*-symmetry LBF's. The coordinates are relative to the basis vectors given in equaequation (4).

Shell	Atom	$[\mu\nu]$
1	(0,0,0)	*ss*, *xx*
2	$(\frac{1}{4},\frac{1}{4},\frac{1}{4})$	*ss*, *sx*, *xx*, *sy*
3	$(\frac{1}{2},\frac{1}{2},0)$	*ss*, *sx*, *sz*
		xx, *zz*, *xy*, *xz*
4	$(\frac{3}{4},\frac{1}{4},-\frac{1}{4})$	*ss*, *sx*, *sy*
		xx, *yy*, *xy*, *yz*
5	(1,0,0)	*ss*,,*sx*, *sy*
		xx, *yy*, *xy*, *yz*
6	$(\frac{3}{4},\frac{3}{4},\frac{1}{4})$	*ss*, *sx*, *sz*
		xx, *zz*, *xy*, *xz*

where n is the band index. The coefficients $C_{n\alpha}{}^{\vec{k}}$ are adjusted to minimize the total energy. The following Hamiltonian and overlap matrices which are block diagonal in $\vec{k}$ are obtained:

$$\sum_{\beta} \mathcal{H}_{\alpha\beta}{}^{\vec{k}} C_{n\beta}{}^{\vec{k}} = \varepsilon_n{}^{\vec{k}} \sum_{\beta} \Delta_{\alpha\beta}{}^{\vec{k}} C_{n\beta}{}^{\vec{k}}, \tag{3.12}$$

$$\Delta_{\alpha\beta}{}^{\vec{k}} = \int \Phi_{\alpha}{}^{\vec{k}\dagger}(\vec{r})\Phi_{\beta}{}^{\vec{k}}(\vec{r})d\vec{r}$$

$$= \sum_{b} \mathcal{U}_{\alpha\beta}(\vec{R}_a,\vec{R}_b)e^{i\vec{k}\cdot(\vec{R}_a-\vec{R}_b)}, \tag{3.13}$$

$$\mathcal{U}_{\alpha\beta}(\vec{R}_a,\vec{R}_b) = \int \phi_{\alpha}(\vec{r} - \vec{R}_a)\phi_{\beta}(\vec{r} - \vec{R}_b)d\vec{r}, \tag{3.14}$$

$$\mathcal{H}_{\alpha\beta}{}^{\vec{k}} = \sum_{b} \mathcal{H}_{\alpha\beta}(\vec{R}_a,\vec{R}_b)e^{i\vec{k}\cdot(\vec{R}_b-\vec{R}_a)} \tag{3.15}$$

$$\mathcal{H}_{\alpha\beta}(\vec{R}_a,\vec{R}_b) = \int d\vec{r}_1\phi_2(\vec{r}_2 - \vec{R}_a) \times\{ - \nabla_1{}^2 - 2 \sum_{c} \frac{Z}{|\vec{r}_1 - \vec{R}_c|} + \int d\vec{r}_2\left[2\,\frac{\rho(\vec{r}_2,\vec{r}_2)}{r_{12}} - \frac{\rho(\vec{r}_1,\vec{r}_2)}{r_{12}}\,P_{12}\right]\}\phi_\beta(\vec{r}_1 - \vec{R}_b). \tag{3.16}$$

Equations (3.13, 15) need only be summed over atoms b owing to translational symmetry. Note that the sum over b is over all atoms of type 1 or of type 2, depending on whether the Bloch function $\Phi_\beta{}^{\vec{k}}$ is for atom type 1 or type 2. In equation (3.16), c sums over all atoms and P_{12} is the permutation operator which interchanges $\vec{r}_1$ with $\vec{r}_2$. Matrix elements where the atom at $\vec{R}_a$ is of type 2 can be obtained from those where the atom at $\vec{R}_a$ is of type 1 by application of the inversion operator about $(\frac{1}{8},\frac{1}{8},\frac{1}{8})$.

The matrix eigenvalue problem (equation (3.12)) is solved by performing a Choleski decomposition [34] on each positive definite overlap matrix $\Delta^{\vec{k}}$:

$$\Delta = LL^{\dagger}; \qquad L_{ij} = 0 \qquad \text{for } i < j. \tag{3.17}$$

After this decomposition is performed, a single matrix diagonalization yields the desired eigenvalues and eigenvectors:

$$[L^{-1}\mathcal{H}L^{-1\dagger}][L^{\dagger}C] = \lambda[L^{\dagger}C]. \tag{3.18}$$

Since in equations (3.14, 16) one has integrals which involve local basis function product pairs as in $\rho(\vec{r},\vec{r}\,')$, one can use the same decomposition into symmetry sets to reduce the number of integrals which need to be done. All of the operators in the Hamiltonian, as well as the overlap matrix (the unit operator) have full crystalline symmetry. Therefore, none of the one-electron operators can change the symmetry of the function it operates on. Thus all of the one-electron integrals may be classified into the same symmetry sets as $\rho(\vec{r},\vec{r}\,')$. Only a single one-electron integral of each type for each different symmetry set need be done. The two-electron integrals in equation (3.16) involve localized basis pairs multiplying $\rho(\vec{r},\vec{r}\,')$. Thus, if K symmetry-independent coefficients characterize $\rho(\vec{r},\vec{r}\,')$ in equation (3.6), then all of the two-electron integrals can be stored in a

K by K array. The two-electron part of equation (3.16) is summed over $\vec{R}_b$, α and β for $\vec{R}_a$ fixed to give the array,

$$A_{IJ} = \frac{1}{N} \sum_{\substack{(\alpha\beta ab) \\ \text{in } I}} \sum_{\substack{(\gamma\delta cd) \\ \text{in } J}} \int d\vec{r}_1 d\vec{r}_2 \phi_\alpha(\vec{r}_1 - \vec{R}_a)\phi_\gamma(\vec{r}_2 - \vec{R}_c)$$

$$\times \frac{2 - P_{12}}{r_{12}} \phi_\delta(\vec{r}_2 - \vec{R}_d)\phi_\beta(\vec{r}_1 - \vec{R}_b). \quad (3.19)$$

Element A_{IJ} is thus proportional to the total Coulomb-plus-exchange integral of all members of symmetry set I with all members of symmetry set J. The constant of proportionality is the reciprocal of the number of cells in the crystal. The symmetry of equation (3.19) guarantees the symmetry of the array, so that $A_{IJ} = A_{JI}$. In practice, the centers for the atoms a,b,c, and d are chosen such that they all are within M shells of each other and atom a is taken at zero. The two-electron integral contribution in equation (3.16) is now given by

$$\frac{1}{W_I} \sum_J P_J A_{IJ}, \quad (3.20)$$

where W_I is the number of members of symmetry set I for fixed $\vec{R}_a$. Note that in evaluating equation (3.19), a single member of symmetry set I, say μ, can be chosen and the integral for that member can be multiplied by the number of members of I with $\vec{R}_a$ at zero. In addition, one can use the set of symmetry operations $\{X\}$ which leave μ invariant, for a further reduction. The members of symmetry set J can be decomposed into subsets which transform into each other under the operations $\{X\}$. Only integrals between μ and one member of each subset need be calculated. Finally, one has the permutational symmetry that the Coulomb integral for $(ab\alpha\beta:cd\gamma\delta)$ is equal to the exchange integral for $(ac\alpha\gamma:bd\beta\delta)$ and $(ad\alpha\delta:bc\beta\gamma)$. All of these considerations greatly reduce the number of integrals which must be calculated.

In order to construct a new first-order density matrix from the self-consistent results, one in principle needs an integral over the occupied Hartree-Fock eigenfunctions in the first Brillouin zone:

$$\rho(\vec{r},\vec{r}') = \int_{BZ} d\vec{k} \sum_{\substack{n \\ \text{(filled)}}} \psi_n^{\vec{k}}(\vec{r})\psi_n^{\vec{k}\dagger}(\vec{r}'). \quad (3.21)$$

In practice, one replaces the Brillouin-Zone integral by a weighted sum over a numerical mesh in an irreducible sector of the Brillouin zone:

$$\rho(\vec{r},\vec{r}\,') = \sum_{\vec{k}} W_{\vec{k}} \sum_{\substack{n \\ \text{(filled)}}} \Psi_n^{\vec{k}}(\vec{r})\Psi_n^{\vec{k}\dagger}(\vec{r}\,'). \tag{3.22}$$

A natural weight to use for a given mesh point is one proportional to that volume of the first zone which is closer to that mesh point than to any other mesh point.

Another important computational simplification that is employed in these calculations is the approximation of some of the less important two-electron integrals involving sets of contracted Gaussians. The core like local basis functions are usually contracted sets of Gaussians with coefficients chosen to help simulate the wave function behavior in the vicinity of the nucleus. Each two-electron integral over four basis functions thus involves many integrals over individual Gaussians. If each basis function consists of m_i Gaussians, $m_1 \times m_2 \times m_3 \times m_4$ Coulomb integrals over individual Gaussian products must be done for each Coulomb integral over four basis functions.

One can, however, view each two-electron integral as representing the Coulomb interaction between two charge distributions.

$$I_{\mu\nu} = \int \frac{n_\mu(\vec{r}_1)n_\nu(\vec{r}_2)}{r_{12}}\, d\vec{r}_1 d\vec{r}_2. \tag{3.23}$$

For a contracted Gaussian lobe-function basis, $n_\mu(\vec{r})$ is given by

$$n_\mu(\vec{r}) = \phi_\alpha(\vec{r} - \vec{R}_a)\phi_\beta(\vec{r} - \vec{R}_b)$$

$$= \sum_{i=1}^{m_1 m_2} C_i e^{-\alpha_i |\vec{r}-\vec{R}_i|^2}; \tag{3.24}$$

where we use the fact that the product of two Gaussians about different centers is again a Gaussian about some third center. The approximation is to replace the sum over $m_1 m_2$ by a single Gaussian

$$n_\mu(\vec{r}) \doteq D e^{-\delta|\vec{r}-\vec{R}_d|^2}, \tag{3.25}$$

such that the total charge of $n_\mu(\vec{r})$ is correct:

$$Q_\delta = D\left(\frac{\pi}{\delta}\right)^{3/2} = \sum_i Q_i; \qquad Q_i = C_i\left(\frac{\pi}{\alpha_i}\right)^{3/2}. \tag{3.26}$$

The center of charge is correct also:

$$Q_\delta \vec{R}_d = \sum_i Q_i \vec{R}_i \qquad (3.27)$$

and δ is adjusted for a best least-squares fit to $n_\mu(\vec{r})$:

$$\zeta(\delta) = \int \left[De^{-\delta|\vec{r}-\vec{R}_d|^2} - \sum_i C_i e^{-\alpha_i|\vec{r}-\vec{R}_i|^2} \right] d\vec{r}, \qquad (3.28)$$

$$\frac{d\zeta}{d\delta} = 0. \qquad (3.29)$$

When p-symmetry functions are involved, each lobe-lobe pair in $n_\mu(\vec{r})$ is treated independently; that is two separate fits are made for s-p and four separate fits for p-p pairs. This procedure maintains the predominant dipole character of the s-p charge density as well as the essential character of the p-p charge density. In practice, all relevant two-electron integrals are first done approximately. Approximate integrals larger than some tolerance (10^{-4} for example) are then recalculated exactly. Molecular calculations suggest that the total energy per cell is uncertain by 10^{-4} a.u. due to this approximation.

Another very useful charge-conserving device is used to reduce the total number of integrals which must be done. The "optimal" total charge of each 2-basis-function product is compared against some tolerance (10^{-4} for example). By optimal total charge we mean that for s-p and p-p functions, the orientation of the p functions is taken to maximize the charge. If both functions are on a single center, the total charge is also assumed larger than the tolerance. If the charge is smaller than the tolerance, the 1-electron nuclear and overlap integrals are zeroed, as are all 2-electron Coulomb integrals involving this charge distribution. Zeroing of the overlap integral guarantees the charge-conserving nature of this approximation. Again, for tolerance of 10^{-4}. molecular studies suggest that the total energy per cell is uncertain by 10^{-4} a.u. due to this approximation.

Coulomb integrals involving far separated 2-basis-function products for which the separation is greater than the computational radius can be treated using Madelung summation techniques for each of the approximated lobes and positions. The Madelung interaction energy is calculated separately for each lobe-lobe and lobe-nuclear interaction.

Preliminary Computational Results for Diamond

Our first Hartree-Fock compuational results on diamond were presented in reference [*13*]. In that paper, the decision as to whether to use exact or approximate two-electron integrals depended upon the number of distinct centers involved for the four basis functions. The better criterion described in section 2 of the present article was introduced in reference [35] in which the total energy per cell was studied as a function of the lattice constant. Reference [36] presents the momentum distribution calculated with a relatively crude basis set. The same programs were used as were used in reference [35], but the computational boundary enclosed only 29 atoms rather than 47, and fewer basis functions were used. In references [35, 36], two-integrals smaller than some tolerance (for example 10^{-8}) were zeroed and not further processed. This procedure, which is not charge-conserving, is less preferable than the charge-conserving procedure described in section 2 in which both one- and two-electron integrals involving certain symmetry sets are systematically zeroed. In all three references, so-called monopole and dipole charge compensation terms were used at the computational boundary. This procedure is less preferable than the Madelung procedure described in section 2.

Hartree-Fock results consist of both eigenvalues of the crystal Fock operator (which by Koopman's theorem correspond to excitation energies of the non-relaxed system), and of various results derived from the first-order density matrix which is constructed from Fock eigenfunctions. The Fourier transforms of the charge density are obtained from

$$\rho_0(\vec{k}) = \frac{1}{\Omega}\int e^{i\vec{k}\cdot\vec{r}}\rho_1(\vec{r},\vec{r})d\vec{r}, \tag{3.30}$$

while the electron momentum distribution is obtained from

$$\rho_1(\vec{k},\vec{k}) = \frac{1}{(2\pi)^3\Omega}\int_\Omega e^{i\vec{k}\cdot\vec{r}}\rho_1(\vec{r},\vec{r}\,')e^{-i\vec{g}\cdot\vec{r}\,'}d\vec{r}d\vec{r}\,'\Big|_{\vec{g}=\vec{k}} . \tag{3.31}$$

The radial momentum distribution is then

$$I_0(k) = \int k^2\rho_1(k,k)d\Omega \tag{3.32}$$

obtained by integrating over solid angle. Within the framework of the impulse approximation,

$$J(g) = \frac{1}{2}\int_g^\infty \frac{I_0(k)}{k}\, dk.$$

The total energy per cell consists of the electronic kinetic energy (the kinetic energy of the lattice is taken as zero),

$$T = \frac{1}{\Omega}\int(-\nabla_r^{\,2})\rho_1(\vec{r},\vec{r}\,')\Big|_{r=r'} d\vec{r}, \tag{3.34}$$

while the potential energy is made up from electron-electron, electron-nuclear and nuclear-nuclear contributions:

$$V = \frac{1}{N\Omega}\left\{\int \sum_{a} \left(\frac{-2Z_a}{|\vec{r} - \vec{R}_a|}\right)\rho_1(\vec{r},\vec{r})d\vec{r} + \frac{N\Omega}{2}\sum_{ab}{}' \frac{2Z_aZ_b}{|\vec{R}_a - \vec{R}_b|}\right.$$

$$+ \frac{1}{2}\int \frac{2\rho_1(\vec{r},\vec{r})\rho_1(\vec{r}',\vec{r}')}{|\vec{r} - \vec{r}'|} d\vec{r}d\vec{r}'$$

$$\left. - \frac{1}{2}\int \frac{\rho_1(\vec{r},\vec{r}')\rho_1(\vec{r}',\vec{r})}{|\vec{r} - \vec{r}'|} d\vec{r}d\vec{r}' \right\} . \qquad (3.35)$$

One can compute the equilibrium lattice constant by computing the total energy for various lattice constants and finding the lattice constant for which the total energy is a minimum. At equilibrium, the virial theorem should be close to satisfied for a good basis set; i.e.,

$$\eta = -\frac{2T}{V} = 1. \qquad (3.36)$$

We shall present results for the four basis sets given in Table VII. The Hartree-Fock results are summarized in Table VIII.

The first striking aspect of Table VIII is that the results are so similar for quite different Gaussian basis. Even with the very crude basis (column 4 of Table VII) used to obtain columns 4 and 5 of Table VIII, the essential physics is obtained. The difference in basis is reflected most directly in the absolute position of the top valence band, and in the total energy per cell. The only difference between columns 4 and 5 of Table VIII is that a 29 atom computational boundary was used for column 4, while a larger 47 atom computational boundary was used for column 5. The Γ_{1v} entry gives the overall width of the top valence band of diamond. The Hartree-Fock widths are expected to be about 1.5 times the experimental widths, and they are. The Γ_{15c} energy is the direct band gap between *p*-symmetry valence and conduction bands. The Hartree-Fock band gaps are about 8 eV wider than experiment. Kunz and co-workers usually get band gaps about 4 eV wider than experiment for alkali halides and rare gas solids. The entries from Γ_{15c} down to and including X_{1c} give the bottom conduction band energies in the (100) direction of the Brillouin zone. Experimentally the absolute minimum of the conduction band is found to be in this direction between (0.5,0,0) and (0.75,0,0). The Hartree-Fock conduction bands in this direction also have their minimum in this direction at roughly (0.5,0,0), although the band from column 4 is quite warped and the dip is very small. None of the dips from Γ_{15c} to the minimum is as large as experiment. Run 2 comes closest.

TABLE VII

Contracted Gaussian Local Function, LBF, exponents (in a.u.) and coefficients, C, for four different cases. The coefficients multiply normalized individual s or p Gaussian.

LBF	1		2		3		4	
	α	C	α	C	α	C	α	C
1s	16371.074	0.00022939	16371.074	0.00022939	16371.074	0.00022939		
	2439.1239	0.00177527	2439.1239	0.00177527	2439.1239	0.00177527	1087.0968	0.00658908
	545.1677	0.00946479	545.1677	0.00946479	545.1677	0.00946479	163.86717	0.04833467
	151.0038	0.03962765	151.0038	0.03962765	151.0038	0.03962765	37.409003	0.2070169
	47.80399	0.131291	47.80399	0.131291	47.80399	0.131291	10.51807	0.4781396
	16.43566	0.32055634	16.43566	0.32055634	16.43566	0.32055634	3.3207801	0.40229784
2s	5.949118	0.7252186	5.949118	0.7252186	5.949118	0.7252186	1.0	1.0
	2.215878	0.3104604	2.215878	0.3104604	2.215878	0.3104604		
3s	1.0	1.0	0.65	1.0	0.85	1.0	0.4	1.0
4s	0.317	1.0	0.40	1.0	0.36	1.0		
1p	40.79042	0.00409698	24.17881	0.04081133	24.17881	0.04081133	1.0	1.0
	9.503463	0.02758477	5.7634925	0.23370951	5.7634925	0.23370951		
	2.940836	0.10635435	1.7994821	0.8158967	1.7994821	0.8158967		
2p	1.0	1.0	0.65	1.0	0.85	1.0	0.4	1.0
3p	0.317	1.0	0.40	1.0	0.36	1.0		

TABLE VIII

Diamond HF Calculation Results

Column 1 gives some cogent experimental information on diamond. Columns 2-6 are calculated HF results for different parameters. The row listed as basis refers to the columns of Table VII. The SCF row gives the top eigenvalues for the top of valence band as it comes from the SCF. All other band eigenvalues are given relative to this level. The Fourier transform of the charge density are in electrons per crystallographic unit cell. The last rows gives the total energy per atom, the binding energy, *BE*, per atom, the virial coefficient, and the pressure-volume product. All energies are in Rydbergs.

	Exp	Calculated Results				
	1	2	3	4	5	6
Basis		1	2	3	4	4
No. of Atoms		47	47	47	29	47
SCF		- 0.42	- 0.41	- 0.42	- 0.11	- 0.18
1*s*		-21.95	-21.97	-21.99	-22.47	-22.38
	- 1.5	- 2.16	- 2.15	- 2.17	- 2.34	- 2.36
	- 0.7	- 1.32	- 1.34	- 1.35	- 1.59	- 1.49
		- 0.58	- 0.61	- 0.62	- 0.77	- 0.67
	0.54	1.10	1.10	1.05	0.92	1.08
		1.07	1.08	1.04	0.93	1.06
		1.03	1.00	1.01	0.94	1.01
	0.4	1.07	1.02	1.05	0.91	1.06
		1.25	1.22	1.23	0.95	1.30
		1.97	1.93	1.91	1.61	1.74
	3.32	3.25	3.29	3.26	3.28	3.28
	1.98	1.94	1.93	1.93	1.83	1.82
	1.66	1.69	1.68	1.69	1.56	1.55
	0.14	0.08	0.09	0.08	0.10	0.10
	-76.23	-75.752	-75.728	-75.715	-75.297	-75.264
	0.56	0.38	0.35	0.34	- 0.08	- 0.11
	1.0	1.0001	1.0006	1.0004	1.0001	0.9999

The Fourier transforms of the charge density are in good agreement with each other for runs 1 through 3, but the disagreement with experiment is disappointing, especially for the crystalline-environment-sensitive (111) and (222). At present we don't know if this disagreement between calculation and experiment is because of generally poor Gaussian basis sets or not. There certainly are reatively large variations in the (111) from one basis to another. Probably longer-range Gaussians are needed as well as more variational freedom in the charge bonding regions.

The binding energies were calculated using the value -75.375 Ry for the Hartree-Fock atomic total energy. It would be better to use the different Gaussian basis sets to calculate different Hartree-Fock atomic total energies. However, the crystal Gaussian basis sets would have to be augmented with long-range Gaussians for the atomic calculations, and there is no clear way of how to do this. The negative binding energies for runs 4 and 5 are probably due to the relatively poor core portion of the basis and would be vastly improved if a comparable basis were used for the atomic total energy. The binding energies for runs 1 through 3 are larger than we expected. The free atom correlation energy for the carbon atom is 0.316 Ry. One would expect an increase in correlation energy in going from atom to crystal. In the C_2 molecule, the correlation energy is 0.415 Ry/atom, an increase of 0.1 Ry/atom. Estimating from this a correlation energy per C-C bond of 0.1 Ry/atom suggests that the Hartree-Fock binding energy should differ from experiment by 0.4 Ry rather than by 0.2 Ry. Better calculations should give more information on this situation.

The Compton profile, derived from the electron momentum distribution, is shown in figure 2 for the (100) crystalline direction. The dashed line gives the result of free-atom Hartree-Fock charge distributions placed in the solid lattice. Even with the crude (3*s*/2*p*) basis (column 4 of Table VII), the Hartree-Fock crystalline Compton profile is strikingly closer to the experimental curve. The overall agreement is very satisfactory. This agreement indicates that the SCF Hartree-Fock calculations have correctly accounted for the crystalline effects that contribute to the Compton profile, i.e., the confinement of electrons in real space compared to the diffuse atomic case. The results also indicate that correlation effects on $J(q)$ are apparently small for diamond, since the Hartree-Fock results are uncorrelated.

Calculations of $J(q)$ for other crystalline directions have been carried out with similar results. The anisotropies of $J(q)_{111}-J(q)_{100}$ observed by experiment (references [*37*,*38*]) are not obtained in the calculations. The charge densities from SCF Hartree-Fock calculations are heavily *s*-like. It is anticipated that a larger basis will correct this in the future. This must be investigated further, particularly with functions in the [111] bonding directions. The bulk modulus, B, and equilibrium lattice constant, a_0, for diamond were calculated using the basis

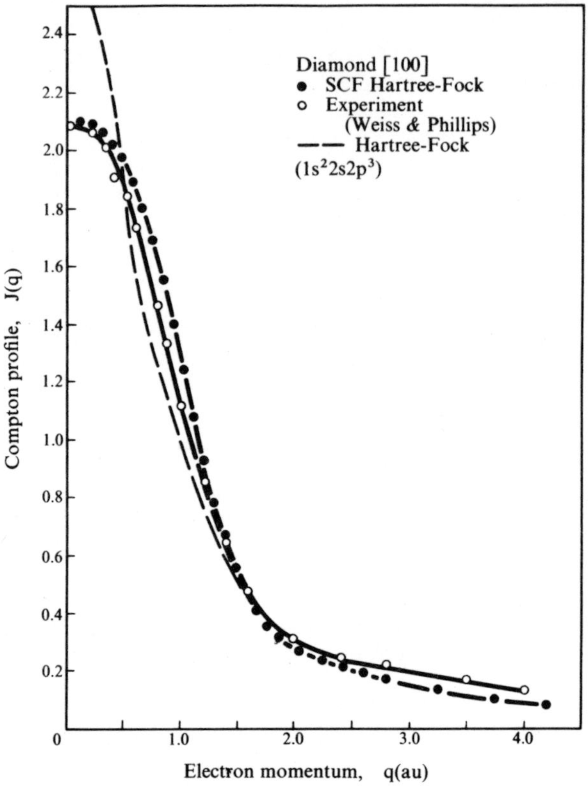

Figure 2 - Comparison of the calculated Compton profile for the [100] direction in diamond with X-ray measurements of reference [*37*]. The dashed line is an atomic Hartree-Fock calculation for carbon ($1s^2 2s2p^3$).

set in column 3 of Table VII. The calculation was done for five lattice constants, and the results are given in figure 3.

These points were fitted to a parabola with and without the point 3.61 A. The resulting curves are labelled 3^5 and 3^4, respectively. The B's and a_0's determined from these curves are given in Table IX. The lattice constant is shorter than experiment, which is the usual Hartree-Fock result. The bulk modulus is reasonably close (within 5%) to experiment. By comparison, (Table IX) the values obtained by Goroff and Kleinman [*39*] are very dependent upon how they fit their calculated points to a parabola. They used an OPW calculation with Slater exchange followed by a correlation correction. Their best correlated value is not as good as Hartree-Fock. If one subtracts the correlation correction and fits to all their calculated points, the Slater exchange values can be determined. These values show that within the model used by Goroff and Kleinman, correlation was needed to bring the results in line with experiment, while

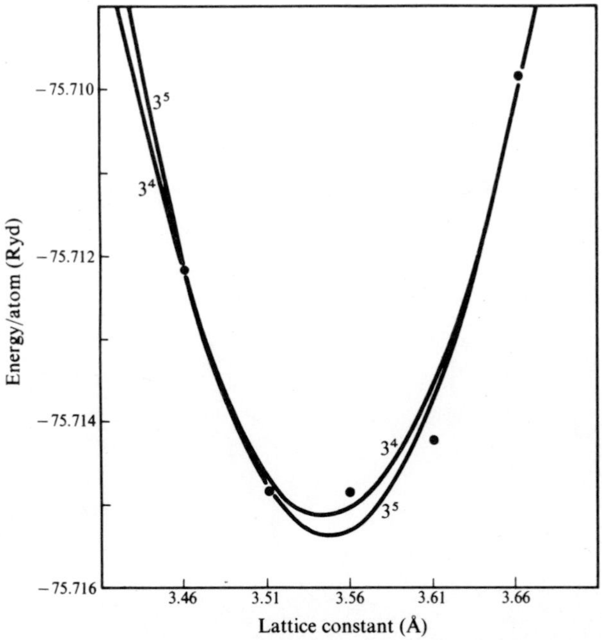

Figure 3 - The least-square fitted parabolas for 3^4 and 3^5 points. The calculated points are denoted by •.

the Hartree-Fock results are by definition uncorrelated.

4. SOME CORRECTIONS TO THE HARTREE-FOCK MODEL

Let's look at a simple derivation of the very difficult problem [42] of formulating correlation operators which will reproduce the total ground state energy, and hopefully lead to operators which will give relaxed excited states of the N particle system. The total energy may be defined as:

$$E = \frac{\langle D|\mathcal{H}_e|\Psi\rangle}{\langle D|\Psi\rangle} \tag{4.1}$$

Assume intermediate normalization, i.e. $\langle D|\Psi\rangle = 1$ where Ψ is the true eigenstate of the system and D is the Slater determinant. Define the operators

$$\Psi = \hat{W}D \qquad \text{and} \qquad \hat{t} = V\hat{W}. \tag{4.2}$$

The first operator, $\hat{W}$, is called the "wave operator" and $\hat{t}$ is called the "reaction operator". Now the Hamiltonian may be written in the following form:

$$\mathcal{H}_e = \sum_i \mathcal{H}_i + \sum_{i<j} \mathcal{H}_{ij} = \sum_i (\mathcal{H}_0 + \mu_i) - \sum_i \mu_i + \sum_{i<j} \mathcal{H}_{ij} = \mathcal{H}_0 + V. \tag{4.3}$$

TABLE IX

Comparison of calculated and experimental lattice constants and bulk moduli for diamond. The Goroff and Kleinman results come from reference [*39*]. The experimental lattice constant comes from reference [*40*] and bulk moduli from reference [*41*].

	Lattice Constant a_0 (Å)	Bulk Moduli (10^{12} dynes/cm^2)
Hartree-Fock		
4-point fit	3.544 ± 0.003	4.38 ± 0.30
5-point fit	3.547 ± 0.002	4.64 ± 0.27
Reference [*39*]		
Slater Exchange	3.515	3.04
plus Nozieres-Pines correlation	3.544	4.69
Reference [*39*]		
Slater Exchange	3.569	3.16
plus Gell-Mann Brueckner correlation	3.576	4.69
Reference [*39*]		
Slater Exchange	3.611	3.74
Experiment	3.567	4.42

In this case,

$$E^{\mathrm{HF}} = \langle D|\mathcal{H}_0 + V|D\rangle, \tag{4.4}$$

and the correlation energy is defined as:

$$E_{\mathrm{corr}} = \langle D|\hat{t} - V|D\rangle \tag{4.5}$$

where one calls $\hat{t} - V$ the correlation operator.

In the electron-gas model there have been a number of approximations to $E_{\mathrm{corr}} = \frac{1}{2}\int \varepsilon_c(\vec{r})\rho(\vec{r})d\vec{r}$. For example, Wigner [*43*] obtained

$$\varepsilon_c = -\frac{0.576}{r_s + 5.1}\,\mathrm{Ry}, \tag{4.6}$$

where r_s is the radius of a unit charge of the particular electron density. More examples are: Wigner and Seitz [*44*] derived

$$\varepsilon_c = \frac{1}{2}\left[-0.005 - \frac{1}{16.2 + 3.07\, r_s}\right] \text{Ry} \tag{4.7}$$

and Brueckner and Gell-Mann [*45*] found

$$\varepsilon_c = [-0.096 + 0.0611 \ln r_s]\ \text{Ry}. \tag{4.8}$$

More recently the results of Singwi, et al. [*46*] were fitted by Hedin and Lundqvist [*47*] to a correlation density functional:

$$\varepsilon_c = -0.045\left[(1 + x^3)\ln\left(1 + \frac{1}{x}\right) + \frac{x}{2} - x^3 - \frac{1}{3}\right] \text{Ry} \tag{4.9}$$

where

$$x = \frac{r_s}{21}.$$

TABLE X

The correlation energy of atomic Ne, Ar, and Fe calculated using the approximations of references [*43*-*46*]. The units are in Rydbergs.

Atom	Experiment	W [*43*]	WS [*44*]	GMB [*45*]	S [*46*]
Ne	0.66 — 0.78	0.99	0.59	2.62	1.50
Ar	1.52	1.79	1.07	5.09	2.84
Fe	2.33	2.63	1.57	8.02	4.33

The results of calculating the correlation energy in the above approximations are given in Table X for Ne and Ar and Fe. One finds the correlation energy obtained by reference [*43*] to be fair. The results obtained from equation (4.7.) are too small and the results of equation (4.8) too large. The results of Singwi, et al. [*46*] seem to be off by a factor of 2. In obtaining the self-consistent results one also obtains an operator in the wave equation called μ_c and an eigenvalue. In Table XI the results of the top eigenvalue obtained in the different approximations are given. One sees very close agreement between experiment and Singwi's approximation for this top state.

TABLE XI

The top eigenvalue of the operator derived using different approximations for the correlation operator. The units are in Rydbergs.

Atom	Experiment	W [*43*]	WS [*44*]	GMB [*45*]	S [*46*]
Ne	1.58	1.52	1.49	1.62	1.56
Ar	1.16	1.15	1.12	1.21	1.18
Fe	0.58	0.57	0.55	0.56	0.58

Hedin and Lundqvist developed a local effective-mass operator using the results of reference [*46*] in the following equation:

$$\left[-\frac{\hbar^2\nabla^2}{2m} + V(\vec{r})\right]\phi_{\vec{K}}(\vec{r}) + \int \Sigma(\vec{r},\vec{r}\,';\varepsilon_{\vec{K}})\phi_{\vec{K}}(\vec{r}\,')d\vec{r}\,' = \varepsilon_{\vec{K}}\phi_{\vec{K}}(\vec{r}), \quad (4.10)$$

where Σ is the mass operator. We used their results to calculate the excitation energies of atomic Ar. These results are given in Table XII. One sees that the results given in column 3 match the results of Kohn-Sham-Gaspar given in column 2. If one adds a gradient term suggested by Herman, et al. [*48*] the results are improved, and are given in the other columns of Table XII for different values of the coefficient of the gradient term as well as for variations of the parameters suggested by reference [*47*]. One general conclusion is that this method does not work for atoms. When tried on crystalline systems without the gradient terms, one obtains Kohn-Sham-Gaspar like results. The crystal used was Si.

Our next attempt to generate excitation energies concerns the fact that in the next approximation beyond Hartree-Fock one obtains a screened exchange plus a Coulomb hole. Now the Coulomb hole is thought to give a constant shift to the energy spectrum, whereas, the screened exchange term gives the shift between the levels. In section 2 it was shown that excitation values obtained from Liberman's exchange approximation match those of Hartree-Fock. Thus we will derive a screened Liberman approximation. Instead of a $1/r$ potential we will use $e^{-k_T r}/r$ [*49*]. This, of course, leads to a dielectric function of the form

$$\varepsilon(k) = 1 + \frac{k_T{}^2}{k^2} . \quad (4.11)$$

TABLE XII

The excitation energies of atomic Ar using the approximation of reference [*47*] for different parameters plus the effects of adding a gradient term [*48*]. The coefficient of the gradient term is β. The units are in Rydbergs.

Level	Experiment	KS [*17*]	HL [*47*] $x = \frac{n_s}{21}$ $\beta = 0$	HL [*47*] $x = \frac{n_s}{15}$ $\beta = 0.008$
$1S^{1/2}$	235.42	228.41	228.57	235.64
$2S^{1/2}$	23.52	21.62	21.75	23.00
$2P^{1/2}$	18.18	16.87	17.00	17.84
$2P^{3/2}$	18.02	16.71	16.84	17.69
$3S^{1/2}$	1.86	1.68	1.79	1.99
$3P^{1/2}$	1.17	0.68	0.78	0.94
$3P^{3/2}$	1.16	0.66	0.76	0.96

The screened exchange differs from Liberman by replacing the $F(\eta)$ function by $S(\eta)$, where

$$S(\eta) = \frac{1}{2} + \frac{\tau^2 + 1 - \eta^2}{8\eta} \ln\left[\frac{\tau^2 + (\eta + 1)^2}{\tau^2 + (\eta - 1)^2}\right] - \frac{\tau}{2}\left[\tan^{-1}\left(\frac{1 + \eta}{\tau}\right) + \tan^{-1}\left(\frac{1 - \eta}{\tau}\right)\right] . \qquad (4.12)$$

$\tau = k_T/k_F$, and is of the form of a constant times $r_s^{\frac{1}{2}}$. For example, the Thomas-Fermi screening is $\tau_{TF} = 0.82\ r_s^{\frac{1}{2}}$. Equations similar to equation (4.12) have been obtained by Ferreira [*50*].

In figure 4 is a plot of $S(\eta)$ along with $F(\eta)$ for different r_s values. τ was taken to be the Thomas-Fermi value. Note that for $r_s \approx 2$, it becomes constant up to the Fermi surface although it is very small. Another important point is that the self-energy term must be taken out. If one screens the Liberman exchange approximation, one is screening the self-energy term as well, and there is no longer the right correlation between the

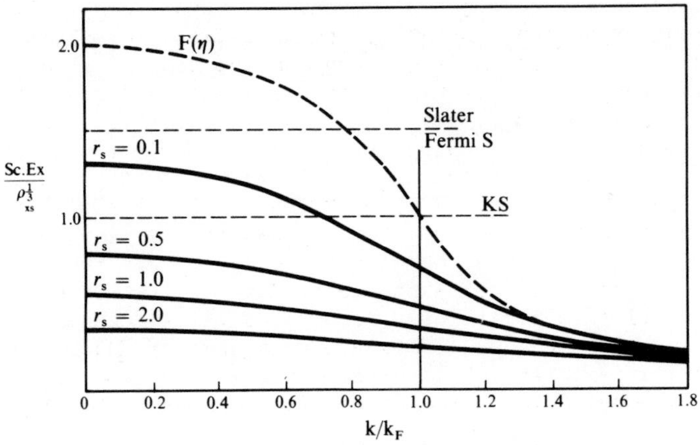

Figure 4 - $S(\eta)$ vs η for different values of r_s. The screening length was taken to be the Thomas Fermi value.

Coulomb and exchange terms. The results of the calculation for atomic Ar are given in Table XIII for various screening lengths. The most striking observation is that when τ_{TF} is used, the results are worse than any of the previous results. However, when $\tau = 0.04\, r_s^{\frac{1}{2}}$, a fair fit to experiment is obtained. Thus one

TABLE XIII

The excitation energies of atomic Ar using the approximation of reference [*49*] using different screening lengths. The units are in Rydbergs.

Level	Experiment	$\tau = 0.82\, r_s^{\frac{1}{2}}$	$\tau = 0.39\, r_s^{\frac{1}{2}}$	$\tau = 0.04\, r_s^{\frac{1}{2}}$
$1S^{1/2}$	235.42	223.45	223.32	235.73
$2S^{1/2}$	23.52	20.52	21.00	24.16
$2P^{1/2}$	18.18	15.15	15.45	18.48
$2P^{3/2}$	18.02	15.01	15.31	18.32
$3S^{1/2}$	1.86	1.84	2.08	2.57
$3P^{1/2}$	1.17	0.83	0.93	1.18
$3P^{3/2}$	1.16	0.82	0.91	1.17

needs long screening lengths which would be beyond simple first term gradient expansions for atoms and most likely crystals. A discussion of this problem is given by Ma and Brueckner [*51*] and also by B.Y. Tong [*52*]. In fact, the screening is done by the outer electrons with the inner electrons being tightly bound. Thus one needs a long screening length to let these electrons rearrange themselves. We conclude from this study that a simple dielectric function described by equation (4.11) is not sufficient to describe the electronic system of atoms or crystals.

We are led to a study of more reliable dielectric functions with which to screen the exchange to produce excitation energies. In this we will show preliminary results of time-dependent Hartree (or Random-Phase Approximation) and simple extensions of this approximation. The first extension will be to add the Slater or X_α exchange term to the Hamiltonian in the derivation [*53*]:

$$\mathcal{H} = h + \int V(\vec{r},\vec{r}\,')\rho(\vec{r}\,',t)d\vec{r}\,' + C\rho^{1/3}(\vec{r},t). \tag{4.13}$$

Here, h is the kinetic energy and the interaction of the electron with the nuclei, and $C = -\alpha\cdot 6(3/8\pi)^{1/3}$ with α being the exchange parameter. In order to derive an expression for the dielectric response function, consider a small external potential $V_{ext}(\vec{r},t)$ which is treated as a perturbation, and look at the response of the system. V_{ext} produces an induced charge density $\rho_1(\vec{r},t)$ so that the total charge density is

$$\rho = \rho_0(\vec{r}) + \rho_1(\vec{r},t), \tag{4.14}$$

where $\rho_0(\vec{r})$ is the unperturbed charge density. Our approximation to the exchange term becomes

$$C[\rho_0(\vec{r}) + \rho_1(\vec{r},t)]^{1/3} \approx C\rho_0^{1/3}(\vec{r})\left[1 + \frac{1}{3}\frac{\rho_1(\vec{r},t)}{\rho_0(\vec{r})} + \ldots\right]. \tag{4.15}$$

The Hamiltonian can now be written

$$\mathcal{H}_0 = \mathcal{H}_0 + \delta V, \tag{4.16}$$

where

$$\delta V(\vec{r},t) = \int V(\vec{r},\vec{r}\,')\rho_1(\vec{r}\,',t)d\vec{r}\,' + \frac{C}{3}\rho_0^{-2/3}(\vec{r})\rho_1(\vec{r},t) + V_{ext}(\vec{r},t). \tag{4.17}$$

Taking the Fourier transform with respect to time given

$$\delta V(\vec{r},\omega) = \int V(\vec{r},\vec{r}\,')\rho_1(\vec{r}\,',\omega)d\vec{r}\,'$$

$$+ \frac{C}{3}\rho_0^{-2/3}(\vec{r})\rho_1(\vec{r}\,',\omega) + V_{ext}(\vec{r},\omega). \qquad (4.18)$$

The dielectric function, ε, is defined by

$$V_{ext}(\vec{r},\omega) = \int \varepsilon(\vec{r},\vec{r}\,',\omega)\delta V(\vec{r}\,',\omega)d\vec{r}\,', \qquad (4.19)$$

and the polarization response function by

$$\rho_1(\vec{r},\omega) = \int P(\vec{r},\vec{r}\,',\omega)\delta V(\vec{r}\,',\omega)d\vec{r}\,'. \qquad (4.20)$$

Substituting equation (4.20) into equation (4.18) and comparing with equation (4.19) gives

$$\varepsilon(\vec{r},\vec{r}\,',\omega) = \delta(\vec{r},\vec{r}\,') - \int V(\vec{r},\vec{r}\,'')P(\vec{r}\,'',\vec{r}\,',\omega)d\vec{r}\,''$$

$$- \frac{C}{3}\rho_0^{-2/3}(\vec{r})P(\vec{r},\vec{r}\,',\omega). \qquad (4.21)$$

The dielectric function above is a two-body function, depending on the coordinates $\vec{r}$ and $\vec{r}\,'$, and if Fourier transform of the coordinates is taken now, one has ε as a function of $\vec{q}$ and $\vec{q}\,'$. At this point, assume a translationally invarient system so that ε and P will be functions of $|\vec{r} - \vec{r}\,'|$ only. Then the Fourier transform becomes

$$\varepsilon(\vec{q},\omega) = 1 - \left[V(\vec{q}) + \frac{C}{3}\overline{\rho_0^{-2/3}}\right]P(\vec{q},\omega) \qquad (4.22)$$

where

$$V(\vec{q}) = \frac{4\pi e^2}{q^2}\,; \qquad \overline{\rho_0^{-2/3}} = \frac{1}{\Omega}\int \rho_0^{-2/3}(\vec{r})d\vec{r}.$$

Ω is the volume of the system. To calculate $P(\vec{q},\omega)$, look at equation (4.20). The induced charge density can be obtained from the density matrix, and the density matrix satisfies the time-dependent Hartree-Fock-Slater equation. So solving the time-dependent equation leads to

$$P(\vec{q},\omega) = \frac{2}{\Omega}\sum_{\vec{k},\ell\ell'} \frac{(n_{\vec{k}+\vec{q},\ell} - n_{\vec{k},\ell'})|\langle \vec{k},\ell'|e^{-i\vec{q}\cdot\vec{r}}|\vec{k}+\vec{q},\ell\rangle|^2}{\varepsilon_{\vec{k}+\vec{g},\ell} - \varepsilon_{\vec{k},\ell'} - \omega - i\delta}\,, \qquad (4.23)$$

where $\vec{k}$ goes over values in the first Brillouin zone, the sums on ℓ and ℓ' are over all bands, $n_{\vec{k},\ell}$ is the occupation number of the state $|\vec{k},\ell\rangle$ and $\varepsilon_{\vec{k},\ell}$ is the eigenvalue. Thus for ε one gets

$$\varepsilon(\vec{q},\omega) = 1 - \frac{2}{\Omega}\left[\frac{4\pi e^2}{q^2} + \frac{C}{3}\overline{\rho_0^{-2/3}}\right]$$

$$\times \sum_{\vec{k},\ell\ell'} \frac{(n_{\vec{k}+\vec{q},\ell} - n_{\vec{k}\ell'})|\langle \vec{k},\ell'|e^{-i\vec{q}\cdot\vec{r}}|k+q,\ell\rangle|^2}{\varepsilon_{\vec{k}+\vec{q},\ell} - \varepsilon_{\vec{k}\ell'} - \omega - i\delta}. \quad (4.24)$$

Without the $\overline{\rho_0^{-2/3}}$ term this is the random-phase approximation for the dielectric function.

Now

$$\varepsilon(\vec{q},\omega) = \varepsilon_1(\vec{q},\omega) + i\varepsilon_2(\vec{q},\omega), \quad (4.25)$$

where

$$\varepsilon_1(\vec{q},\omega) = 1 + \frac{2}{\Omega}\left[\frac{4\pi e^2}{q^2} + \frac{C}{3}\overline{\rho_0^{-2/3}}\right]$$

$$\times \sum_{\vec{k},cv} \left\{\frac{|\langle\vec{k},v|e^{-i\vec{q}\cdot\vec{r}}|\vec{k}+\vec{q},c\rangle|^2}{\varepsilon_{\vec{k}+\vec{q},c} - \varepsilon_{\vec{k},v} - \omega}\right.$$

$$\left. + \frac{|\langle\vec{k},c|e^{-i\vec{q}\cdot\vec{r}}|\vec{k}+\vec{q},v\rangle|^2}{\varepsilon_{\vec{k},c} - \varepsilon_{\vec{k}+\vec{q},v} + \omega}\right\}. \quad (4.26)$$

The sums on c and v are over conduction and valence bands,

$$\varepsilon_2(\vec{q},\omega) = \frac{2\pi}{\Omega}\left[\frac{4\pi e^2}{q^2} + \frac{C}{3}\overline{\rho_0^{-2/3}}\right]$$

$$\times \sum_{\vec{k},cv} |\langle\vec{k},v|e^{-i\vec{q}\cdot\vec{r}}|\vec{k}+\vec{q},c\rangle|^2 \delta(\varepsilon_{\vec{k}+\vec{q},c} - \varepsilon_{\vec{k},v} - \omega). \quad (4.27)$$

The wave functions used in the matrix elements in this work are LCAO wave functions:

$$|\vec{k},\ell\rangle = \frac{1}{\sqrt{N}} \sum_{i,n} b_{\ell i}(\vec{k}) e^{i\vec{k}\cdot\vec{R}_n} \phi_i(\vec{r} - \vec{R}_n), \quad (4.28)$$

where N is the number of unit cells, ℓ is the band index, the sum on n is over all direct lattice sites $\vec{R}_n$, the sum on i is over atomic orbitals, and the atomic orbitals are Gaussian orbitals. The crystal used for this calculation was NaF.

Figure 5 is a graph of the random phase ε_1 along the [100] axis. Now the largest difference between random phase and the

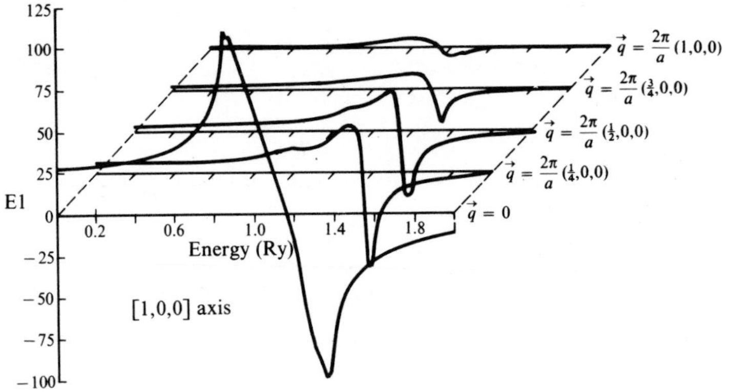

Figure 5 - Random phase dielectric function ε_1, along the [100] axis of NaF.

dielectric function derived here (time-dependent Hartree-Fock-Slater) will be at the zone boundaries, because for larger $\vec{q}$ the term $4\pi e^2/q^2$ is smaller compared to $(C/3)\overline{\rho_0^{-2/3}}$. The difference is shown in figure 6. The upper curve is the random phase and the lower curve is the Hartree-Fock-Slater. One observes very little difference between the two. Figure 7 shows a comparison between the Hartree-Fock and the random-phase approximations. One obtains the Hartree-Fock approximation by expanding the random-phase approximation in term of the polarization and retaining only the first term. One observes that there is a large difference between the two. In order for the two to be close, the polarization term must be small. For NaF this is not the case.

As soon as we have a good approximation to the dielectric function we will return to obtaining screened exchange operators. With this screened exchange operator, we hope to generate more correct excitation energies for the electronic systems.

5. CONCLUSION

The basic overall conclusion that can be made from this work is that simple free-electron approximations (local exchange) do not work well in generating excitation spectra. There does not appear to be any simple extension of the local potential theory which will lead to more correct results. One has to return to

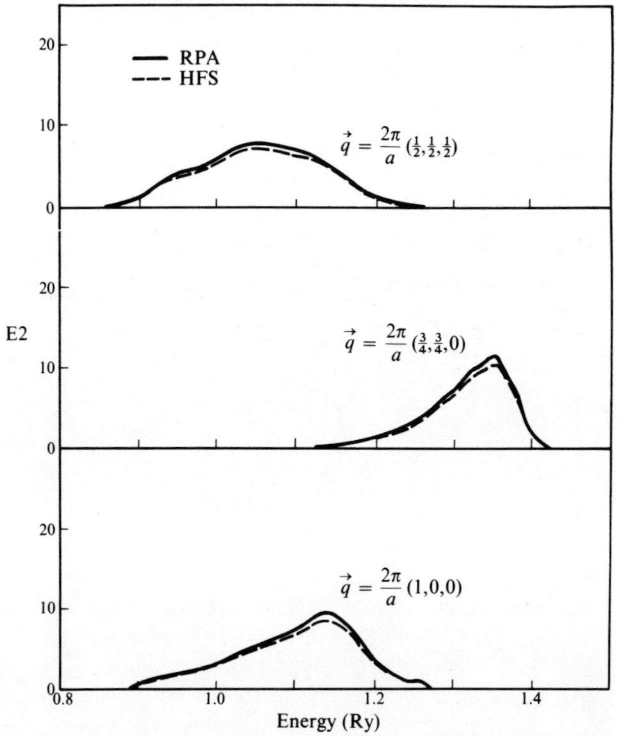

Figure 6 - Comparison of ε_2 in the random phase and Hartree-Fock-Slater approximation. Upper curve is the random phase and lower curve is the Hartree-Fock-Slater.

more fundamental foundations, namely the Hartree-Fock Atom and Crystal electronic systems. Then one obtains excitation Hamiltonians which include relaxation and correlation effects. As can be seen from this study, the crystal as well as the atom has local excitations. These excitations break symmetry if only a single Slater determinant is used in the crystal. However, the symmetry can be restored if one uses a configuration-interaction picture to describe the system.

There are many ways to build excitation Hamiltonians similar to the ones given here. In other talks at this NATO school we will hear about the electronic polaron [*54*] and the extension of this idea to obtain excitation energies of atoms and molecules [*55*].

Another very important point is that the Hartree-Fock ground state values match experiment quite well. And this is really the foundation on which to develop the excitation spectra and correlated ground state values. It is a large step from the free electron gas approximation to the Quantum chemistry point of view in obtaining theoretical solid state models.

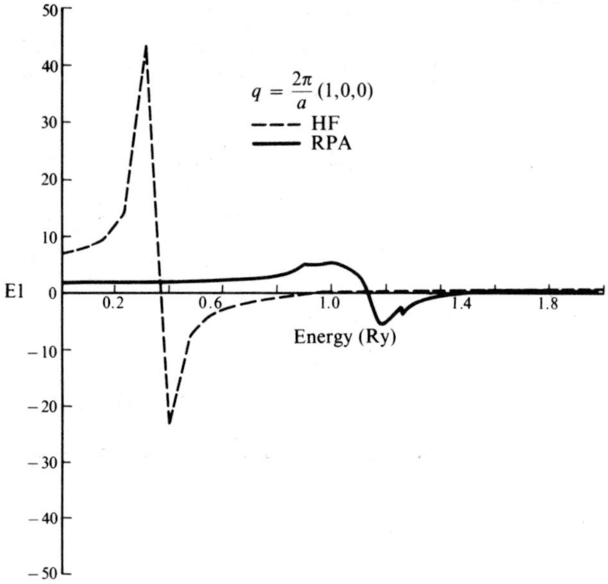

Figure 7 - Comparison of ε_2 in the random phase and the Hartree-Fock approximation.

REFERENCES

1. Born, M. and Huang, K. (1954). *Dynamical Theory of Crystal Lattices*, (Oxford University Press).
2. Sham, L.J. *Proceedings of NATO Advanced Study Institute on Elementary Excitations in Solids, Molecules and Atoms.*
3. Fock, V. (1930). *Z. Phys.*, **61**, 126. See also the next reference.
4. Löwdin, P.O. (1955). *Phys. Rev.*, **97**, 1490.
5. Löwdin, P.O. (1969). *Advances in Chemical Physics, Vol.XIV*, (eds. Lefebvre, R. and Moser, C.), (Interscience Publishers), p. 283.
6. Slater, J.C. (1960). *Quantum Theory of Atomic Structure*, (McGraw-Hill Book Company, Inc.), chapter 17, pp. 1-30.
7. Koopman, T. (1933). *Physica*, **1**, 104.
8. Herring, C. (1940). *Phys. Rev.*, **57**, 1169.
9. Slater, J.C. (1930). *Phys. Rev.*, **36**, 57.
10. Euwema, R.N. (1971). *Phys. Rev.*, **B4**, 4332.
11. Kelly, H.P. (1969). *Adv. Chem. Phys.*, **14**, 129.
12. Scofield,D.F., Dutta, N.C. and Dutta, C.M. (1972). *Int. J. Quan. Chem.*, **6**, 9.
13. Euwema, R.N., Wilhite, D.L. and Surratt, G.T. (1973). *Phys. Rev.*, **B7**, 818.
14. Kunz, A.B. (1969). *Phys. Stat. Sol.*, **36**, 301.

15. Slater, J.C. (1951). *Phys. Rev.*, **81**, 385.
16. Gaspar, R. (1954). *Acta Phys. Sci. Hung.*, **3**, 263.
17. Kohn, W. and Sham, L.J. (1965). *Phys. Rev.*, **A140**, 1133.
18. Slater, J.C. (1971). *Computational Methods in Band Theory*, (eds. Marcus, P., Janak, J. and Williams,A.),(Plenum, N.Y.).
19. Liberman, D.A. (1966). *Phys. Rev.*, **171**, 1.
20. Raccah, P.M., Euwema, R.N., Stukel, D.J. and Collins, T.C. (1970). *Phys. Rev.*, **B1**, 756.
21. Brillouin, L. (1933). *Actual. Sci. Ind.*, No. 71; (1934). *Actual. Sci. Ind.*, No. 159.
22. Hedin, L. and Lundqvist, S. (1969). *Solid State Physics*, *Vol. 23*, (eds. Seitz, F., Turnbull. D. and Ehrenreich, H.), (Academic Press, New York).
23. Löwdin, P.O. (1962). *Rev. Mod. Phys.*, **34**, 520.
24. Löwdin, P.O. (1955). *Phys. Rev.*, **97**, 1474.
25. By a contracted set of Gaussians, we mean that the $\{A_\alpha\}$ in equation (3.1) are fixed and only one coefficient multiplying the basic function ϕ is allowed to have variational freedom in the SCF calculation.
26. Pruess, H. (1956). *Z. Naturforsch*, **11**, 823.
27. Whitten, J.L. (1963). *J. Chem. Phys.*, **39**, 349; (1966). *J. Chem. Phys.*, **44**, 359.
28. Petke, J.D., Whitten, J.L., and Douglas, A.W. (1969). *J. Chem. Phys.*, **51**, 256.
29. Huzinaga, S. (Unpublished).
30. Slater, J.C. (1965). *Quantum Theory of Molecules and Solids*, *Vol. 2*, (McGraw-Hill, New York).
31. Dirac, P.A.M. *Proc. Camb. Phil. Soc.*, **26**, 376.
32. The atom at zero is the first shell, the four nearest neighbors comprise the second shell, etc..
33. Roothan, C.C.J. (1960). *Rev. Mod. Phys.*, **32**, 179.
34. Wendroff, B. (1966). *Theoretical Numerical Analysis*, (Academic Press, New York).
35. Surratt, G.T., Euwema, R.N. and Wilhite, D.L. (1973). *Int. J. Quan. Chem.*, (to be published).
36. Wepfer, G.G., Euwema, R.N., Surratt, G.T. and Wilhite, D.L. (1973). *Int. J. Quan. Chem.*, (to be published).
37. Weiss, R.J. and Phillips, W.C. (1968). *Phys. Rev.*, **176**, 900.
38. Reed, W.A. and Eisenberger, P. (1972). *Phys. Rev.*, **B6**, 4596.
39. Goroff, I. and Kleinman, L. (1970). *Phys. Rev.*, **131**, 2574.
40. Thewlis, J. and Davey, A.R. (1958). *Phil. Mag.*, **1**, 409.
41. McSkimin, H.J. and Bond, W.L. (1957). *Phys. Rev.*, **105**, 116.
42. Löwdin, P.O. (1971). Treatment of Exchange and Correlation Effects in Crystals, in *Proceedings of the Wilbad Conference*, (IBM).
43. Wigner, E.P. (1934). *Phys. Rev.*, **46**, 1002.
44. Wigner, E.P. and Seitz, F. (1933). *Phys. Rev.*, **43**, 804.
45. Gell-Mann, M. and Brueckner, K. (1957). *Phys. Rev.*, **106**, 364.
46. Singwi, K.S., Sjölander, A., Tosi, M.P. and Land, R.H. (1970). *Phys. Rev.*, **B1**, 1044.
47. Hedin, L. and Lundqvist, B.I. (1971). *J. Phys.*, **C4**, 2064.

48. Herman, F., Van Dyke, J.P. and Ortenburger, I.B. (1969). *Phys. Rev. Lett.*, **22**, 325.
49. Collins, T.C., Euwema, R.N. and Kunz, A.B. (1972). *Int. J. Quan. Chem.*, **6**, 459.
50. Leite, José R. and Ferreira, Luiz G. (1971). *Phys. Rev.*, **A3**, 1224.
51. Ma, S. and Brueckner, K. (1968). *Phys. Rev.*, **165**, 18.
52. Tong, B.Y. (1971). *Phys. Rev.*, **A4**, 1375.
53. Brener, N.E. (1973). *J. Quan. Chem.*, (to be published).
54. Kunz, A.B. (1972). *Phys. Rev.*, **B2**, 606; Devreese, J.T., Kunz, A.B. and Collins, T.C. (1972). *Solid State Commun.*, **11**, 673.
55. Collins, T.C., Devreese, J.T., Evrard, R. and Kunz, A.B. (To be published).

AB INITIO ENERGY BAND METHODS†

A. BARRY KUNZ

Department of Physics and Materials Research Laboratory, University of Illinois, Urbana, Illinois 61801

1. INTRODUCTION

The beginning of this work was in 1964, when I was a graduate student. At that point, any detailed experimental data relating to the band structure of the alkali-halide crystals was lacking. The only generally accepted results for these materials was the identification of the shoulder in the optical absorption of these substances on the high energy side of the exciton absorption as being due to a band to band transition at Γ, the center of the Brillouin Zone. These transitions were identified using the usual BSW notation as being from the *p*-like valence band (Γ_{15}) to an *s*-like conduction band (Γ_1) [*1*]. There appeared, at this time, an attempt by Phillips to identify virtually every spectra feature of these materials in terms of excitons at critical points and in terms of van Hove singularities [*2*]. The attempts of Phillips were shortly shown to be in error [*3*,*4*,*5*].

Therefore, due to the lack of experimental understanding of the optical absorption spectra of the Alkali Halides (and also the Rare Gas solids for that matter) in terms of band models, any attempt to construct a band structure of these materials based upon any empirical potential model seemed doomed to failure. In addition, such techniques as cyclotron resonance which might yield band curvatures were of no use in these substances, since due to the ionic nature of the host, the effective mass

† Work supported in part by Contract No. 33-615-72-C-1506 issued by the Aerospace Research Laboratory, Air Foce Systems Command, USAF Wright-Patterson AFB, Ohio, and by the National Science Foundation under Grant GH-33634.

of the carriers is severely modified by polaron processes.

Based upon such considerations and also upon personal preference for techniques without parameters, the decision was made to attempt *ab initio* studies. The problem then becomes: (1) define *ab initio*, (2) develop computationally realistic techniques, (3) look at experiments critically.

In the present context, *ab initio* band calculations are minimally said to imply solution to the Hartree-Fock equation for the solid to an accuracy greater than the errors introduced by the neglect of correlation.

Let us briefly review the important features of Hartree-Fock Theory as it applied to solids. In many cases, the solid's ground state is non-degenerate. We, therefore, postulate that (neglecting motion of the nuclei and their mutual repulsion) the Hamiltonian is given as

$$\mathcal{H} = -\sum_{i} \nabla_i^2 - 2\sum_{i,I} \frac{Z_I}{|\underset{\sim}{R}_I - \underset{\sim}{r}_i|} + \sum_{i,j}{}' \frac{1}{|\underset{\sim}{r}_i - \underset{\sim}{r}_j|} . \tag{1}$$

Here we use upper case letters to denote nuclear properties and lower case letters to denote electronic coordinates. $\hbar = 1$, $e = \sqrt{2}$, $m = 0.5$, the unit of energy is the Rydberg and the unit of length is the Hydrogen atom Bohr radius. We postulate that the solution to

$$\mathcal{H}\psi_0 = E_0\psi_0, \tag{2}$$

the equation defining the ground state of the system may be approximated as

$$\psi_0 \approx \frac{1}{\sqrt{N!}} \det\{\phi_i(\underset{\sim}{r}_\alpha)\}, \tag{3}$$

N is the total number of electrons. We may require

$$\int \phi_i^*(\underset{\sim}{r}_\alpha)\phi_j(\underset{\sim}{r}_\alpha)d\underset{\sim}{r}_\alpha = \delta_{ij}.$$

The imposition of the requirement on orthonormality is a convenience in some cases. It is, however, non-essential. At a later point, the author will exploit this factor. In this case, we find that the ϕ_i's may be labelled by quantum numbers; α for the irreducible representation of the point group in question, $\underset{\sim}{k}$ denoting a reciprocal lattice vector and n a principal quantum number. One finds by employing the variational theorem for the wavefunction of form equation (3),

$$F\phi_n^\alpha(\underset{\sim}{K},\underset{\sim}{r}) = \varepsilon_n^\alpha(\underset{\sim}{K})\phi_n^\alpha(\underset{\sim}{K},\underset{\sim}{r}). \tag{4}$$

If we define

$$\rho(\underset{\sim}{r},\underset{\sim}{r}') = \sum_{\substack{\alpha n\underset{\sim}{K} \\ \text{occupied}}} \phi_n^\alpha(\underset{\sim}{K},\underset{\sim}{r})\phi_n^{\alpha\dagger}(\underset{\sim}{K},\underset{\sim}{r}'), \tag{5}$$

one has

$$F = -\nabla^2 - \sum_I \frac{2Z_I}{|\underset{\sim}{R}_I - \underset{\sim}{r}|} + 4\int\frac{\rho(\underset{\sim}{r}',\underset{\sim}{r}')}{|\underset{\sim}{r} - \underset{\sim}{r}'|}\,d\underset{\sim}{r}' - \frac{2\rho(\underset{\sim}{r},\underset{\sim}{r}')}{|\underset{\sim}{r} - \underset{\sim}{r}'|}\,. \tag{6}$$

It is clear from equations (4-6) that the equation to be solved for a given occupied orbital involved a knowledge of all N occupied orbitals. This equation is clearly non-local (i.e., state dependent). We adopt the usual procedure introduced by Koopman and identify the eigenvalues of (4), $\varepsilon_n^\alpha(K)$, as being the negative of the amount of energy required to remove the electron from the crystal [*6*]. We also identify the virtual solutions (i.e., the unoccupied orbitals) of (4) with the excited states of the solid. It is noted that whatever computational deficiencies are introduced by this, at least the problem of the so-called Koopman's corrections is avoided [*7*]. It is also noted that all correlation other than that which comes from having an antisymmetric wavefunction is omitted.

2. TECHNICAL DIFFICULTIES

A SELF-CONSISTENCY

There are a number of immediate and obvious difficulties associated with operators of the form (4-6). One is that for the usual infinite solid, one may not be able to solve for the infinite number of occupied solutions to (4) since these equations are all coupled. There is a tendency to solve this dilemma by constructing a mesh in $\underset{\sim}{K}$-space and solving (4) for a few finite points in $\underset{\sim}{K}$-space. The wavefunctions obtained are then weighted by the volume of K-space surrounding the points solved and are used to reconstruct the density operator ρ for the crystal. Such techniques are used [*7,8*]. In our work, we have chosen to solve this problem by avoiding it.

The technique of solving the dilemma is in principle quite simple. One first allows for the orbitals to be overlapping, i.e. non-orthogonal, then one finds (using a new set of quantum numbers Ai, where A is a site label and i stands for the n_α of before).

$$\rho(\underset{\sim}{r},\underset{\sim}{r}') = \sum_{Ai,Bj} \tilde{\phi}_{Ai}(\underset{\sim}{r}) S_{AiBj}^{-1} \tilde{\phi}_{Bj}^\dagger(\underset{\sim}{r}'), \tag{7}$$

$$S_{Ai,Bj} = \langle Ai|Bj\rangle . \tag{8}$$

The summation is over all occupied orbitals. Let us define an operator A^F,

$$A^F = \rho A \rho, \tag{9}$$

where A is an arbitrary Hermitian operator. Then if one solves the equation

$$[F + \rho A \rho]\tilde{\phi}_{Bi} = \varepsilon'_{Bi}\tilde{\phi}_{Bi}, \tag{10}$$

on has using the $\tilde{\phi}$'s to form ρ by equation (7) the same ρ as if one had solved (4) [*9,10*]. Equation (10) is the Adams-Gilbert Local Orbitals Equation. Note the virtual orbitals of (10) and of (4) are identical. The process is simple in principle. For the crystal, pick a unit cell, and pick an A for that cell. Solve (10) for the charge density in that cell. By symmetry, this is the charge density in all the other cells when translated. It can now be shown that the set of solutions to (10) for all sites B, where B now denotes the unit cell in question form a complete set for the ground state Fock problem. One proves this by assuming it to be untrue. This implies that a band state is given as

$$\phi_n^\alpha(\underset{\sim}{K},\underset{\sim}{r}) = \sum_{\substack{A,i \\ =\text{occupied}}} c_{Ai}^{n\alpha\underset{\sim}{K}}\tilde{\phi}_{Ai}(\underset{\sim}{r}) + bX(\underset{\sim}{r}).$$

Here $X(r)$ is an arbitrary linear combination of the virtual solution to any of the equation (10) which is then constrained to be orthogonal to all the ϕ's in ρ with no loss of generality. Consider the matrix of F and 1 to see this:

$$\langle Ai|1|Bj\rangle = S_{Ai,Bj},$$

$$\langle Ai|1|X\rangle \equiv 0 \text{ for all } A \text{ and } i,$$

$$\langle Ai|F|Bj\rangle = \varepsilon^1_{Bj}S_{AiBj} - \langle Ai|\rho A \rho|Bj\rangle,$$

$$\langle Ai|F|X\rangle = \varepsilon_X S_{AiX} - \langle Ai|\rho A \rho|X\rangle,$$

but $S_{AiX} \equiv 0$, and

$$\langle Ai|\rho A \rho|X\rangle \equiv 0,$$

therefore $\langle Ai|F|X\rangle \equiv 0$. Thus the $\tilde{\phi}$'s are complete on the occupied space of equation (4). Therefore we have chosen to concentrate on (10). The reason for this is clear: By use of symmetry one has only a few distinct orbitals to find, one for each electron on the average in a given unit cell. Clearly the sum over the Brillouin Zone is avoided. Further if the orbitals are spatially local, and we have found them to be so, only a finite number of unit cells surrounding the one in question need be considered. In fact, due to our relaxing the orthogonality requirement, the local orbitals are found to be substantially more local than Wannier functions or even the free atomic orbitals.

Another advantage to the employment of localized orbitals is the possibility of systematic approximations to the equations. The author has studied this question and found in the limit of small overlap, S, one may retain terms in (10) which are linear only in S and thus find a 'first order' local orbitals equation [*11*]

$$[F_B + U_B - \epsilon_{Bi}']\tilde{\phi}_{Bi} = - \left[\sum_{j,k} \tilde{\phi}_{Bj}\langle Bj|A|Bk\rangle\tilde{\phi}^{\dagger}_{Bk}\right]\tilde{\phi}_{Bi}, \qquad (11)$$

with

$$F_B = -\nabla^2 - \frac{2Z_B}{|\underset{\sim}{r} - \underset{\sim}{R}_B|} + 2\int\frac{\rho_B(\underset{\sim}{r}',\underset{\sim}{r}')}{|\underset{\sim}{r}' - \underset{\sim}{r}|}\,d\underset{\sim}{r}' - \frac{2\rho_B(\underset{\sim}{r},\underset{\sim}{r}')}{|\underset{\sim}{r} - \underset{\sim}{r}'|}, \qquad (12)$$

$$U_B = -\sum_{c}{}' \frac{2Z_C}{|\underset{\sim}{r} - \underset{\sim}{R}_C|} + 2\int\frac{\rho_C(\underset{\sim}{r}',\underset{\sim}{r}')}{|\underset{\sim}{r} - \underset{\sim}{r}'|}\,d\underset{\sim}{r}', \qquad (13)$$

and

$$\rho_C(\underset{\sim}{r},\underset{\sim}{r}') = \sum_{i=\text{occupied}} \tilde{\phi}_{Ci}(\underset{\sim}{r})\tilde{\phi}^{\dagger}_{Ci}(\underset{\sim}{r}'). \qquad (14)$$

In general, it is the system of equation (11-14) which we chose to solve in order to define a charge density, equation (7), which in turn defines the Fock operator. In doing this, we use F of equation (6) with ρ of equation (7). Excepting for the choice of using a first order equation (11) everything so far is rigorous. It is possible to obtain local orbital equations to other powers of overlap. We have done so for second order. The author notes that the only approximation in the set of equations (11-14) is in the form of U_B. The reduction of the term $\rho A\rho$ on the right-hand side of equation (11) is exact [*12*]. It is often convenient to let $A = -U_B$.

B THE ENERGY BAND MODEL

Given the nature of the problem, obtaining solutions to (4), and the nature of the local orbitals, band theoretic methods fall into two classes — those which work and those which don't. The former class is further subdivided into two classes —

methods which have been used and those which haven't been used.

Essentially, there are two models which work and which the author has used. These are the LCAO method [*13*] and the MB (mixed basis) method [*14*]. For LCAO calculations, the local orbitals form an ideal expansion set. One has then trial orbitals of the form $\xi_j(\underset{\sim}{k},\underset{\sim}{r})$,

$$\xi_j(\underset{\sim}{k},\underset{\sim}{r}) = \sum_{\mu} e^{i\underset{\sim}{k}\cdot\underset{\sim}{R}_\mu}\tilde{\phi}_j(\underset{\sim}{r} - \underset{\sim}{R}_\mu),$$

and of the form $\eta_j(\underset{\sim}{k},\underset{\sim}{r})$,

$$\eta_j(\underset{\sim}{k},\underset{\sim}{r}) = \sum_{\mu} e^{i\underset{\sim}{k}\cdot\underset{\sim}{R}_\mu}(\underset{\sim}{r} - \underset{\sim}{R}_\mu)^{A_j}\exp[-\alpha_j(\underset{\sim}{r} - \underset{\sim}{R}_\mu)]. \tag{15}$$

One could in principle include ξ's formed form the local orbitals equations virtual orbitals in the basis, however, the use of free Slater Type orbitals to describe the virtual bands has been found to be most accurate and efficient. In terms of the new basis set, one finds

$$\phi_n^\alpha(\underset{\sim}{k},\underset{\sim}{r}) = \sum_j c_{j\alpha}^n\xi_j(\underset{\sim}{k},\underset{\sim}{r}) + \sum_j d_{j\alpha}^n\eta_j(\underset{\sim}{k},\underset{\sim}{r}). \tag{16}$$

In principle, using F and the basis given by (16), one can solve for the energy bands.

The second model is the mixed basis method. Here one forms a set of ξ's from *some* of the occupied local orbitals. These are similar to (15) in form. One then proceeds to form symmetrized linear combinations of plane waves to complete the basis set. Thus we define a set

$$S_j^\alpha(\underset{\sim}{k},\underset{\sim}{r}) = \sum_\ell b_{\ell j}^\alpha(k)e^{i(\underset{\sim}{k}+\underset{\sim}{K}_\ell)\cdot\underset{\sim}{r}}. \tag{17}$$

Here we chose K_ℓ for a given set such that $|(\underset{\sim}{k} + \underset{\sim}{K}_\ell)|$ is a constant. In terms of this basis, the block orbitals become

$$\phi_n^\alpha(\underset{\sim}{k},\underset{\sim}{r}) = \sum_j c_j^n\xi_{j\alpha}(\underset{\sim}{k},\underset{\sim}{r}) + \sum_\ell d_\ell^n S_j^\alpha(\underset{\sim}{k},\underset{\sim}{r}). \tag{18}$$

Again, in principle, using F, the band problem is standard.

There are two common band models which could be used in principle, but which the author does not use. These are the APW and KKR methods. We do not use them because (1) prejudice against muffin tin potentials, (2) for ionic solids, muffin tin potentials are not obviously useful due to the strong long range ionic potentials, (3) it is difficult to completely eliminate muffin tin potentials, (4) one hasn't had time to try everything. The APW method has been used successfully for *ab initio* studies by Dagens and Perrot [*15*], and by Perrot [*16*].

Finally, I must mention two common band models which are not useful for our purposes. These are the OPW method and the Pseudo-Potential method. The reason one is unable to use these methods is that the local orbital core states are not eigenfunctions of F and the usual procedures for OPW or Pseudo-Potential simply don't apply.

C CORRELATION METHODS

There are three models for including the effects of electron correlation which the author has used: (1) Semi-classical Mott-Littleton method [*17,18*], (2) Screened Exchange plus Coulomb Hole [*19,20*], (3) Electronic Polaron [*21,22,23*]. We briefly sketch the physical ideas behind each model and note its limitations and features. No derivations are given and the interested reader is referred to the appropriate literature.

In the semi-classical Mott-Littleton type calculation, for an alkali halide, the conduction electron is assumed to be a stationary point charge situated on an alkali ion site, while the hole is assumed to be a stationary point of charge on an halogen ion site. One takes the point charge on a given site and calculates in a self-consistent manner, the amount of energy associated with deforming the charge cloud on the surrounding ion sites. This is done separately for the electron and the hole. In this model as in the other two, the motion of the ion cores is neglected. One finds the valence band moves up by about two eV while the conduction band moves down by about two eV. Thus this effect is large and naturally $\underset{\sim}{K}$-independent. The direct effect of electron-hole interaction is neglected as is the polarization of the central cell ion.

In the second method, that of screened exchange plus Coulomb hole, one proceeds by replacing the Fock operator with a new operator which contains the effect of correlation. This operator is [*24,25*]

$$(T + V_{\mathrm{N}} + V_{\mathrm{c}} + M - \varepsilon^{\alpha}_{n\underset{\sim}{k}})\psi^{\alpha}_{n\underset{\sim}{k}}(\underset{\sim}{r}) = 0. \tag{19}$$

This operator is the same as F except that M replaces the exchange operator. M is the self-energy operator and is

$$M(\underset{\sim}{p},\underset{\sim}{p} + \underset{\sim}{K}) = V^{\mathrm{core}}_{\mathrm{ex}}(\underset{\sim}{p},\underset{\sim}{p} + \underset{\sim}{K}) + V^{\mathrm{valence}}_{\substack{\mathrm{screened}\\ \mathrm{exchange}}}(\underset{\sim}{p},\underset{\sim}{p} + \underset{\sim}{K})$$

$$+ E_{\mathrm{ck}}(\delta_{0,\underset{\sim}{K}} - \langle \underset{\sim}{p}|\bar{P}_{\mathrm{c}}|\underset{\sim}{p} + \underset{\sim}{K}\rangle). \tag{20}$$

M was determined from the random phase approximation. One must first solve the Hartree-Fock bands in order to obtain the parameters needed to construct (20). No one has solved (19) self-consistently. Again the valence levels move up some two eV while the conduction levels move down two eV. However, in this case there is considerable dispersion in the self-energies of the valence electrons and the valence bands narrow appreciably. There is no obvious dispersion in the self-energy for the conduction bands.

The third model is that of the electronic polaron. This model was proposed by Toyozawa for conduction electrons in ionic solids in 1954. This lay dormant until it was second quantized by Inoue et al. in 1970. Inoue et al. gave a great number of studies of this operation based on a free electron model. Recently, the author has been able to derive Toyozawa's model from first principles and to extend this model to the case of occupied levels (hole bands) [23]. The physical model is simple. The true wave function of the solid is constructed from expansions in terms of Slater Determinants as follows. Let

$$|\mathrm{vac}\rangle = \frac{1}{\sqrt{N!}} \det\{\phi_i(r_\alpha)\}$$

Let ψ_0^T be the true ground state wave function, for example. Now if we label so $1 < - i < j \leqslant N < a < b \ldots$ and i labels occupied orbital and a the virtual orbitals, we let

$$\psi_0^T(r_1,\ldots,r_N) \cong |\mathrm{vac}\rangle + \sum_{j=1}^{N} \sum_{a=N+1}^{\infty} A_{0ja} C_a^\dagger C_j |\mathrm{vac}\rangle$$

$$+ \sum_{j,k=1}^{N} \sum_{a,b=N+1}^{\infty} B_{0jkab} C_a^\dagger C_b^\dagger C_j C_k |\mathrm{vac}\rangle .$$

Here the C's are the fermion creation and annihilation operators and the A's and B's are variational parameters. In practice, we allow two possible types of excitation: (1) scattering in the bands, (2) formation of longitudinal excitons. This is discussed by Toyozawa. One finds that the electron in the conduction band is given by the Hamiltonian

$$h = \sum_{\underset{\sim}{k}} \varepsilon_{\underset{\sim}{k}} C_{\underset{\sim}{k}}^\dagger C_{\underset{\sim}{k}} + \varepsilon \sum_{\underset{\sim}{K}} b_{\underset{\sim}{K}}^\dagger b_{\underset{\sim}{K}}$$

$$+ \sum_{\underset{\sim}{k}\underset{\sim}{K}} V_{\underset{\sim}{K}}(0)\{b_{-\underset{\sim}{K}}^\dagger - b_{\underset{\sim}{K}}\} C_{\underset{\sim}{k}+\underset{\sim}{K}}^\dagger C_{\underset{\sim}{k}} . \tag{21}$$

Here the b's are the creation and annihilation operators for the excitons which are considered to be bosons. The energy of the exciton is ε. A similar Hamiltonian is applied to the hole in a filled shell. One finds

$$V_{\underset{\sim}{K}}(0) = e\left[\frac{2\pi\varepsilon}{V}\left(1 - \frac{1}{\varepsilon_\infty}\right)\right]^{\frac{1}{2}}\frac{i}{|\underset{\sim}{K}|}\int\phi_n{}^2(\underset{\sim}{r})e^{i\underset{\sim}{K}\cdot\underset{\sim}{r}}d\underset{\sim}{r}. \qquad (22)$$

V is the crystal volume, ε_∞ is the optical dielectric constant, and ϕ_n is the Wannier function for the band in question. In terms of such a model, the self energies are (to second order)

$$E^{se}_{electron}(\underset{\sim}{K}) = \sum_{\underset{\sim}{k}} \frac{|V_{\underset{\sim}{k}}(0)|^2}{\varepsilon(\underset{\sim}{K}) - \varepsilon - \varepsilon(\underset{\sim}{K} - \underset{\sim}{k})},$$

$$E^{se}_{hole}(\underset{\sim}{K}) = \sum_{\underset{\sim}{k}} \frac{|V_{\underset{\sim}{k}}(0)|^2}{\varepsilon + \varepsilon(\underset{\sim}{K}) - \varepsilon(\underset{\sim}{K} - \underset{\sim}{k})}. \qquad (23)$$

Again actual calculation shows that the effect of this operator is to lower the conduction band by two eV and raise the valence bands by two eV. Here there is seen to be dispersion in valence self-energies. If one permits scattering to all bands of wave-vector $\underset{\sim}{k}$, the dispersion in the self-energy of the condution electrons tends to vanish.

3. A SUMMARY OF THE MOST PROMISING TECHNIQUES AND RESULTS

In this section, I shall try to give what amounts to an overview of the past efforts of my group. I intend to state which efforts are the most productive in terms of reasonable effort for further development. In principle, I believe that local orbitals methods, especially when studied to first order in interatomic overlap are most productive for many problems. There are two basic ways to divide the charge density which seem reasonable. Let us stress first that it is best to use units which consist of complete shells. Firstly, in most ionic cases, it seems reasonable to divide the charge into two units, that associated with the anion and that associated with the cation. We have done this for a number of cases. In general, our method of solving (11-14) is to expand the ϕ_{Ai}'s in terms of STO (Slater Type Orbitals). The solution then reduces to repetitive matrix formation and diagonalization. We have used this method to study He, Ne, Ar, Kr, LiH, NaH, LiF, LiCl, LiBr, LiI, NaF, NaCl, NaBr, KF, KCl, KI, MgO, ZnSe, C (diamond), Si, Al, Ca.

More recently, the author has programmed both (7-10) and (11-14) in terms of Gaussian Type Orbitals (GTO's). In many ways, we find the GTO's form an excellent basis set for our purposes. I have not embarked upon an extensive program of study with our GTO basis, but I have obtained both exact and first order solutions for LiH and solid He. One point must be made. The first order method is two orders of magnitude faster than the exact method. In obtaining solutions with GTO's, I do not necessarily divide the system into atoms or ions, but work instead with the basic unit cell as a whole. I feel this is an especially useful feature when studying molecular crystals. This is so because a given molecule interacts weakly with its surroundings but atoms within a given molecule may interact strongly. This feature I feel overcomes the principal objections to the extension of the method of local orbitals to a wide class of systems.

It is of some interest then to study the accuracy of solutions to (11-14), our approximate method as a function of overlap. Our test system is solid fcc He. We have solved (11-14) and (7-10) for five values of lattice constant using GTO's. In Table I, the solutions, basis and overlaps, are given. Please note the overlap given is the total overlap with the surrounding twelve unit cells of He. It is this measure of total overlap which is most important. It is seen from Table I that the first order solution remains reasonable as long as $\sum_B S_{AB} \leqslant 0.6$. This would essentially include a large number of non-metallic solids. In fact, even a substance as bad as ZnSe has been found to lie at about this limit. It is further noted, as we shall see the error here comes from a poor choice of basis rather than from the first order expansion. In our GTO program, we choose

$$A = \delta(\underset{\sim}{r}). \tag{24}$$

We have also studied the effect of various choices of A on the local orbitals. The test system used was LiF. In Table II, we show the dominant overlaps, $S_{pp\sigma}$ and $S_{pp\pi}$ as a function of S for the $F^- 2p$ orbitals. We see the orbitals are only weakly dependent upon the choice of A. Thus one may as well choose an A which is to his liking. We also see that the local orbitals overlap less than the greatest of all the alkali halides.

The test which is easiest to perform is to study the behavior of simple molecules using (1) canonical restricted Hartree-Fock techniques, (2) exact local orbitals equations, (3) first order theory. For this study, the H_4 molecule and the He_2 molecule was chosen. Gaussian basis functions were placed on all nuclei and varied along with basis coefficients. The same basis set was used for each type of calculation for a given system. This flexible basis permits one to replicate cusp conditions for the local orbitals about sites other than that about which the orbital is sited. This was the chief deficiency in the study for solid He [*26*].

In the case of H_4, the atoms were in a planar rectangular array with sides of 1.4 and 2.0 Bohr radii. The basic unit for the local orbital was the H_2 molecule. The electronic energy of the system using the exact Hartree-Fock methods was -10.60 Ry whereas the energy using the first order wavefunctions was -10.56 Ry. This error is very small compared with the correlation energy. The overlap of the local orbitals approaches unity here and is clearly not small.

TABLE I

In this Table, we specify the basis used for the solid He calculation. We specify the coefficients of the exact and the first order in overlap local orbitals, and the one electron parameters as a function of overlap. We also give the free atom coefficients. The total overlaps of the central cell atom with the 12 nearest neighbors is also given for the exact local orbitals, first order local orbitals and free atom wave functions. The orbitals are of the form $\sum b_{cj}\exp(-\alpha_i(r - r_j))$. When a = 2.0, we found the exact solution to oscillate between two extremes which are given here. Whether these extremes are indicative of a poor basis for this lattice constant or of some other effect is unclear.

Helium

j	α	r_j	b (free atom) E = -1.830 Ry
1	32.0	0	0.286
2	4.6	0	0.452
3	0.9	0	0.383
4	0.2	0	0.077

a	E_{1s}		S			j	b_j	
	exact	first order	exact	first order	atomic		exact	first order
9.54	-1.903	-1.905	0.0188	0.0188	0.0188	1	0.286	0.286
						2	0.452	0.452
						3	0.383	0.383
						4	0.078	0.078
8.0	-1.904	-1.905	0.0841	0.0840	0.0840	1	0.286	0.286
						2	0.452	0.452
						3	0.383	0.383
						4	0.078	0.078
6.0	-1.923	-1.910	0.468	0.466	0.457	1	0.284	0.285
						2	0.452	0.452
						3	0.377	0.381
						4	0.080	0.078
4.0	-2.593	-2.085	2.85	2.25	1.85	1	0.246	0.271
						2	0.437	0.449
						3	0.275	0.341
						4	0.118	0.094
2.0	-6.41 -6.40 (oscillatory solution)	-4.47				1	0.302† 0.309¶	0.297
						2	0.246† 0.259¶	0.245
						3	0.457† 0.481¶	0.446
						4	0.065† 0.054¶	0.069

† First solution.
¶ Second solution.

TABLE II

Here the results of sixteen choices for U_A are given with respect to the overlap parameters $S_{pp\sigma}$ and $S_{pp\pi}$. Results using free ion wave functions are given by way of comparison. If the choice of A is a square well then its radius, R, is given in terms of the lattice constant a and its depth, D, in Rydbergs is given.

A	$S_{pp\sigma}$	$S_{pp\pi}$
free ion	-0.0658	0.0218
U_A	-0.0588	0.0172
$-U_A$	-0.0585	0.0170
$\lim_{\alpha\to 0} \alpha U_A$	-0.0588	0.0172
Point ion potential	-0.0590	0.0173
Well $R = \frac{1}{2}a$ $D = -10.0$	-0.0613	0.0187
Well $R = \frac{1}{4}a$ $D = -10.0$	-0.0609	0.0185
Well $R = \frac{1}{2}\sqrt{2}a$ $D = -10.0$	-0.0613	0.0187
Well $R = \frac{1}{4}a$ $D = -1.0$	-0.0589	0.0173
Well $R = \frac{1}{2}a$ $D = -1.0$	-0.0590	0.0174
Well $R = \frac{1}{2}\sqrt{2}a$ $D = -1.0$	-0.0591	0.0174
Well $R = \frac{1}{4}a$ $D = 1.0$	-0.0585	0.0170
Well $R = \frac{1}{2}a$ $D = 1.0$	-0.0583	0.0169
Well $R = \frac{1}{2}\sqrt{2}a$ $D = 1.0$	-0.0584	0.0170
$-1/r$	-0.0591	0.0174
$-1/r^2$	-0.0593	0.0175
$1/r$	not local	not local

As a second test, the He_2 molecule was studied as a function of internuclear separation. These results are summarized in Table III. The quality of the first order results is excellent for all values of internuclear separation and the error is an order of magnitude, or more, better than the Hartree-Fock error in neglecting correlation.

Using the MB method, the author and his associates have computed band structures for Ar, Kr, LiCl, LiBr, NaCl, NaBr, and

TABLE III

The basis set employed in the study of the He_2 molecule is given. The total energies and splitting between the 1σ and 2σ orbital are shown for both the restricted Hartree-Fock case and for the first order local orbitals solution. The conditions of the calculation are presented in the text. The overlap of the first order local orbitals is also given. The Gaussian basis functions are of the form $\exp(-\alpha_j r^2)$. Energies are in Rydbergs and the α_j in $(\text{bohr})^{-2}$. The basis functions are sited about each nucleus.

j	1	2	3	4	5
α_j	150.0	32.0	4.6	0.9	0.2

Inter-atomic separ-ation (bohr)	Restricted Hartree-Fock		First order local orbitals		
	total energy (Ry)	$\varepsilon_{2\sigma} - \varepsilon_{1\sigma}$	total energy (Ry)	$\varepsilon_{2\sigma} - \varepsilon_{1\sigma}$	over-lap
10.0	-11.42	0.000	-11.42	0.000	0.0000
8.0	-11.42	0.001	-11.42	0.001	0.0002
6.0	-11.42	0.005	-11.42	0.004	0.0044
4.0	-11.41	0.07	-11.41	0.08	0.050
2.0	-11.16	0.79	-11.14	0.78	0.33

KCl. Of these calculations, all but Kr are self-consistent. Self-consistency in its initial stages is achieved by iterating (11-14) to self-consistency. In the cases of Kr, LiCl, and LiBr correlation is included by means of the Mott-Littleton method and in the other cases by means of the screened exchange plus Coulomb hole. More recently, the author and his associates have used the LCAO method to study the energy bands of He, Ne, Ar, Kr, LiH, NaH, LiF, NaF, KF, and Ca. All calculations here are self-consistent and this is achieved by iterating (11-14) except for He where the exact local orbitals equations are iterated. As of this writing, correlation effects have been incorporated into the calculations for He, Ne, Ar, Kr, and LiF by means of the electronic polaron. The numerical details of these calculations are given in the literature and need not be given here [*27-30*].

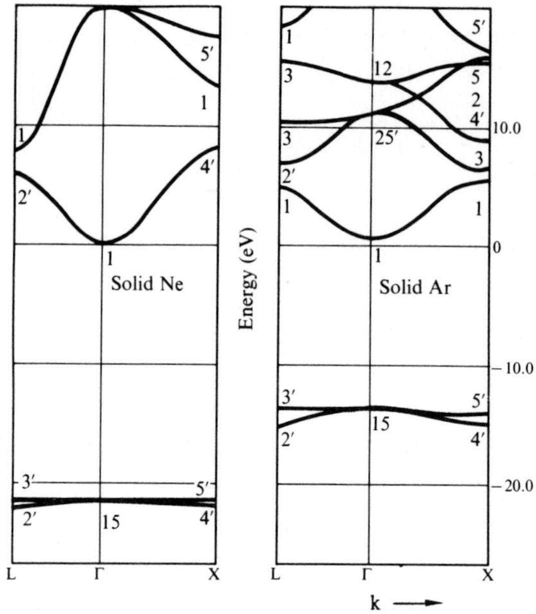

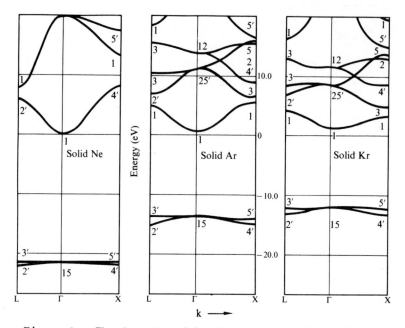

Figure 1 - The Correlated band structures for solid Ne, Ar, and Kr obtained by the author and D.J. Mickish are shown [*30*].

In figure 1, we show the correlated band structures of Ne, Ar, and Kr as computed by the present authors. These are in good agreement with other *ab initio* results for these materials. In figure 2, we show the Hartree-Fock band structure for LiF and in figure 3 for LiH. Finally, in figure 4, we show the Hartree-Fock band structure for Ca at the normal lattice constant and at a 20% reduced lattice constant where Ca is found both theoretically and experimentally to be a semimetal. An appendix giving a short introduction to available *ab initio* band structures is given.

4. EXPERIMENTAL COMPARISON

We shall consider here five types of experiments. These are (1) X-ray emission spectroscopy, (2) ESCA, (3) Optical Absorption and optical constants in the U.V., (4) Soft X-ray absorption spectra, (5) Photoemission spectroscopy. This largely covers the available field of experiments for the insulators, at least it covers the classes for which reasonably unambiguous results may be obtained.

The first type of available data is X-ray emission data. Some of the earliest data of interest comes from O'Brian and Skinner [*31*]. This data is for a large number of alkali hal-

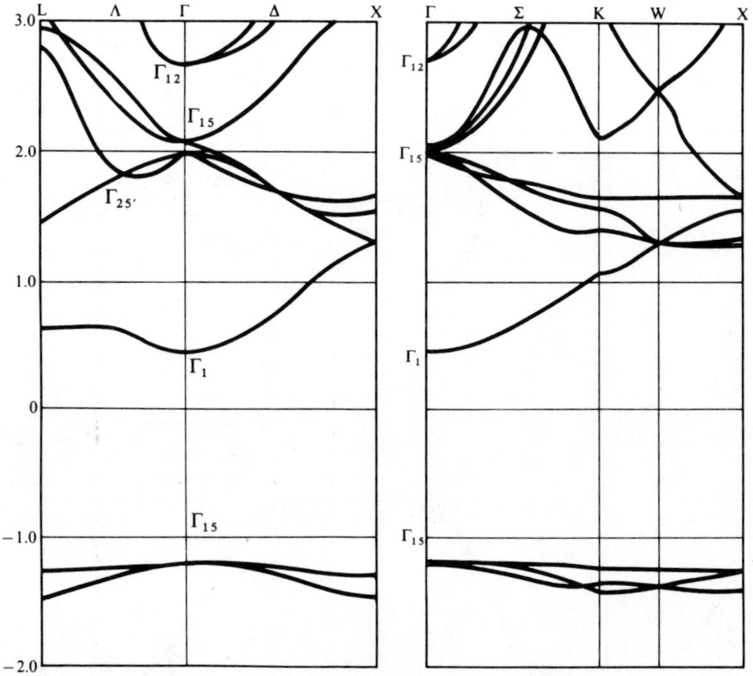

Figure 2 - The Hartree-Fock band structure obtained by the author and D.J. Mickish for LiF is seen [*29*].

ides as well as for other solids. In general, O'Brian finds rather broad emission spectra for these solids. In figure 5, we see the results for the $2p$ valence band in LiF along with our predictions. It has been noted by Parratt and Jossen [*32*] that the interpretation of X-ray emission data in terms of band structure may be (1) ambiguous, (2) misleading. This point of view has recently been elucidated theoretically by Best [*33*]. Best's point is essentially related to the localization of the valence hole after emission. Keeping the possible difficulties in mind with respect to X-ray emission spectroscopy we pass on to ESCA.

ESCA (Electron Spectroscopy for Chemical Analysis) is another tool which permits us to examine the structure of the occupied levels. Basically one hits the solid with a high energy X-ray such as the AlK_{α} line. This photo-ionizes the solid. One measures the range of kinetic energies of the emitted electrons. From this, it is possible to deduce the shape and width of the occupied levels. In general, the available ESCA data seem to favor the interpretations of O'Brian and Skinner rather than the very narrow band predictions of Parratt and Jossen. In figure 6, we show ESCA data from Thomas [*34*] for NaCl $3p$ and $3s$ valence band along with our theoretical prediction. There is obvious

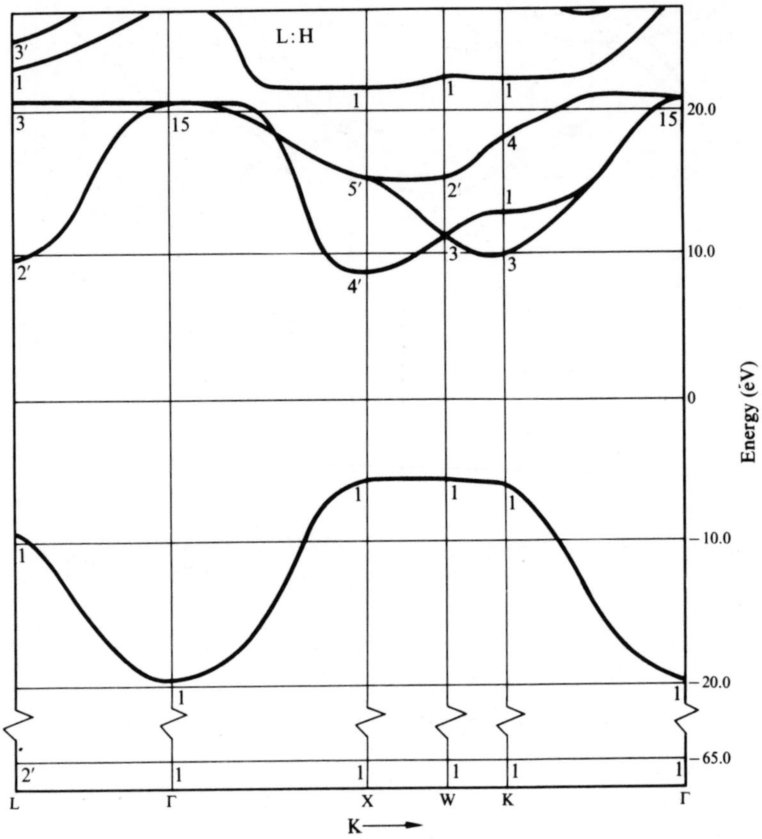

Figure 3 - The Hartree-Fock band structure for LiH as obtained by the author and D.J. Mickish is seen [*30*].

similarity in width, but not of structure. This is likely due to the poor resolution of the ESCA experiment (0.5 - 1.0 eV typically). We also show essentially equivalent type data of Krolokowski and Spicer for the $3p$ valence band of CsCl and the $5p$ valence band of KI [*35*]. These are in figure 7. It is clear from all this that the valence bands are broad, 3 - 4 eV. These predictions are not found using $x - \alpha$ type exchange unless $\alpha >> 1$. For $\alpha = 1$ typically one has widths of 0.5 - 1.0 eV. Please note, Hartree-Fock also is quite wrong here. It over-estimates the band widths. Correlation here is essential.

The next class of experiments is that of optical absorption in the UV region. For our present purposes, we are, in general, interested in the region from say 5 to 27 eV. This covers the region from the first exciton to the plasmon in all alkali-halides. In general, the optical absorption data is available for the alkali-halides [*36*]. A detailed study of the optical con-

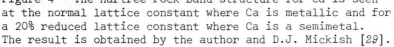

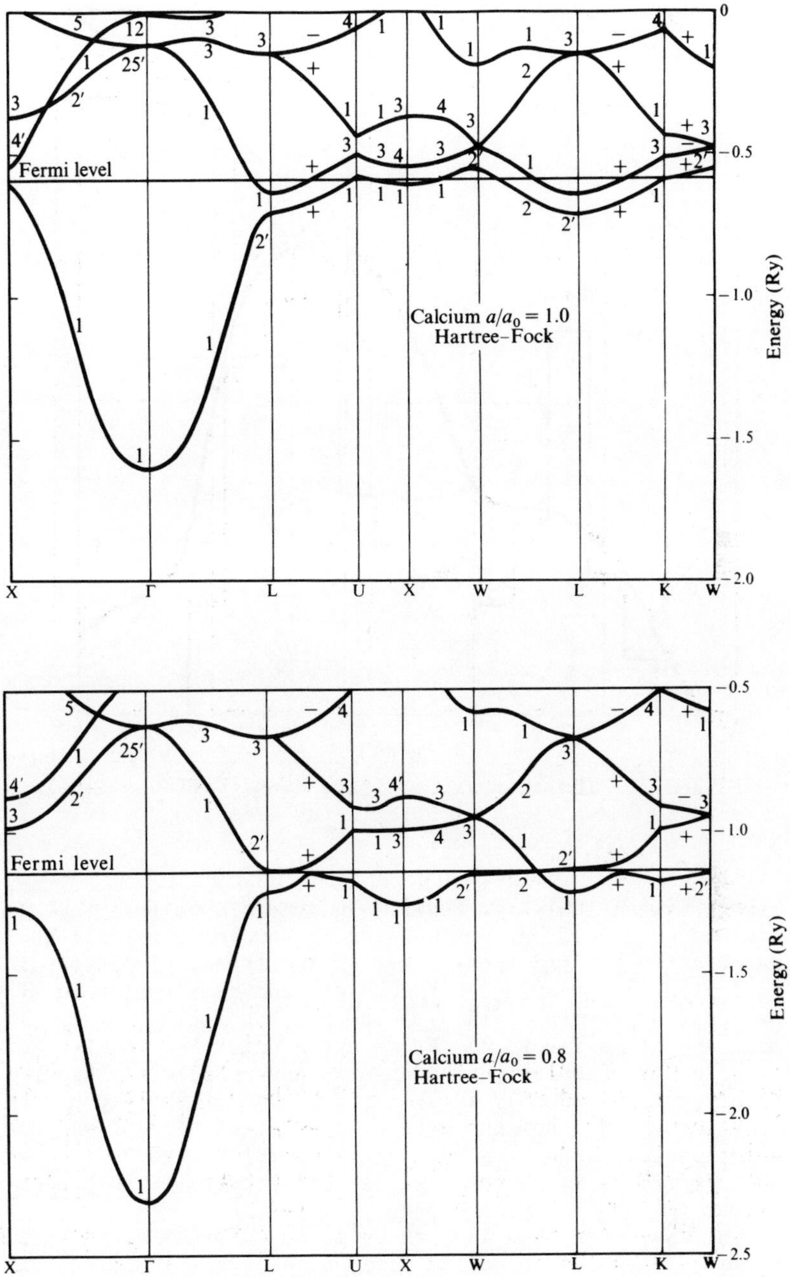

Figure 4 - The Hartree-Fock band structure for Ca is seen at the normal lattice constant where Ca is metallic and for a 20% reduced lattice constant where Ca is a semimetal. The result is obtained by the author and D.J. Mickish [*29*].

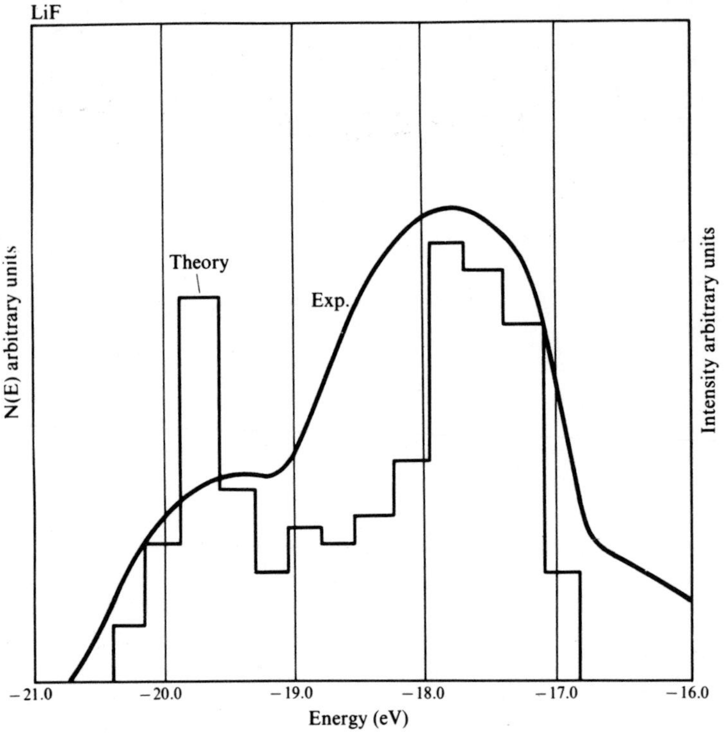

Figure 5 - The theoretical state density of the LiF 2*p* band is shown along with the X-ray emission spectrum of O'Brian and Skinner [*31*].

stants is available for fewer. In general, we feel that as a minimum we should have values for ε_2 and Im1/ε for all these materials. Exciton spectra are of two types, simple (as in NaCl with only a spin-orbit doublet) and complex (as in KI with at least 5 principal lines). The exciton region is shown for NaCl and KI in figure 8. We do not predict the exciton absorption in our calculations because we have explicitly ignored it. We are interested only in the band to band transitions here. In figure 9, we show the optical absorption of NaBr and our calculation for it using a constant transition element. Basically this agreement is acceptable. As one progresses to higher energies 'plasma' effects become of interest.

Miyakawa has seen that there is an interesting feature associated with the so-called plasmon resonance in Im1/ε. That is, that it always lies on the high energy side of a peak in ε_2 by a small amount. This is seen in the data for KCl [*37*]. We show this in figure 10. It happens that this phenomenon may be explained using our electronic polaron concept. Using our Hamil-

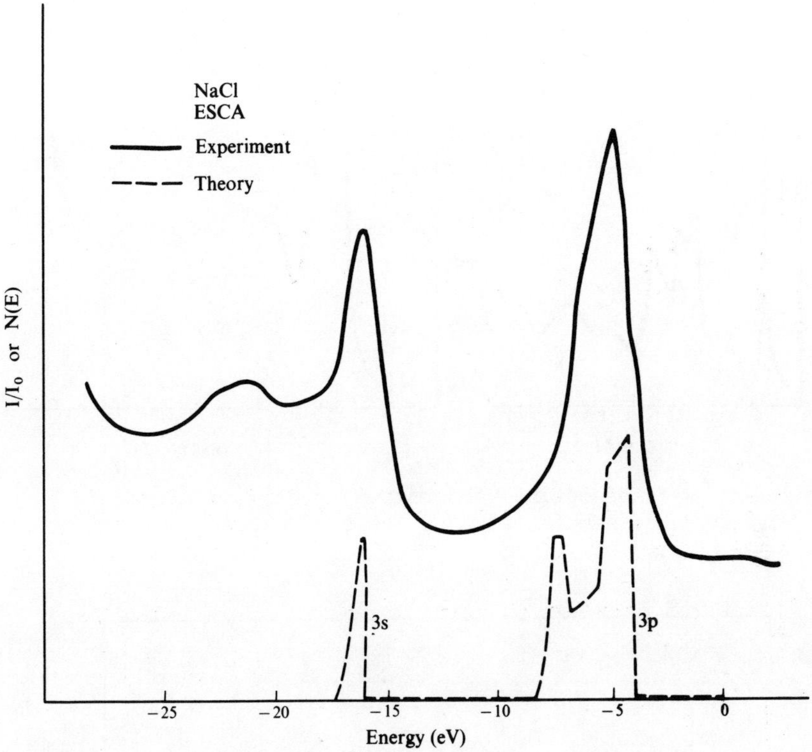

Figure 6 - The valence density of NaCl is shown along with the ESCA data of Thomas [*34*].

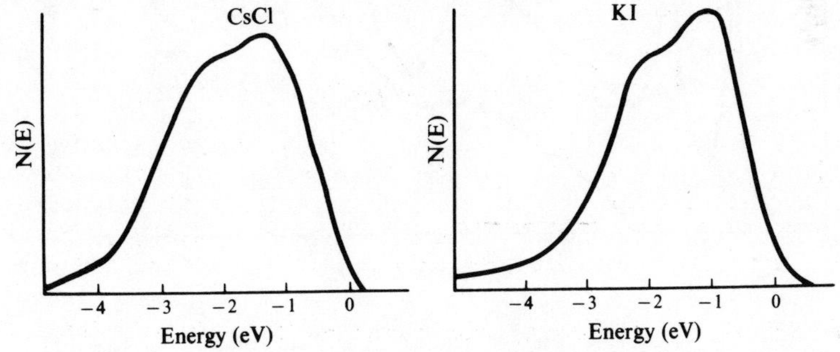

Figure 7 - The valence band photoemission state densities are seen for CsCl and KI. These are measured by Krolokowski [*35*].

tonian equation (21) we have evaluated the contribution to ε_2 and Im1/ε analytically for free electronic polaron complexes.

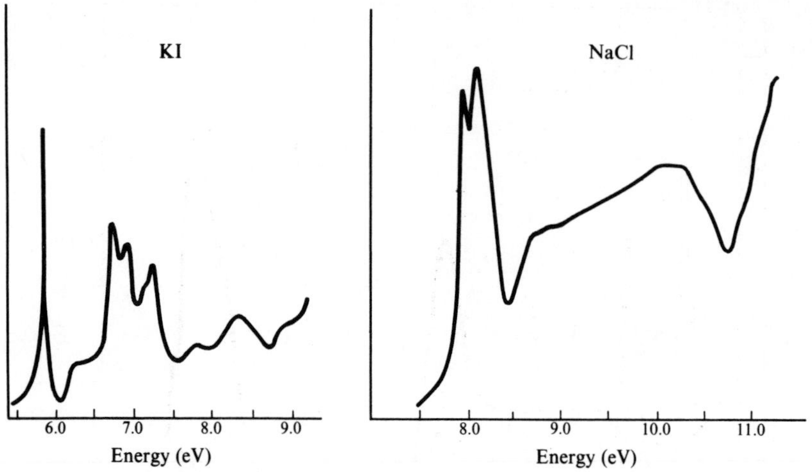

Figure 8 - The optical absorption spectra of KI and NaCl are shown to demonstrate the types of exciton spectra seen [*36*].

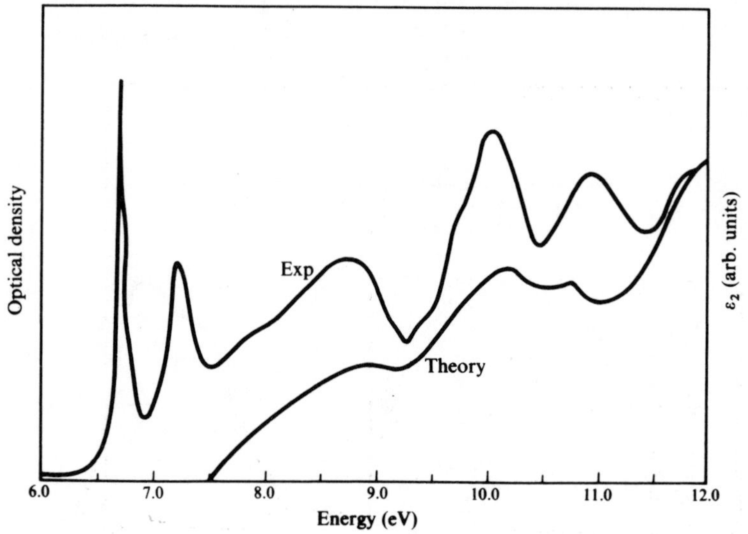

Figure 9 - The optical absorption of NaBr is shown [*36*] along with the theoretical contribution due to band theory [*25*].

Devreese, Kunz and Collins have applied the results on the optical absorption of Fröhlich polarons to the electronic polaron and find [*38*]

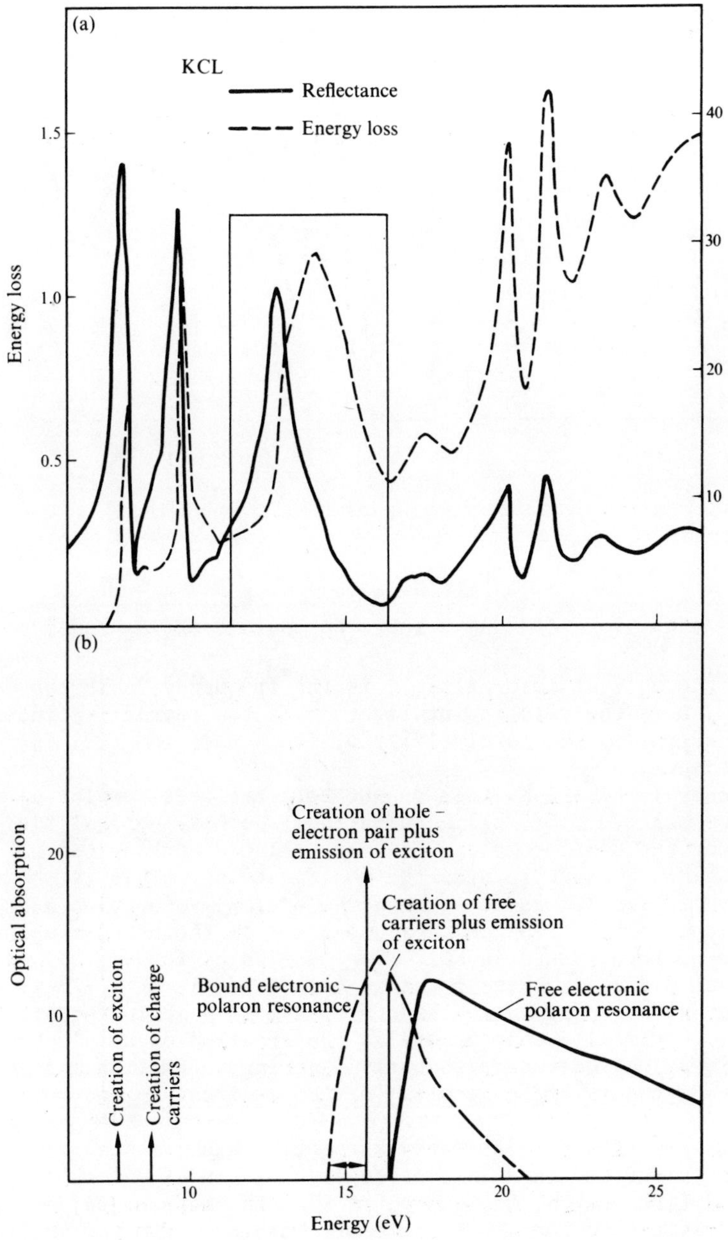

Figure 10 - The optical constants of KCl are shown [37] along with the optical effects due to the many body resonance contained in the electronic polaron theory, Devreese, Kunz and Collins [38].

$$\Gamma(\Omega) = \frac{1}{\varepsilon_0 cn} \times \frac{\Omega \operatorname{Im} X((\Omega - \Delta),\alpha)}{(\Omega - \Delta)^4 - 2(\Omega - \Delta)^2 \operatorname{Re} X((\Omega - \Delta),\alpha) + |X(\Omega - \Delta,\alpha)|^2}, \tag{25}$$

with

$$X(\Omega) = \Omega^2 + iZ(\Omega),$$

$$\alpha = \frac{e^2}{2\dfrac{\hbar}{2m\omega_{ex}}} \frac{1}{\hbar\omega_{ex}}\left(1 - \frac{1}{\varepsilon_\infty}\right),$$

and

$$-\operatorname{Im}\frac{1}{\varepsilon} = \frac{\operatorname{Im} X((\Omega - \Delta),\alpha)}{((\Omega - \Delta)^2 + 1)^2 - 2((\Omega - \Delta)^2 + 1)\operatorname{Re} X((\Omega - \Delta),\alpha) + |X((\Omega - \Delta,\alpha)|^2}. \tag{26}$$

$\Gamma(\Omega)$ is the optical absorption, Ω is the frequency, Δ is the band gap, c is the velocity of light, n is the refractive index of the medium, ε_0 the permittivity of free space and $Z(\Omega)$ is the impedance.

In general, alkali halides do not form the free complex although we believe LiBr, LiI, and NaI do. We have no analytic result for the bound complex; however, related work on the bound exciton phonon dictates the modifications. This is shown in figure 10 for KCl where we show the optical absorption and energy loss for the free complex and also for the bound complex which we believe exists in KCl. The peak in ε_2 for a free complex is at $\Delta + 1.2\ \varepsilon$ whereas the peak in $\operatorname{Im} 1/\varepsilon$ is at $\Delta\ 1.2\ \varepsilon$. This then is our explanation of the 'plasmon' peak in the alkali halides. Physically what occurs is the creation of an electron hole pair which scatters resonantly emitting an exciton and an electron hole pair (free complex) or two excitons (bound complex).

Let us now turn to soft X-ray absorption experiments. We shall be interested here in four edges, the LiK edge in LiF, LiCl, and LiBr and the $M_{IV,V}$ edge in KI. In these cases, exciton transitions to the $N = 1$ state are forbidden and our analysis is less ambiguous.

In figure 11, we show data for KI$M_{IV,V}$ edge along with our calculation [*39*]. In figure 12, we show data for LiF from DESY [*40*]. Here we also show the contribution to the absorption due to the electronic polaron as given by Kunz, Devreese and Collins.

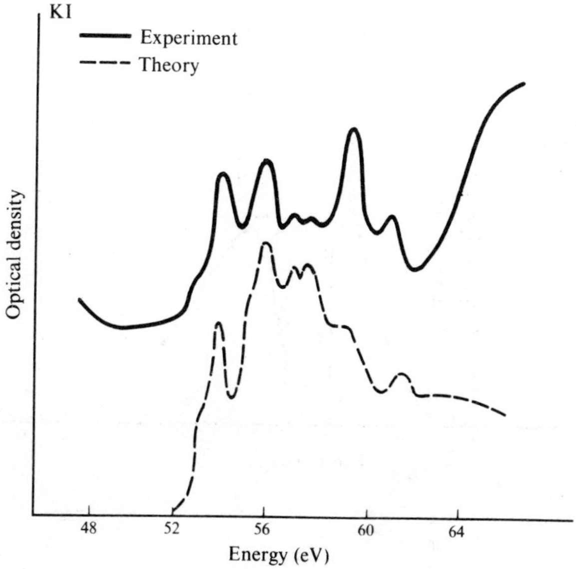

Figure 11 - The soft X-ray spectrum of KI along with the theoretical prediction is seen [*39*].

We note that Lynch has also found Im1/ε for LiF and finds a peak in Im1/ε to lie about 0.6 eV on the high energy side of the peak we identify as our absorption. This agrees well with our theory. In figure 13, we show LiCl and LiBr. We show the experiment, band theory contribution and electron polaron contribution. LiCl has a bound complex and LiBr a free complex. Unfortunately, Im1/ε is unknown for these substances, Devreese, Kunz and Collins [*41*].

Finally, we turn to photoemission spectroscopy. Krolokowski and Spicer have deduced the conduction band structure of KI from their data. This is shown in figure 14 with our calculations. The agreement is fantastic. I would like to encourage further studies of this kind.

5. APPLICABILITY TO OTHER PROBLEMS

Basically the method of local orbitals should apply to a wide class of solids. We intend to extend our efforts to the molecular crystals. We shall try CH_4 solid. It is also intended to study H_2O solid. Of course, He and Ne are molecular solids of a very simple type. In the ionic crystal range, we shall like to try our hand at CaF_2. We feel the LCAO method is excellent here also.

It is clear the electronic polaron model has consequences far

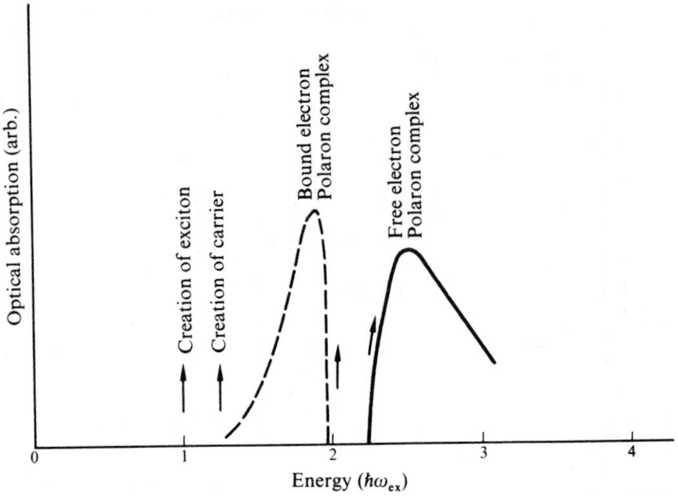

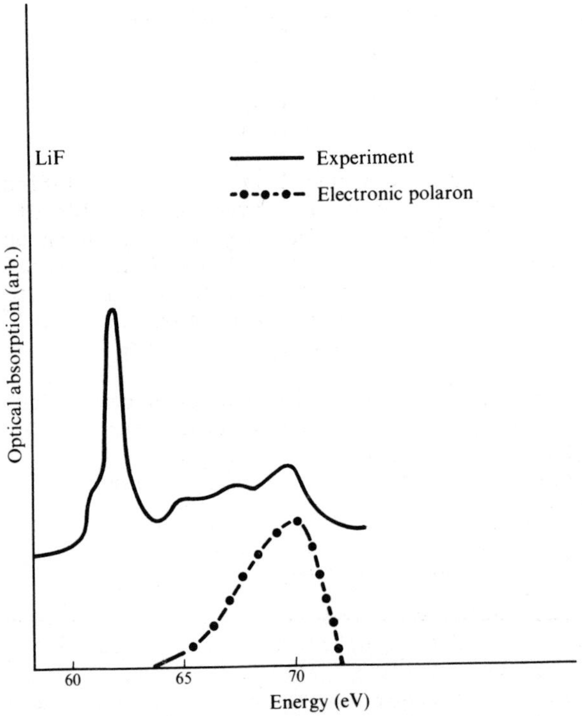

Figure 12 - The soft X-ray spectrum of LiF [*40*] and the contribution to the absorption due to the electronic polaron is shown, Devreese, Kunz and Collins [*41*].

studies of self-energies of electrons and holes. It is hoped to probe fully the consequences of this model.

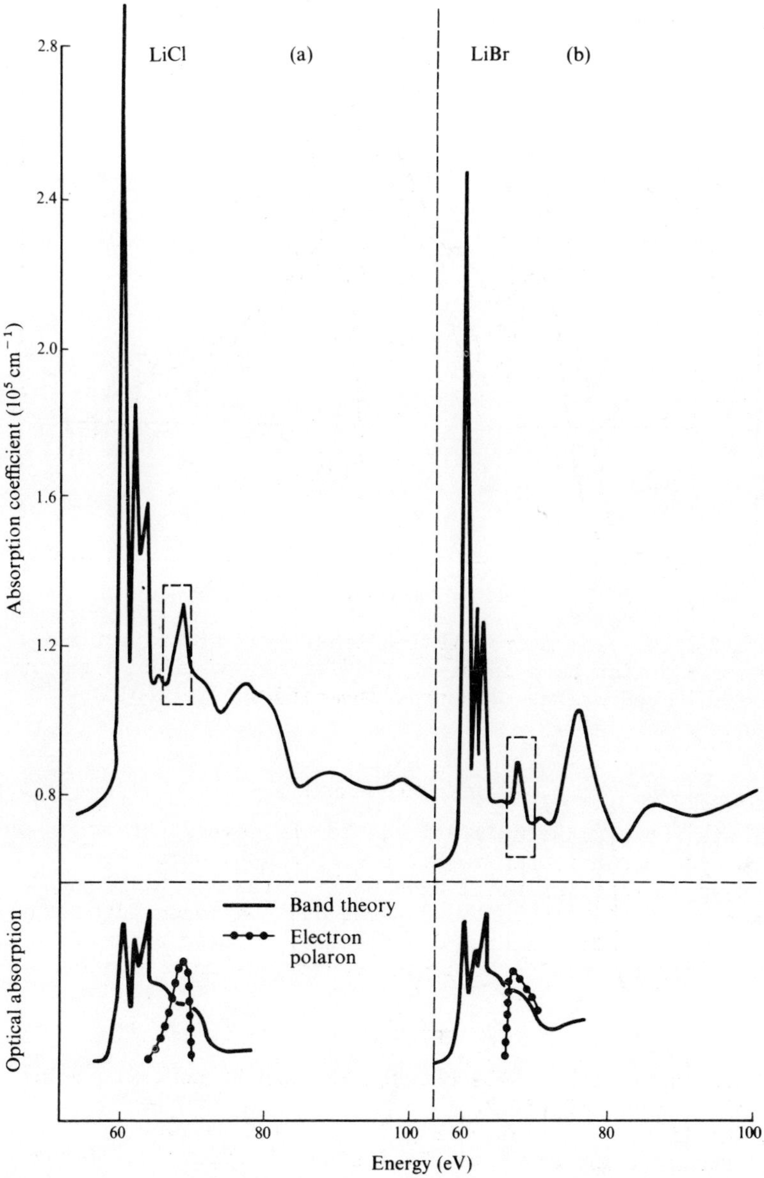

Figure 13 - The soft X-ray spectrum of LiCl and LiBr is shown along with the contributions to the absorption due to band theory and the electronic polaron, Devreese, Kunz and Collins [*41*].

Clearly, the local orbitals model is useful for studies of molecules and of solid defect structures. I hope these models will find full employment in these fields.

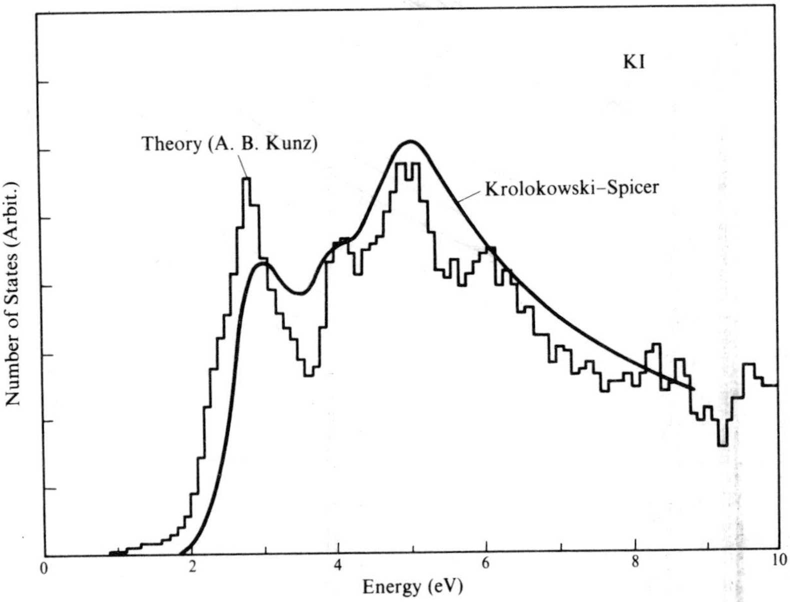

Figure 14 - The photoemission density of states for the KI conduction band is seen along with the author's theoretical prediction. The experiment is due to Krolokowski and Spicer [*35*].

ACKNOWLEDGMENT

I wish to express my gratitude to the members of my research group and in particular to Dr. D.J. Mickish, whose work is heavily used in this paper. I also wish to thank F.C. Brown, R. Haensel, T.C. Collins, R.N. Euwema, J.T. Devreese, John Bardeen, and B. Sonntag for helpful discussions. I also especially thank R.J. Maurer for his interest and support of this project.

REFERENCES

1. Bouckaert, L.P., Smoluchowski, R. and Wigner, E. (1936). *Phys. Rev.*, **50**, 58.
2. Phillips, J.C. (1964). *Phys. Rev.*, **136**, A1705.
3. Oyama, S. and Miyakawa, T. (1966). *J. Phys. Soc. (Japan)*, **21**, 868.
4. Onodera, Y., Okazaki, M. and Inoue, T. (1966). *J. Phys. Soc. (Japan)*, **21**, 2229.
5. Kunz, A.B. (1966). *Phys. Rev.*, **151**, 620.
6. Koopman, T. (1933). *Physica*, **1**, 104.
7. Herman, F., Kortun, R.L., Kuglii, C.D. and Short, R.A. (1966). In *Quantum Theory of Atoms, Molecules, and The Solid State*, (ed. Löwdin, P.O.), (Academic Press).

8. Stukel, D.J. and Euwema, R.N. (1970). *Phys. Rev.*, **B1**, 1635.
9. Adams, W.H. (1961). *J. Chem. Phys.*, **34**, 89.
10. Gilbert, T.L. (1964). In *Molecular Orbitals in Chemistry, Physics, and Biology*, (eds. Löwdin, P.O. and Pullman, B.), (Academic Press).
11. Kunz, A.B. (1966). *Phys. Stat. Sol.*, **36**, 301.
12. Kunz, A.B. *J. Phys.*, **B**, (to be published).
13. Slater, J.C. and Koster, G.F. (1954). *Phys. Rev.*, **94**, 1498.
14. Kunz, A.B. (1969). *Phys. Rev.*, **180**, 934.
15. Dagens, L. and Perrot, F. (1972). *Phys. Rev.*, **B5**, 641.
16. Perrot, F. (1972). *Phys. Stat. Sol.*, **52**, 163.
17. Mott, N.F. and Littleton, M.J. (1938). *Trans. Faraday Soc.*, **34**, 485.
18. Fowler, W.B. (1966). *Phys. Rev.*, **151**, 657.
19. Hedin, L. (1965). *Phys. Rev.*, **A139**, 796.
20. Brinkman, W.F. and Goodman, B. (1966). *Phys. Rev.*, **149**, 597.
21. Toyozawa, Y. (1954). *Prog. Theor. Phys.*, **12**, 421.
22. Inoue, M., Mahutte, C. and Wang, S. (1970). *Phys. Rev.*, **B2**, 539.
23. Kunz, A.B. (1972). *Phys. Rev.*, **B6**, 606.
24. Lipari, N.O. and Kunz, A.B. (1971). *Phys. Rev.*, **B3**, 491.
25. Kunz, A.B. and Lipari, N.O. (1971). *Phys. Rev.*, **B3**, 1374.
26. Kunz, A.B. *Phys. Rev.*, (to be published).
27. Kunz, A.B. (1970). *J. Phys.*, **C3**, 1542.
28. Kunz, A.B. (1970). *Phys. Rev.*, **B2**, 5015.
29. Mickish, D.J. and Kunz, A.B. *J. Phys.*, **C**, (to be published); Others (to be published).
30. Kunz, A.B. and Mickish, D.J. *J. Phys.*, **C**, (to be published); *Phys. Rev.* (to be published); Others (to be published).
31. O'Brian, H.M. and Skinner, H.N.B. (1940). *Proc. Roy. Soc.*, **A176**, 229.
32. Parratt, L.G. and Jossen, E.M. (1955). *Phys. Rev.*, **97**, 916.
33. Best, P. (1971). *Phys. Rev.*, **B3**, 4377.
34. Thomas, J.M. (Unpublished report).
35. Krolokowski, W. (1967). (Thesis), (Stanford University), (unpublished); Krolokowski, W. and Spicer, W. (Unpublished data).
36. Teegarden, K.J. (1967). *Phys. Rev.*, **155**, 896.
37. Rossler, D.M. and Walker, W.C. (1968). *Phys. Rev.*, **160**, 599.
38. Devreese, J.T., Kunz, A.B. and Collins, T.C. (1972). *Solid State Commun.*, **11**, 673.
39. Brown, F.C., Gahwiller, C., Fujita, H., Kunz, A.B., Scheifley, W. and Carrera, N. (1970). *Phys. Rev.*, **B2**, 2126.
40. Haensel, R., Kunz, C. and Sonntag, B. (1968). *Phys. Rev. Lett.*, **20**, 262.
41. Kunz, A.B., Devreese, J.T. and Collins, T.C. (1972). *J. Phys.*, **C5**, 3259.

APPENDIX: TABULATION OF *AB INITIO* BAND CALCULATIONS

In this appendix, a brief tabulation of the available *ab intio* band structure calculations is given. This tabulation is complete to the best of the author's knowledge. All calculations use to some degree the non-local Fock operator.

SOLID RARE GASES

He — Kunz, A.B. *Phys. Rev.*, (to be published).

Ne — Dagens, L. and Perrot, F. (1972) *Phys. Rev.*, **B5**, 641.
Kunz, A.B. and Mickish, D.J. *Phys. Rev.*, (to be published).

Ar — Lipari, N.O. and Fowler, W.B. (1970). *Phys. Rev.*, **B2**, 3354.
Dagens, L. and Perrot, F. (1972). *Phys. Rev.*, **B5**, 641.
Lipari, N.O. (1972). *Phys. Rev.*, **B6**, 4071.
Mickish, D.J. and Kunz, A.B. *J. Phys.*, **C**, (to be published).
Kunz, A.B. and Mickish, D.J. *Phys. Rev.*, (to be published).

Kr — Lipari, N.O. (1970). *Phys. Stat. Sol.*, **40**, 691.
Kunz, A.B. and Mickish, D.J. *Phys. Rev.*, (to be published).

MOLECULAR CRYSTALS

CH_4 — Piela, L., Pietronero, L. and Resta, R. *Phys. Rev.*, (to be published).

ALKALI HALIDE CRYSTALS

LiF — Perrot, F. (1972). *Phys. Stat. Sol.*, **52**, 163.
Mickish, D.J. and Kunz, A.B. *J. Phys.*, **C**, (to be published).

NaF — Perrot, F. (1972). *Phys. Stat. Sol.*, **52**, 163.
Mickish, D.J. and Kunz, A.B. (to be published).

KF — Perrot, F. (1972). *Phys. Stat. Sol.*, **52**, 163.
Mickish, D.J. and Kunz, A.B. (to be published).

LiCl — Kunz, A.B. (1970). *J. Phys.*, **C3**, 1542.
Kunz, A.B. (1970). *Phys. Rev.*, **B2**, 5015.
Perrot, F. (1972). *Phys. Stat. Sol.*, **52**, 163.
Kucher, T.I. (1958). *Soviet Physics, JETP*, **34**, 274.

NaCl — Lipari, N.O. and Kunz, A.B. (1973). *Phys. Rev.*, **3**, 491.
Perrot, F. (1972). *Phys. Stat. Sol.*, **52**, 163.
Kucher, T.I. (1958). *Soviet Physics, JETP*, **34**, 274.

KCl — Lipari, N.O. and Kunz, A.B. (1971). *Phys. Rev.*, **4**, 4639.
Perrot, F. (1972). *Phys. Stat. Sol.*, **52**, 163.
Kucher, T.I. (1958). *Soviet Physics, JETP*, **34**, 274.
Howland, L.P. (1958). *Phys. Rev.*, **109**, 1927.

LiBr — Kunz, A.B. and Lipari, N.O. (1971). *J. Phys. Chem. Sol.*, **32**, 1141.

NaBr — Kunz, A.B. and Lipari, N.O. (1971). *Phys. Rev.*, **B4**, 1374.

HYDRIDES

LiH — Kunz, A.B. and Mickish, D.J. *J. Phys.*, **C**, (to be published).

NaH — Kunz, A.B. and Mickish, D.J. (to be published).

PLANAR COMPOUNDS

C (graphite)
— Zupan, J. (1972). *Phys. Rev.*, **6**, 2477.
BN — Zupan, J. (1972). *Phys. Rev.*, **6**, 2477.

SEMICONDUCTORS

C (diamond)
— Euwema, R.N., Wilhite, D.L. and Surratt, G.T. (1973). *Phys. Rev.*, **B7**, 818.
MgO — Pantelides, S., Mickish, D.J. and Kunz, A.B. (to be published).

METALS

H — Harris, F. and Monkhorst, H. *Phys. Rev.*, (to be published).
Li — O'Keefe, P.M. and Goddard, W.A., III. (1969). *Phys. Rev. Lett.*, **23**, 300.
Be — Shankland, D. *Bull. Amer. Phys. Soc.*.
Ca — Mickish, D.J. and Kunz, A.B. (to be published).

EFFECTS OF EXCHANGE AND CORRELATION IN THE ELECTRON BANDSTRUCTURE PROBLEM

LARS HEDIN

Department of Theoretical Physics, University of Lund, Sweden

A. QUALITATIVE DISCUSSION OF ELEMENTARY EXCITATIONS IN SOLIDS

1 The Concept of an Elementary Excitation

The typical example of an elementary excitation in a solid is the one-electron quasi-particle, that is the type of excitation that we are concerned with in a bandstructure calculation. We call it *quasi*-particle to remember that the excitation is something more general and complicated than just an electron moving in some periodic potential. To bring out the essential points in the simplest way, let us discuss the electron gas. The Hamiltonian in second quantization is

$$H = \sum_k \varepsilon_k c_k^\dagger c_k + \frac{1}{2\Omega} \sum_q v(q) \rho_q \rho_{-q}, \tag{1}$$

where $\varepsilon_k = h^2 k^2 / 2m$ is the kinetic energy, Ω the volume of the electron gas, $v(q) = 4\ e^2/q^2$ the Coulomb interaction and ρ_q the density Fourier component

$$\rho_q = \sum_k c_{k+q}^\dagger c_k . \tag{2}$$

We will not write out spin indices explicitly nor use fat letters for vectors, but leave that to the imagination of the reader. We have further written the electron-electron interaction on the form $\rho_q \rho_{-q}$ which differs from the correct expression $c_{k+q}^\dagger c_{k'-q}^\dagger c_{k'} c_k$ by the (infinite) constant $N/2\Omega \sum_q v(q)$. Those who objects to infinite constants may flatten the Coulomb potential for $r < r_c$ which gives a convergence factor $\sin(qr_c)/qr_c$. In the final expressions we can then take $r_c \to 0$.

The simplest approximate description of an electron quasi-particle is obtained if we neglect the electron-electron interactions. We then have the quasi-electron and quasi-hole states

$$|N+1,k\rangle_0 = c_k^\dagger |N\rangle_0, \qquad |N-1,k\rangle_0 = c_k |N\rangle_0, \tag{3}$$

where $|N\rangle_0$ stands for a filled Fermi sea

$$|N\rangle_0 = \prod^{occ} c_k^\dagger |0\rangle. \tag{4}$$

These states clearly satisfy the Schrödinger equation

$$\begin{aligned} H_0|N\rangle_0 &= E_N|N\rangle_0 \\ H_0|N+1,k\rangle_0 &= (E_N + \varepsilon_k)|N+1,k\rangle_0 \\ H_0|N-1,k\rangle_0 &= (E_N - \varepsilon_k)|N-1,k\rangle_0 \end{aligned} \tag{5}$$

where $H_0 = \sum \varepsilon_k c_k^\dagger c_k$ and $E_N = \sum^{occ} \varepsilon_k$. By the quasi-particle energies one does not understand the total energies $E_{N\pm1,k} = E_0 \pm \varepsilon_k$ but only the excitation energies

$$\begin{aligned} E_{N+1,k} - E_N &= \varepsilon_k, \qquad k > k_F, \\ E_N - E_{N-1,k} &= \varepsilon_k, \qquad k < k_F. \end{aligned} \tag{6}$$

Let us now study the term in H, which we neglected, namely the electron-electron interaction. We first try to estimate the life-time of the states in equation (3) by using the Golden Rule. The probability for decay per unit of time for the state $|N+1, k\rangle_0$ is, to lowest order

$$W = \frac{2\pi}{\hbar} \sum_f |\langle \Phi_f | \frac{1}{2\Omega} \sum_q v(q) \rho_q \rho_{-q} c_k^\dagger |N\rangle_0|^2 \delta(E_f - E_{N+1,k}). \tag{7}$$

The simplest type of final state that gives a non-vanishing matrix-element of the operator $c_{k_1+q}^\dagger c_{k_1} c_{k_2-q}^\dagger c_{k_2} c_k^\dagger$ contains two electrons and one hole and is reached when k_1 or k_2 equals k. These two cases give the same contribution and we thus obtain

$$W = \frac{2\pi}{\hbar} \frac{1}{\Omega^2} \sum_{qk_1} v^2(q) \delta(\varepsilon_{k_1+q} + \varepsilon_{k-q} - \varepsilon_{k_1} - \varepsilon_k), \tag{8}$$

(Contd) $\quad k - q > k_F, \quad k_1 + q > k_F, \quad k_1 < k_F.$ (8)

We study this expression for k close to k_F. Replacing $v(q)$ by an average value v_0 and introducing some new variables; $k = k_F(1 + \delta)$, $k_1 = k_F(1 - x)$, $q = k_F\eta$, $\xi = \hat{k}\hat{q}$ and $\xi_1 = -\hat{k}_1\hat{q}$, we have

$$W = \frac{2\pi}{\hbar}\frac{(2\pi)^2}{(2\pi)^6} v_0{}^2 k_F{}^6 \int_0^\infty \eta^2 d\eta \int_0^1 (1-x)^2 dx \int_{-1}^1 d\xi \int_{-1}^1 d\xi_1 \delta[2\varepsilon_F\eta(\eta - \xi - \xi_1)]$$

$$\eta^2 - 2\eta\xi > -2\delta; \qquad \eta^2 - 2\eta\xi_1 > 2x,$$

$$= \frac{v_0{}^2 k_F{}^6}{\hbar(2\pi)^3}\frac{1}{2\varepsilon_F}\int_0^1 (1-x)^2 dx \int_{-1}^1 d\xi \int_{-1}^1 d\xi_1 (\xi + \xi_1), \tag{9}$$

$$2\delta > \xi^2 - \xi_1{}^2 > 2x; \qquad \xi + \xi_1 > 0.$$

The triple integral is easily evaluated by noting that $\delta > x$ and that in $\int \xi d\xi \int d\xi$ only $\xi > 0$ contributes while in $\int d\xi \int \xi_1 d\xi_1$ only $\xi_1 < 0$ contributes. The result is simply $2\delta^2$ giving

$$W = \frac{1}{(2\pi)^3}\frac{v_0{}^2 k_F{}^6}{\hbar\varepsilon_F}\frac{1}{4}\left(\frac{\varepsilon_k - \varepsilon_F}{\varepsilon_F}\right)^2. \tag{10}$$

We estimate v_0 by $4\pi e^2/k_F{}^2$ and obtain

$$W = \frac{1}{2\pi}\frac{(e^2 k_F)^2}{\hbar\varepsilon_F}\left(\frac{\varepsilon_k - \varepsilon_F}{\varepsilon_F}\right)^2. \tag{11}$$

For metallic densities $e^2 k_F$ is comparable to ε_F and hence the life-time $\tau \approx 1/W$ becomes

$$\tau \cong \frac{\hbar}{\varepsilon_F}\left(\frac{\varepsilon_F}{\varepsilon_k - \varepsilon_F}\right)^2. \tag{12}$$

Our calculation thus indicates that the life-time of the quasi-particle is greatly increased above the value we might have expected on dimensional grounds, namely $\hbar/\varepsilon_F \approx 10^{-16}$ sec, and actually becomes infinite when k approaches the Fermi surface. The reason for this enhancement is clear, namely the sever restriction of phase space available for final states having a prescribed energy and momentum and being built only from such one-electron states that are not blocked by Fermi statistics.

The estimate we have made of the life-time seriously lacks in rigour; the integral in equation (8) actually diverges due to the singular behaviour of $v(q)$ for small q. Thus our first attempt to define a quasi-particle, namely with equation (3), leads to a life-time that is zero! If we, however, define our quasi-particles with more care we will find that their interaction is more like a screened Coulomb potential $4\pi e^2/(k^2 + k_F{}^2)$ and thus our estimate for the life-time τ should actually be somewhat too small. This indeed turns out to be the case, a more careful estimate gives a result about a factor of 10 larger for τ than that given in equation (12). The full analysis also verifies the quadratic dependence on the energy difference $\varepsilon_k - \varepsilon_F$.

We now must look for better ways than equation (3) to define a quasi-particle, and try the following. Consider a time dependent Hamiltonian

$$H(t) = \sum \varepsilon_k c_k^\dagger c_k + e^{\lambda t}\, \tfrac{1}{2} \sum v(q)\rho_q\rho_{-q}, \tag{13}$$

chosen such that $H(-\infty) = H_0$ and $H(0) = H$. We define a state $|N + 1,k;0\rangle$ which is the $t = 0$ solution of the time dependent Schrödinger equation

$$i\hbar \frac{\partial}{\partial t}|N + 1,k;t\rangle = H(t)|N + 1,k;t\rangle, \tag{14}$$

obtained with the boundary condition

$$|N + 1,k;-\infty\rangle = c_k^\dagger|N\rangle_0. \tag{15}$$

In the limit that $\lambda \to 0$ such a procedure gives the true eigenstates of the time independent Hamiltonian $H = H(0)$ provided the 'starting' state $|N + 1,k;-\infty\rangle$ is non-degenerate (the adiabatic theorem). Our starting state in equation (15) is, however, degenerate, except when $k = k_F$. But in our discussion of decay we have just seen that the number of degenerate states rapidly goes to zero when k approaches k_F. Thus at least for this case $(k \to k_F)$ we could expect the adiabatic theorem 'almost' to work so that the life-time τ of our quasi-particle $|N+1,k;0\rangle$ is long. To succeed in obtaining a large τ we must be careful in our choice of λ. We cannot on one hand allow λ to become much smaller than $1/\tau$ since otherwise the state we try to create will have decayed before the interaction has reached its full strength (at $t = 0$). On the other hand λ cannot be allowed to become very large since we then would approach the 'sudden' limit and obtain a state with a substantial amount of high energy components and thus necessarily having a rapid decay. Somewhere between Scylla and Charybdis there should be a possible course to sail.

The definition of quasi-particle that we have attempted to give, clearly is not very precise, and in the next section we will do a better job. We emphasize, however, that there is always a certain degree of arbitrariness in any definition of a quasi-particle *per se*; the precisely defined concepts which are introduced in the next section and which are chosen to have a close connection with experiment, only have a qualitative relation to the quasi-particles.

Whatever our choice of definition for a quasi-particle, we can always consider it as built from a cloud of virtual excitations

$$|N + 1,k\rangle = (\alpha c_k^\dagger + \beta c^\dagger c^\dagger c + \gamma c^\dagger c^\dagger c^\dagger cc + \ldots)|N\rangle_0. \quad (16)$$

We truly have a cloud with a large number of particle-hole pairs since it can be shown that any individual coefficient in the expansion tends to zero with the number N of particles, and that the number of particle-hole pairs needed for convergence is of order N.

To summarize, the quasi-electron and the quasi-hole are complicated many-body states built from a cloud of virtual excitations. They have sharply defined momenta but a certain spread in energy Γ and thus a finite life-time $\tau = \hbar/\Gamma$. The spread in energy can be precisely defined. If we expand the quasi-particle state in terms of the exact eigenstates

$$|N + 1,k\rangle = \sum_s \alpha_s |N + 1,k,s\rangle, \quad (17)$$

then

$$\Gamma^2 = \sum_s \alpha_s^2 E_s^2 - (\sum_s \alpha_s^2 E_s)^2. \quad (18)$$

Note that Γ has nothing to do with the extension of the cloud of virtual excitations, thus, e.g., at the Fermi surface $\Gamma \to 0$ while there is little change in the size of the virtual cloud.

We have only discussed one particular type of quasi-particle: electrons and holes. Other types of quasi-particles are, e.g., plasmons, polarons and excitons. They are all characterized by having a sharp (crystal) momentum and a finite life-time. We also sometimes use the concept quasi-particle for states that are well defined in real space like electronic states on an F-center or a decaying core-hole state. For localized energetically more or less sharp states we, however, often use the term resonance. Thus we may speak of a d-band resonance in the bandstructure of transition metals, and a $2s2p$ state of a free Helium atom is called a Fano resonance.

2 THEORETICAL TOOLS FOR DESCRIBING ELEMENTARY EXCITATIONS

We will in this section outline a theoretical framework for

discussing quasi-particles or elementary excitations as well as more general relations of importance to describe experiments in a systematic way. The framework builds on a set of closely related functions: Green's functions or propagators, correlation functions and linear response functions. Our discussion will not in any way be exhaustive, we will instead try to sketch a bird's eye view of the motives for introducing the different functions, their inter-relations and some of their important general features such as spectral resolutions and sum rules. We will also illustrate the general shape of some functions by discussing approximate results for the electron gas.

The one-electron Green's function or the Feynman propagator is defined as

$$G(rt,r't') = -i\langle T(\psi(rt)\psi^{\dagger}(r't'))\rangle . \tag{19}$$

Here $\psi(rt)$ is the Heisenberg representation, $\psi(rt) = \exp(iHt)\times \psi(r)\exp(-iHt)$, of the field operator

$$\psi(r) = \sum_n c_n\phi_n(r), \tag{20}$$

$\{\phi_n\}$ is a complete set of orthonormal functions and $\{c_n\}$ are the corresponding operators. The symbol T stands for the time ordering operator; explicitly equation (19) becomes

$$G(rt,r't') = \begin{cases} -i < \psi(rt)\psi^{\dagger}(r't')\rangle, & t > t', \\ \\ +i < \psi^{\dagger}(r't')\psi(rt)\rangle, & t' > t. \end{cases} \tag{21}$$

The brackets stand for an expectation value with respect to the ground state of the considered N-particle system, or for a finite temperature, an ensemble average. G is called a propagator since it describes the probability that an electron injected into the system at time t' at the point r', will propagate to the point r so that it arrives there at time t. Using for simplicity just the index number of r and t as argument, the one, two and three-electron Green's functions are

$$G(1,2) = (-i)\langle T(\psi(1)\psi^{\dagger}(2))\rangle,$$

$$G(1,2,1',2') = (-i)^2\langle T(\psi(1)\psi(2)\psi^{\dagger}(2')\psi^{\dagger}(1'))\rangle, \tag{22}$$

$$G(1,2,3,1',2',3') = (-i)^3\langle T(\psi(1)\psi(2)\psi(3)\psi^{\dagger}(3')\psi^{\dagger}(2')\psi^{\dagger}(1'))\rangle .$$

Another important propagator describes density fluctuations

$$F(rt,r't') = -i\langle T(\rho'(rt)\rho'(r't'))\rangle,$$

$$\rho'(r) = \psi^\dagger(r)\psi(r) - \langle\psi^\dagger(r)\psi(r)\rangle,$$

F is also called a density-density correlation function. The time ordered function F is very closely related to the commutator expression H

$$H(rt,r't') = -i\theta(t - t')\langle[\rho(rt),\rho(r't')]\rangle, \tag{24}$$

as we will see shortly when we discuss spectral resolutions. H is the basic ingredient in the linear response or dielectric function

$$\varepsilon^{-1}(1,2) = \delta(1,2) + \int H(1,3)v(3,2)d(3), \tag{25}$$

where

$$v(1,2) = \frac{e^2}{|r_1 - r_2|}\delta(t_1 - t_2). \tag{26}$$

The dielectric function gives the linear response of the system to an external perturbation. Thus if we introduce a 'classical' charge distribution $\rho^{ext}(rt)$ the charge density of the system is given by the expression

$$\langle\rho'(1)\rangle = \int\varepsilon^{-1}(1,2)\rho^{ext}(2)d(2). \tag{27}$$

One motive for introducing Green's functions is obvious, with their help we can calculate the expectation value of any operator like, e.g., the Coulomb interaction, $\frac{1}{2}\int\psi^\dagger(1)\psi^\dagger(2)v(1,2)\psi(2)\times\psi(1)\delta(t_1)d(1,2)$. But to this end we only need the equal time limit of the Green's functions. The full time-dependent form is, however, very useful for other purposes. First we may mention the systematic perturbation expansions with Feynman graphs. For the one-electron Green's function these expansions lead to a particularly attractive 'equation of motion'

$$\left[i\frac{\partial}{\partial t_1} + \frac{\hbar^2}{2m}\frac{\partial^2}{\partial r_1{}^2} - V(r_1)\right]G(1,2) - \int\Sigma(1,3)G(3,2)d(3) = \delta(1,2). \tag{28}$$

Here $\Sigma(1,2)$ is the 'self-energy' operator which itself has a systematic Feynman graph expansion, the lowest order term being the usual Hartree-Fock exchange operator.

The full time dependence of our functions also allows us to make spectral resolutions. Thus we have for, e.g., the $T = 0$

one-electron Green's function

$$G(r,r';\omega) \equiv \int e^{i\omega(t-t')}G(rt,r't')dt$$

$$= \sum_s \frac{f_s(r)f_s^*(r')}{\omega - \omega_s - i\delta} + \sum_t \frac{g_t(r)g_t^*(r')}{\omega - \omega_t + i\delta}, \qquad (29)$$

where $f_s(r) = \langle N - 1,s|\psi(r)|N\rangle$ and $g_t(r) = \langle N|\psi(r)|N + 1,t\rangle$ are direct generalizations of the one-electron functions of the independent particle approximation and the indices s and t label *all* states of the system. The energies $\varepsilon_s = E_N - E_{N-1,s}$ and $\varepsilon_t = E_{N+1,t} - E_N$ are also straightforward generalizations of the 'orbital' energies (cf. equation (6)). The Fourier transform in equation (29) can be obtained after a very elementary calculation starting from equation (21), the infinitesimals δ coming from convergence factors. From the spectral resolution we see that the Green's function has poles at the excitation energies of the $(N \pm 1)$-particle system.

For a large system the energy levels generally form a continuum and one then usually writes the spectral resolution as an integral

$$G(r,r';\omega) = \int_C \frac{A(r,r';\omega')}{\omega - \omega'} d\omega', \qquad (30)$$

where the integration contour C has the form

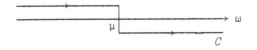

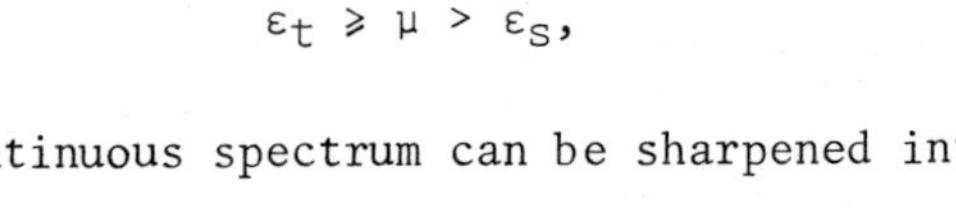

and the cross-over of the real axis occurs at the chemical potential $\mu = E_{N+1,0} - E_N$. In general we have the inequalities

$$\varepsilon_t \geqslant \mu > \varepsilon_s, \qquad (31)$$

which for a continuous spectrum can be sharpened into $\varepsilon_t \geqslant \mu \geqslant \varepsilon_s$.

The spectral function obeys the sum rule

$$\int A(r,r';\omega)d\omega = \delta(r - r'), \qquad (32)$$

and as is immediately clear from equation (29), it has the Hermitean property

$$A^*(r,r';\omega) = A(r',r;\omega). \qquad (33)$$

For an electron gas $G(r,r';\omega)$ depends only on $r - r'$ and we can make a new Fourier transform to

$$G(k\omega) \equiv \int e^{ik(r-r')}G(r,r';\omega)d^3r$$

$$= \int_C \frac{A(k\omega')}{\omega - \omega'}\,d\omega', \tag{34}$$

Equation (32) implies that

$$\int A(k\omega)d\omega = 1. \tag{35}$$

The explicit expression for A is

$$A(k\omega) = \sum_s |\langle N|a_k|N - 1,s\rangle|^2\delta(\omega - \varepsilon_s)$$

$$+ \sum_t |\langle N + 1,t|a_k^\dagger|N\rangle|^2\delta(\omega - \varepsilon_t). \tag{36}$$

In the independent particle approximation where $|N\rangle$ is a single Slater determinant of plane waves inside the Fermi sphere, only one state gives a non-vanishing matrix element and A becomes a delta function

$$A(k\omega) = \delta(\omega - \varepsilon_k). \tag{37}$$

In the previous section we defined a quasi-particle by making a smooth switching-on of the electron-electron interactions. Starting form the one-electron spectral function A we can now give a more satisfactory definition: 'If for a given momentum k, the spectral function $A(k\omega)$ has a reasonably well-defined peak, we call that peak a quasi-particle'. This definition still does not tell us the precise form of the state vector, but the energy of the peak at least gives a well-defined quasi-particle energy and the HWHM (half width at half maximum) gives us the energy spread Γ and the decay time $\tau = \hbar/\Gamma$.

We do not yet know very well the shape of even the electron gas spectral function. We do know, however, that $\Gamma_k \sim (\omega_k - \mu)^2$ for k close to k_F, and reasonable estimates show that out of the total oscillator strength 1 (cf. equation (35)) about 2/3 resides in the quasi-particle. Results from numerical estimates of the width Γ are given in figure 1. We note that the quasi-particle seems to be quite well-defined for all k since Γ_k always is appreciably smaller than ε_k. The mean free path, however, becomes quite small for k-values in the range 2-3 k_F. If we take, say, $r_s = 2$ (which is about the value appropriate for aluminium) and

$k = 2k_F$, the mean free path is predicted to be (see figure 1) $\ell = v\tau = 2(k/k_F)(\varepsilon_F/\Gamma)(1/k_F) = 2 \times 2 \times 2.9 \times 0.55$ Å $= 6.5$ Å. Thus at an energy of $4\varepsilon_F = 47$ eV the width of the quasi-particle state is about 4 eV while the mean free path is as short as only about two nearest neighbour distances.

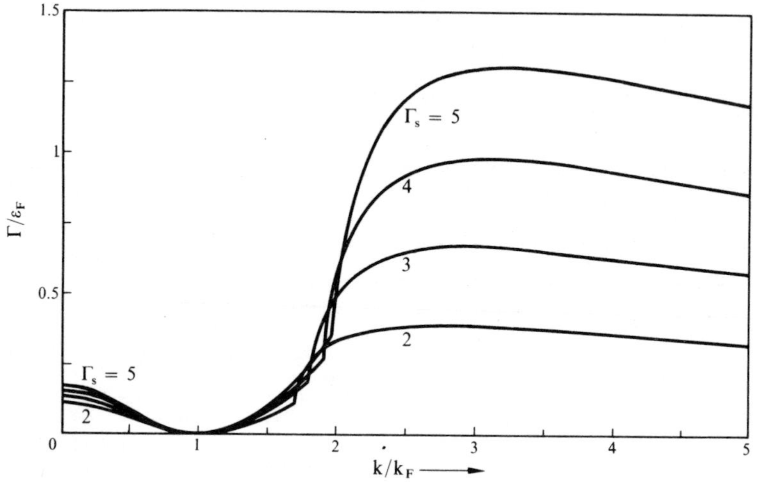

Figure 1 - The HWHM value Γ for quasi-particles in an electron gas as calculated by Lundqvist, B.I. (1969). *Phys. Stat. Sol.*, 32, 276.

Before leaving our discussion of the one-electron Green's function we will elaborate a little more on the decay process of quasi-particles. For $t > 0$ we have

$$G(kt) = -i\langle a_k \exp[-i(H - E_0)t] a_k^{\dagger}\rangle$$

$$= -i\sum_t |\langle N + 1, t|a_k^{\dagger}|N\rangle|^2 \exp(-i\varepsilon_t t). \tag{38}$$

If there is a group of states t with nearly the same energy, they will contribute coherently to $G(k,t)$ and give a component with long lifetime. To show that in more detail we assume that $A(k\omega)$ has a Lorentzian peak

$$A(k\omega) = \frac{Z_k}{\pi} \frac{\Gamma_k}{(\omega - E_k)^2 + \Gamma_k^2} + A_{inc}(k\omega). \tag{39}$$

From equation (34) we then obtain for G

$$G(k\omega) = \frac{Z_k}{\omega - E_k - i\Gamma_k \mathrm{sgn}(\mu - \omega)} + G_{inc}(k\omega). \tag{40}$$

We study the case when $E_k > \mu$ and transfer part of the pole-expression to G_{inc}

$$G(k\omega) = \frac{Z_k}{\omega - E_k + i\Gamma_k} + G_{inc}'(k\omega). \tag{41}$$

A Fourier transform from ω to t gives

$$G(kt) = - iZ_k e^{-iE_k t} e^{-\Gamma_k t} \theta(t) + G_{inc}'(kt). \tag{42}$$

Thus when the spectral function has a Lorentzian peak with a HWHM value of $2\Gamma_k$ centered at an energy E_k the Green's function has a decay time of $1/\Gamma_k$ and oscillates with a frequency E_k (please add or subtract $\hbar$ at the right places!). The quasiparticle part in G dominates over the G_{inc}' only for an intermediate time range when t is large enough to make the contribution of the smooth and extended function G_{inc} become small, and before t becomes so large that the sharp step in $G_{inc}' - G_{inc}$ starts to dominate the Fourier transform.

Spectral resolutions for the F and H functions connected with the linear response expression can be obtained in a similar way as for G. Thus we obtain from equation (23)

$$F(r,r';\omega) = \sum_n \frac{2\omega_n \rho_{0n}'(r)\rho_{n0}'(r')}{\omega^2 - \omega_n^2}, \ \mathrm{Im}\omega_n < 0, \tag{43}$$

where $\omega_n = E_{N,n} - E_{N,0}$ is the excitation energy of state n and $\rho_{0n}'(r)$ a matrix element, $\rho_{0n}'(r) = \langle N,0|\rho'(r)|N,n\rangle$. Clearly the prime on ρ has importance only for $n = 0$, making ρ_{0n}' equal to zero. In deriving equation (43) we have made use of the fact that since it always is possible to work with real wavefunctions (in the absence of magnetic fields), we have $\rho_{0n}(r) = \rho_{n0}(r)$. The infinitesimal imaginary parts in ω_n come from the convergence factors. Similarly we obtain from equation (24)

$$H(r,r';\omega) = \sum_n \frac{2\omega_n \rho_{0n}(r)\rho_{n0}(r')}{\omega^2 - \omega_n^2}, \ \mathrm{Im}\omega > 0. \tag{44}$$

Clearly F and H are very closely related and we have $\mathrm{Re}F(\omega) = \mathrm{Re}H(\omega)$ and $\mathrm{Im}F(\omega) = \mathrm{sgn}(\omega)\mathrm{Im}H(\omega)$.

For an electron gas we can make further simplifications and from equation (25) we obtain

$$\frac{1}{\varepsilon(q\omega)} = 1 + v(q)H(q\omega) \tag{45}$$

The Fourier transform of H in equation (44) is

$$H(q\omega) = \int_0^\infty \frac{2\omega_1 B(q\omega_1)}{\omega^2 - \omega_1^2} d\omega_1, \tag{46}$$

where $B(q\omega)$ is real and non-negative

$$B(q\omega) = \frac{1}{\Omega} \sum_n (\rho_q)_{0n}(\rho_{-q})_{n0}\delta(\omega - \omega_n). \tag{47}$$

From equations (45,46) we find that

$$\operatorname{Im} \frac{1}{\varepsilon(q\omega)} = -\pi v(q)B(q\omega), \tag{48}$$

and thus $1/\varepsilon(q\omega)$ is entirely determined by its imaginary part. (The Kramers-Kronig relations). By evaluating the double commutator $[[\rho_q,H],\rho_{-q}]$ the generalized f-sum rule follows

$$\int_0^\infty \omega B(q\omega)d\omega = \frac{nq^2}{2m}, \tag{49}$$

where $n = N/\Omega$ is the particle density. From equations (48,49) we obtain the well-known sum rule

$$\int_0^\infty \omega \operatorname{Im} \frac{1}{\varepsilon(q\omega)} d\omega = -\frac{\pi}{2}\omega_p^2, \tag{50}$$

where ω_p is the plasmon frequency

$$\omega_p^2 = \frac{4\pi ne^2}{m}. \tag{51}$$

Using the general analytic properties of the spectral resolution for H and equation (49), a sum rule for $\varepsilon(q\omega)$ can be obtained

$$\int_0^\infty \omega \operatorname{Im}\varepsilon(q\omega)d\omega = \frac{\pi}{2}\omega_p^2. \tag{52}$$

A comparatively simple and often used approximation for the dielectric function is obtained from the linearized time-dependent Hartree approximation

$$\varepsilon(q\omega) = 1 - v(q)P(q\omega),$$

$$P(q\omega) = \frac{2}{(2\pi)^3}\int \frac{n_{k+q} - n_k}{\varepsilon_{k+q} - \varepsilon_k - \hbar\omega} d^3k, \qquad \operatorname{Im}\omega > 0. \tag{53}$$

This approximation for $\varepsilon(q\omega)$ is also called the Lindhard function or the Random Phase Approximation (RPA). We immediately see that for $\omega > 0$

$$\mathrm{Im}\varepsilon(q\omega) = v(q)\,\frac{1}{(2\pi)^2}\int(1 - n_{k+q})n_k\delta(\varepsilon_{k+q} - \varepsilon_k - \hbar\omega)\mathrm{d}^3k, \qquad (54)$$

and thus $\mathrm{Im}\varepsilon(q\omega)$ has contributions only from free particle-hole excitations. The excitation energy falls between two parabolas

$$\frac{\hbar^2}{2m}(q^2 + 2k_Fq) > \varepsilon_{k+q} - \varepsilon_k > \frac{\hbar^2}{2m}(q^2 - 2k_Fq) \qquad (55)$$

as sketched in figure 2. With q in units of k_F and ω in units

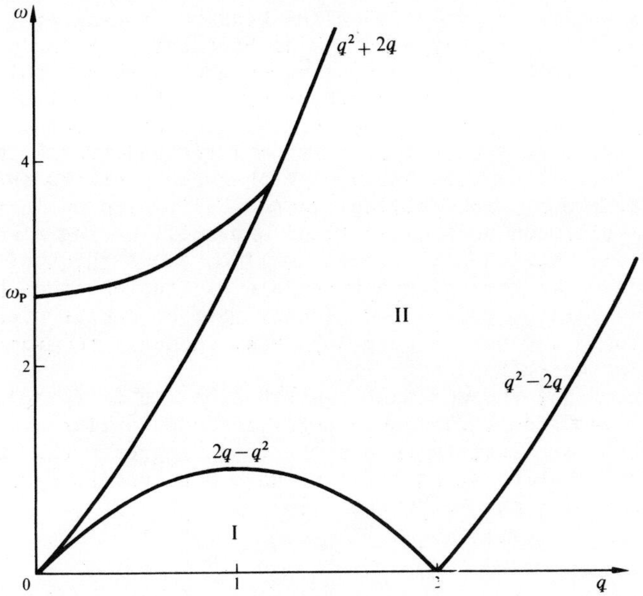

Figure 2 - $\mathrm{Im}\varepsilon(q\omega)$ is different from zero only in regions I and II while $\mathrm{Im}\,1/\varepsilon(q\omega)$ has a further delta function contribution along the plasmon line starting at ω_P. Here q is in units of k_F and ω in units of ε_F.

of ε_F we have

$$\mathrm{Im}\varepsilon(q\omega) = \frac{\alpha\Gamma_S}{q^3}\begin{cases}\omega & \text{in I,}\\ 1 - \frac{1}{4}\left(q - \frac{\omega}{q}\right)^2 & \text{in II}\end{cases} \qquad (56)$$

where $\alpha = 0.521$ and r_s is the usual electron gas parameter. For small q, the sum rule of equation (52) is exhausted in the very small interval $0 < \omega < h^2 k_F q/m$ and thus $\mathrm{Im}\varepsilon(q\omega)$ must be very large as we see explicitly in equation (56). One may then ask how the sum rule involving

$$\mathrm{Im}\,\frac{1}{\varepsilon(q\omega)} = -\,\frac{\mathrm{Im}\varepsilon(q\omega)}{|\mathrm{Re}\varepsilon(q\omega)|^2 + |\mathrm{Im}\varepsilon(q\omega)|^2} \tag{57}$$

could possible be satisfied. The answer is that almost all oscillator strength is shifted away from the particle-hole excitations to a collective resonance at the plasmon frequency ω_p. In the present simple approximation for $\varepsilon(q\omega)$ the plasmons are not damped until q reaches the critical momentum where the curve $\omega(q)$ obtained from the solution of the equation $\varepsilon(q\omega) = 0$ cuts the parabola $\hbar^2/2m(q^2 + 2k_F q)$. Even when the electron-electron interactions are fully included the plasmon damping is zero for $q = 0$ and increases only as q^2. The broadening of the plasmon loss peak of fast electrons passing through a thin metal foil in the $q \to 0$ limit is due to other sources than electron-electron interactions, mainly the Drude term from electron-phonon and electron-impurity scattering and from interband transitions.

It is actually a remarkable fact that there exists an excited state like the plasmon falling energetically high up in the continuum of electron-hole excitations and still having a vanishingly small damping (when q becomes small). If we think of the big diffuse cloud of virtual electron-hole excitations that build the plasmon state, this cloud clearly must be constructed in a very delicate way not to have any decay channels to ordinary electron-hole excitations.

The fact that the dominant contributions to $\mathrm{Im}1/\varepsilon(q\omega)$ for small q comes from the plasmon-pole line and for large q from a narrow strip of electron-hole excitations suggests the following approximation which seems first to have been put forward by Hedin, Lundqvist and Lundqvist (1967)

$$\mathrm{Im}\,\frac{1}{\varepsilon(q\omega)} = \frac{\pi}{2}\,\frac{\omega_p^2}{2\omega(q)}\,\delta(\omega - \omega(q)), \tag{58}$$

where the coefficient is chosen to satisfy the sum rule in equation (50). Equation (58) leads to the simple expression

$$\frac{1}{\varepsilon(q\omega)} = 1 + \frac{\omega_p^2}{\omega^2 - \omega(q)^2}\,, \tag{59}$$

which with a reasonable choice of $\omega(q)$ has proved very useful in different types of approximate theories. For a more detailed account of the theoretical machinery and for further references we refer to Hedin and Lundqvist (1969).

3 EXPERIMENTAL MEASUREMENTS OF ELEMENTARY EXCITATIONS

Calculated electron bandstructres are used for comparison with a wide set of different experimental data. The traditional testing ground, optical spectroscopy, holds its place and still provides the bulk of good experimental information. More recently methods like UV photoemission (UPS), X-ray photoemission (XPS) and electron energy loss measurements have, however, been increasingly applied, and old soft X-ray spectroscopy has been succesfully reactivated. These are perhaps the most well-known methods, but there are many others which rely on bandstructure calculations for their interpretation. For recent reviews we refer to Bennett (1971), Azaroff (1973) and Hedin (1973).

We have chosen in this section to concentrate on optical absorption, and will only briefly discuss other methods. We will first derive the independent particle approximation for the imaginary part of the macroscopic dielectric function that connects the macroscopic (averaged) fields. The perturbation from the electromagnetic field on the electrons is $(e/mc)\bar{A}\bar{p}$. The probability that this perturbation causes a transition is given by the Golden Rule expression

$$w = \frac{2\pi}{\hbar}\left(\frac{eA_0}{mc}\right)^2 \sum_{j}^{empty} \sum_{i}^{occ} |\langle\psi_j|\hat{e}\bar{p}|\psi_i\rangle|^2 \delta(\varepsilon_j - \varepsilon_i - \hbar\omega), \qquad (60)$$

where sum 'i' runs over occupied and sum 'j' over empty one-electron states, and where the vector potential has been written as $\bar{A}(t) = \hat{e}A_0[\exp(i\omega t) + \exp(-i\omega t)]$, $\hat{e}$ being a unit vector (the polarization direction). For a perfect crystal the matrix element of $\bar{p}$ vanishes unless the two Bloch functions involved have the same crystal momentum. Introducing band indices and using the reduced zone scheme we have

$$w = \frac{2\pi}{\hbar}\left(\frac{eA_0}{mc}\right)^2 \sum_{n'}^{empty} \sum_{n}^{occ} \frac{2\Omega}{(2\pi)^3}\int d^3k |\langle\bar{k}n'|\hat{e}\bar{p}|\bar{k}n\rangle|^2$$

$$\times\delta(\varepsilon_{n'}(k) - \varepsilon_n(\bar{k}) - \hbar\omega), \qquad (61)$$

where we have inserted an extra factor of 2 from summing over spin states. Equation (61) can be simplified into a two-dimensional integration over the surface $\varepsilon_{n'}(\bar{k}) - \varepsilon_n(\bar{k}) = h\omega$

$$w = \frac{2\pi}{\hbar}\left(\frac{eA_0}{mc}\right)^2 \sum_{n'}^{empty} \sum_{n}^{occ} \frac{2\Omega}{(2\pi)^3}\int \frac{|\langle\bar{k}n'|\hat{e}\bar{p}|\bar{k}n\rangle|^2 dS_k}{|\nabla(\varepsilon_{n'}(\bar{k}) - \varepsilon_n(k))|} . \qquad (62)$$

It remains to connect the transition probability w to the imaginary part of the dielectric function ε_2. The latter is related to the absorption coefficient α by the well-known relation $\varepsilon_2 = nc\alpha/\omega$. The absorption coefficient gives the energy absorbed per unit time per unit volume, $w\hbar\omega/\Omega$ divided by the energy flux, that is the energy density $u = n^2 A_0{}^2 \omega^2/2\pi c^2$ times the velocity of the photons c/n. We thus have

$$\varepsilon_2(\omega) = \frac{2\pi\hbar c^2 w}{\omega^2 \Omega A_0{}^2}$$

$$= \frac{1}{\pi}\frac{e^2}{m^2}\frac{1}{\omega^2}\sum_{n'}^{empty}\sum_{n}^{occ}\int\frac{|\langle\bar{k}n'|\hat{e}\bar{p}|\bar{k}n\rangle|^2 dS_k}{|\nabla(\varepsilon_{n'}(\bar{k}) - \varepsilon_n(\bar{k}))|} . \tag{63}$$

This expression for ε_2 also follows from the linearized time-dependent Hartree approximation for the microscopic dielectric function if 'local field effects' are neglected. The microscopic dielectric function connects the microscopic fields, which have variations over the atomic scale. This approximation was given in equation (53) for the electron gas case. Its general form is (see e.g. Hedin and Lundqvist (1969), p. 16)

$$\varepsilon(\bar{r},\bar{r}';\omega) = \delta(\bar{r} - \bar{r}') - \int v(\bar{r} - \bar{r}'')P(\bar{r}'',\bar{r}';\omega)d^3r'',$$

$$P(\bar{r},\bar{r}';\omega) = \sum_{ij}\frac{n_j - n_i}{\varepsilon_j - \varepsilon_i - \hbar\omega}\phi_i(\bar{r})\phi_j{}^*(\bar{r})\phi_i{}^*(\bar{r}')\phi_j(\bar{r}'). \tag{64}$$

It is convenient to introduce the Fourier transform

$$\varepsilon(\bar{q},\bar{q}';\omega) \equiv \frac{1}{\Omega}\int e^{i\bar{q}\bar{r}}\varepsilon(\bar{r},\bar{r}';\omega)e^{-i\bar{q}'\bar{r}'}d^3r d^3r'$$

$$= \delta_{\bar{q}\bar{q}'} - v(q)P(\bar{q},\bar{q}';\omega), \tag{65}$$

which due to the translational symmetry of the crystal vanishes unless $\bar{q}$ and $\bar{q}'$ differ by a reciprocal lattice vector. Neglect of off-diagonal elements is, as we will discuss later, equivalent to neglect of 'local field effect'. Keeping only the diagonal element $\varepsilon(\bar{q},\bar{q};\omega) \equiv \varepsilon(q\omega)$ we have

$$\varepsilon(q\omega) = 1 - v(q)\frac{1}{\Omega}\sum_{ij}\frac{n_j - n_i}{\varepsilon_j - \varepsilon_i - \hbar\omega}\left|\int\phi_j{}^*(\bar{r})e^{i\bar{q}\bar{r}}\phi_i(\bar{r})d^3r\right|^2. \tag{66}$$

The 'optical limit' when $q \to 0$ gives

$$\varepsilon(q\omega) = 1 - \frac{4\pi e^2}{\Omega}\sum_{ij}\frac{n_j - n_i}{\varepsilon_j - \varepsilon_i - \hbar\omega}|\langle j|\hat{q}\bar{r}|i\rangle|^2. \tag{67}$$

Using the commutator relation $[H,\bar{r}] = -i(\hbar/m)\bar{p}$ to change the matrix element in equation (67) from one over $\bar{r}$ to one over $\bar{p}$, performing the spin summation and taking the imaginary part of ε, we have for $\omega > 0$

$$\varepsilon_2(\omega) = \frac{4\pi e^2}{\Omega} 2\pi \sum_{ij} (1 - n_j)n_i \frac{1}{m^2\omega^2} |\langle j|\hat{q}\bar{p}|i\rangle|^2 \delta(\varepsilon_j - \varepsilon_i - \hbar\omega)$$

$$= \frac{1}{\pi}\frac{e^2}{m^2}\frac{1}{\omega^2} \sum_{n'}^{empty} \sum_{n}^{occ} \int d^3k |\langle n'\bar{k}|\hat{q}\bar{p}|n\bar{k}\rangle|^2$$

$$\times\delta(\varepsilon_{n'}(\bar{k}) - \varepsilon_n(k) - \hbar\omega), \qquad (68)$$

which is precisely the same result as in equation (63). For a crystal with cubic symmetry $\varepsilon_2(\omega)$ is independent of the direction q and a more symmetric expression can be used

$$|\langle n'\bar{k}|\hat{q}\bar{p}|n\bar{k}\rangle|^2 = \frac{1}{3}|\langle n'\bar{k}|\bar{p}|n\bar{k}\rangle|^2. \qquad (69)$$

There are still very few complete calculations of $\varepsilon_2(\omega)$ from equation (63). Most calculations only give the bands along the symmetry directions but an increasing number now give the density of states

$$N(\omega) = \frac{1}{\Omega}\sum_i \delta(\omega - \varepsilon_i) = \frac{1}{(2\pi)^3}\int \frac{dS_k}{|\nabla\varepsilon_k|} . \qquad (70)$$

Such calculations require knowledge of the energies over a large number of non-symmetry k-points and in addition a rather complicated interpolation scheme. Evaluation of $\varepsilon_2(\omega)$ also requires knowledge of the wavefunctions and summation over band indices.

In figures 3 and 4 we show theoretical and experimental results for the metal Copper. We first note that the three experimental curves (two in figure 3b, and one in figure 4) show a good overall agreement except for a scaling factor; e.g., the 5 eV maximum is 5.2, 6.9 and 9.3 in the three cases. This discrepancy seems to stem from the uncertainty in the high energy reflectivity data used in a Kramers-Kronig calculation of ε_2. The theoretical curve labelled 'Mueller and Phillips' (figure 3a) was obtained with very rough estimates of the matrix elements (oscillator strengths) while the two curves 'Theory Chodorow potential' (figure 3a) and 'present theory' (figure 3b) can be considered as faithful representations of equation (63) but calculated with different crystal potentials. The curve 'Fong et al.' is from an eight parameter pseudopotential calculation. The conclusions from looking at figures 3 and 4 may be summarized in the following way:

(i) The oscillator strengths are absolutely necessary in

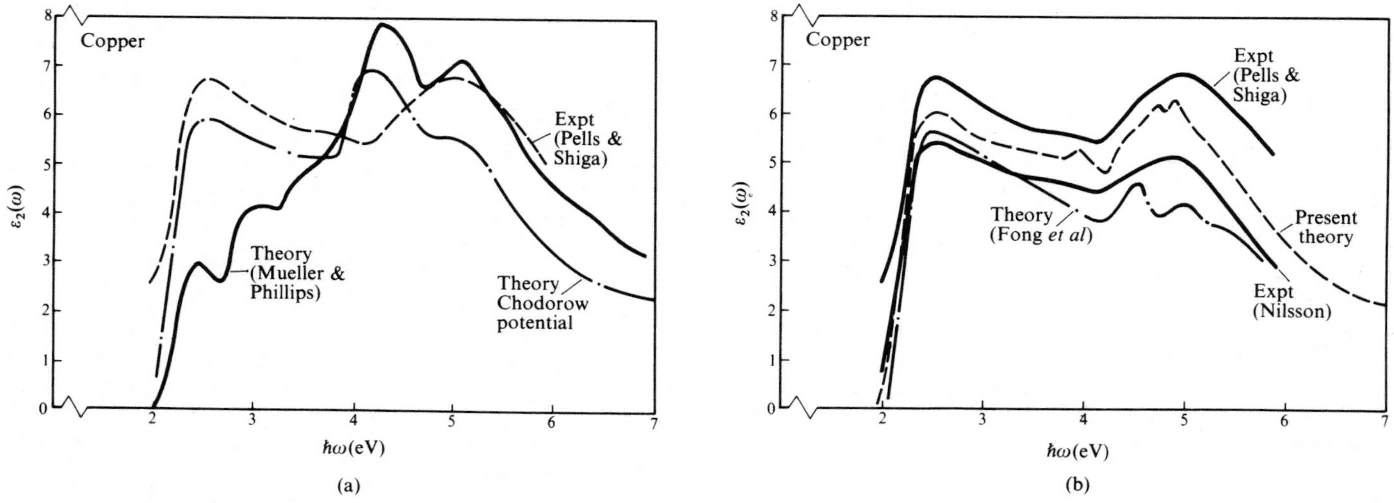

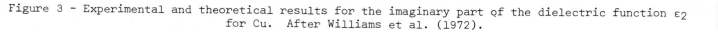

Figure 3 - Experimental and theoretical results for the imaginary part of the dielectric function ε_2 for Cu. After Williams et al. (1972).

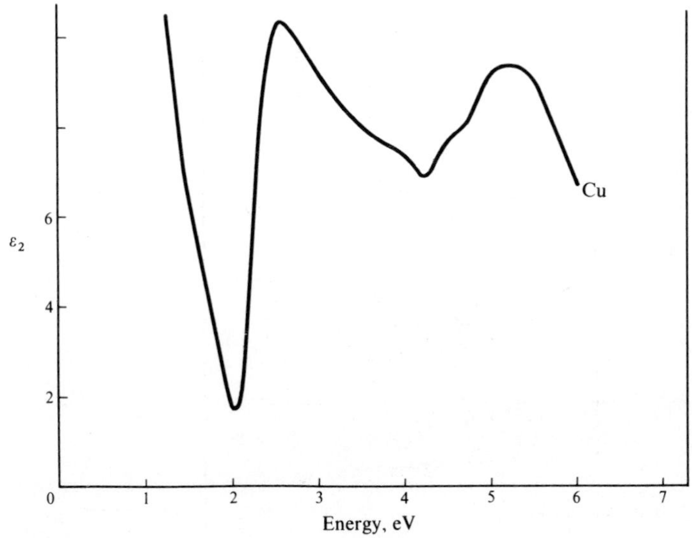

Figure 4 - Experimental results for the imaginary part of the dielectric function ε_2 for Cu. After De Reggi and Rea (1973).

order to have anything like a quantitative agreement with experiment.

(**ii**) The choice of crystal potential is very important, when the Chodorow potential was empirically slightly adjusted to change the main L_2 - $L_{2'}$ gap from 4.6 to 4.9 eV, the 'Chodorow' curve in figure 3a changed into the 'present theory' curve in figure 3b, which essentially agrees with experiment (within present uncertainties of the latter).

In figure 5 we show theoretical and experimental results for the optical conductivity $\sigma = \omega\varepsilon_2/4\pi$ for potassium metal. The theoretical calculation was made by Ching and Callaway (1973) who evaluated equation (63) using a self-consistent LCAO method with Kohn-Sham exchange. The two dominant peaks at 6.5 and 8 eV are expected to be largely washed out if a lifetime broadening with $\Gamma \approx 0.5$ eV (see figure 1) is taken into account. Still some humps should remain, however, and it will be interesting to see if improved experimental accuracy will show up such a structure.

The most important effects which are left out in the simple result of equation (63) can be roughly described as local field effects, dressing of the electrons and holes, and electron-hole interactions (vertex corrections). In deriving equation (63) we assumed that the same average macroscopic field acted on an electron irrespective of its position within the unit cell. This approximation is thought to be acceptable when the jumping

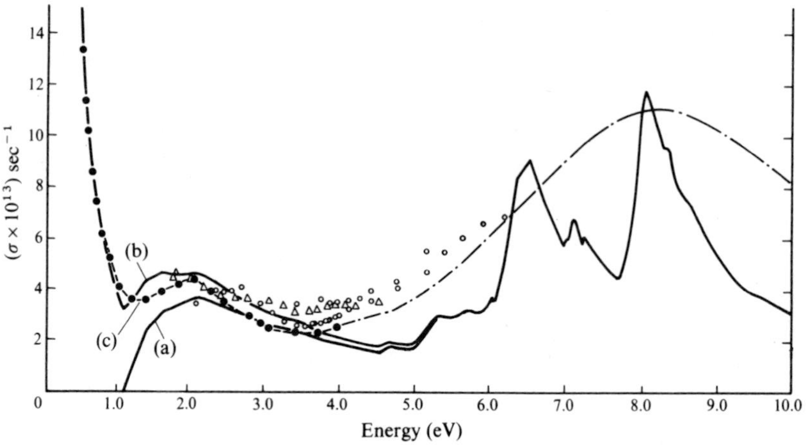

Figure 5 - The optical conductivity for potassium after Ching and Callaway (1973). Curve (a) is the interband contribution (equation (63)), while curve (b) is interband plus Drude. The dashed and dashed-dotted curve is from experimental observations.

electrons have extended wavefunctions like in metals. In the limit of very localized wavefunctions there are on the other hand strong local field effects. For a cubic crystal we then have the well-known Clausius-Mossotti or Lorentz-Lorenz relation $(\varepsilon - 1)/(\varepsilon + 2) = (4/3)\pi N\alpha$ where N is the number of atoms per unit volume and α their polarizability. If we neglected local field effects the dielectric function would be $\tilde{\varepsilon} = 1 + 4\pi N\alpha$ and we thus have the relation

$$\frac{(\varepsilon - 1)}{(\varepsilon + 2)} = \frac{1}{3}(\tilde{\varepsilon} - 1). \tag{71}$$

Bergstresser and Rubloff (1973) have evaluated the real and imaginary parts of ε from experimental data for CsCl and then calculated $\tilde{\varepsilon}$ by using equation (71). The results for ε_2 and $\tilde{\varepsilon}_2$ are shown in figure 6. We see that the two curves are rather different, in particular some of the peaks from the 'simple theory' ($\tilde{\varepsilon}_2$) are dramatically reduced in size. The optical sum rule $\int \omega\varepsilon_2(\omega)d\omega$ is obeyed for both functions and the suppression of the peaks in $\tilde{\varepsilon}_2$ is compensated by a larger high energy tail. Bergstresser and Rubloff make the point that if local field effects can be *that* strong, they may well be non-negligible also for, say, a transition metal like Cu where the d-functions have a certain degree of localization.

By the expression 'dressing of an electron' we mean that the dynamic interaction between a given electron and its surrounding is taken into account. Thus we study the motion of a quasi-electron, that is a bare electron plus a screening or polarization

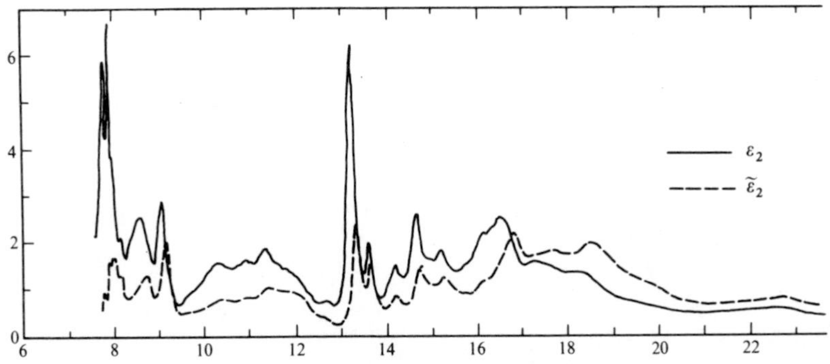

Figure 6 - Experimental results for ε_2 for CsCl in the far ultraviolet at 90°K compared with theoretical results $\tilde{\varepsilon}_2$ without local field corrections. After Bergstresser and Rubloff (1973).

cloud. This leads to correction terms in the crystal potential which if accurately described will give precise values for bandgaps.

We also have the vertex corrections coming from electron-hole interactions. They interfere with the dressing corrections in a delicate manner, in particular both corrections give large scaling effects on $\varepsilon_2(\omega)$ which tend to cancel each other. We will return to this question after having derived the exact expression for optical absorption.

An exact expression for the dielectric function $\varepsilon_2(\omega)$ can be derived starting from the Golden Rule expression (cf equation (60))

$$w = \frac{2\pi}{\hbar}\left(\frac{eA_{\text{ext}}}{mc}\right)^2 \sum_f |\langle\Psi_f| \sum_{i=1}^{N} \hat{e}\bar{p}_i|\Psi_0\rangle|^2 \delta(E_f - E_0 - \hbar\omega), \tag{72}$$

where now the exact many-electron wavefunctions and energies appear. A_{ext} is the amplitude of the vector potential from externally applied sources and should not be confused with the vector potential in the macroscopic Maxwell equations. We want to connect the transition probability w with the linear response function ε^{-1} given in equation (25)

$$\varepsilon^{-1}(\bar{r},\bar{r}';\omega) = \delta(\bar{r} - \bar{r}') + \int H(\bar{r},\bar{r}'';\omega)v(\bar{r}'' - \bar{r}')d^3r''. \tag{73}$$

First we calculate the diagonal part of the Fourier transform (cf equation (65))

$$\varepsilon^{-1}(\bar{q}\omega) \equiv \varepsilon^{-1}(\bar{q},\bar{q};\omega),$$

and obtain using equation (44)

$$\varepsilon^{-1}(q\omega) = 1 + v(q)H(q\omega),$$
$$H(q\omega) = \frac{1}{\Omega}\sum_n \frac{2\omega_n}{\omega^2 - \omega_n^2} |\langle\Psi_0|\rho_q|\Psi_n\rangle|^2. \tag{74}$$

In the 'optical limit' $q \to 0$ we obtain for $n \neq 0$

$$\lim_{q\to 0}\frac{1}{q}\langle\Psi_0|\rho_q|\Psi_n\rangle = \lim_{q\to 0}\frac{1}{q}\langle\Psi_0|\int\psi^\dagger(\bar{r})\exp(i\bar{q}\bar{r})\psi(\bar{r})d^3r|\Psi_n\rangle$$
$$= i\langle\Psi_0|\hat{q}\sum\bar{r}_i|\Psi_n\rangle, \tag{75}$$

where we have used the fact that $\langle\Psi_0|\int\psi^\dagger(r)\psi(r)d^3r|\Psi_n\rangle = N\langle\Psi_0|\Psi_n\rangle = 0$. We again use the commutator relation (cf equation (67)) $[H,\sum\bar{r}_i] = -i\hbar/m\sum\bar{p}_i$ to change the matrix element over $\sum\bar{r}_i$ to one over $\sum\bar{p}_i$ and obtain

$$\lim_{q\to 0}\varepsilon^{-1}(q\omega) = 1 + \frac{4\pi e^2}{\Omega}\sum_n\frac{2\omega_n}{\omega^2 - \omega_n^2}\frac{1}{m^2\omega_n^2}|\langle\Psi_0|\hat{q}\sum\bar{p}_i|\Psi_n\rangle|^2. \tag{76}$$

The imaginary part of this expression (for $\omega > 0$)

$$\mathrm{Im}\,\varepsilon^{-1}(\omega) = -\frac{4\pi^2e^2}{m^2}\frac{1}{\omega^2}\frac{1}{\Omega}\sum_n|\langle\Psi_0|\hat{q}\sum\bar{p}_i|\Psi_n\rangle|^2\delta(\omega - \omega_n) \tag{77}$$

clearly is proportional to the transition probability w in equation (72). Thus by using the relation between $\varepsilon_2(\omega)$ and w in equation (63) we have the simple result

$$\varepsilon_2(\omega) = -\left(\frac{A_{ext}}{A_0}\right)^2\mathrm{Im}\varepsilon^{-1}(\omega). \tag{78}$$

It can further be shown (Adler (1962), Wiser (1963)) that the two amplitudes A_0 and A_{ext}, which both have a smooth variation on the atomic scale, are connected by the relation

$$A_0 = |\varepsilon^{-1}(\omega)|A_{ext}, \tag{79}$$

which leads us to the result wanted, namely a connection between the ordinary dielectric function $\varepsilon_2(\omega)$ associated with the macroscopic Maxwell equations and the microscopic response function $\varepsilon^{-1}(\bar{r},\bar{r}';\omega)$ defined in equations (24,25)

$$\varepsilon_2(\omega) = -\left|\varepsilon^{-1}(\omega)\right|^2 \mathrm{Im}\varepsilon^{-1}(\omega),$$
$$\varepsilon^{-1}(\omega) = \lim_{q\to 0} \varepsilon^{-1}(q,q;\omega), \tag{80}$$

where $\varepsilon^{-1}(\bar{q},\bar{q}';\omega)$ is given by equation (65). We note that if the non-diagonal terms in $\varepsilon^{-1}(\bar{q},\bar{q}';\omega)$ are neglected, the matrix can trivially be inverted giving $\varepsilon^{-1}(\bar{q},\bar{q};\omega) = 1/\varepsilon(\bar{q},\bar{q};\omega)$ and we then have

$$\varepsilon_2(\omega) \cong \lim_{q\to 0} \mathrm{Im}\varepsilon(\bar{q},\bar{q};\omega). \tag{81}$$

Adler and Wiser demonstrated that if the non-diagonal terms are kept in the linearized time dependent Hartree approximation for $\varepsilon(\bar{q},\bar{q}';\omega)$ the Lorentz-Lorenz local field correction was properly reproduced. Thus the non-diagonal terms are associated with local field effects.

To evaluate $\varepsilon_2(\omega)$ we thus need the response function $H \sim \langle[\rho,\rho]\rangle$ or what is essentially the same, $F \sim \langle T\rho'\rho'\rangle$. F in turn is a special case of the two particle Green's function $G_2 \sim \langle T\psi\psi\psi^\dagger\psi^\dagger\rangle$. To calculate optical properties we can thus resort to solving the equation of motion for G_2. It seems, however, to be more rational in many cases directly to calculate the polarization propagator P, $(\varepsilon = 1 - vP)$, by diagram techniques and then to invert $\varepsilon(\bar{q},\bar{q}';\omega)$ if local field effects are considered important. To lowest order P is given by a product of Green's functions

$$P_0(r,r';\omega) = -\frac{i}{2\pi}\int G_0(r,r';\omega')G_0(r',r;\omega' - \omega)d\omega'. \tag{82}$$

Equation (82) gives exactly the same result as the linearized time dependent Hartree approximation. A first attempt to improve on P_0 would be to replace G_0 by a dressed G function where in the spectral function A the strength of the quasi-particle peak has gone down from 1 to, say, 2/3, with the remaining 1/3 of the oscillator strength appearing as a broad background with possibly some plasmon satellites. This would, however, result in a function $P \cong Z^2P_0 + P_{inc}$, i.e. we would have the result P from the simple theory scaled down by about a factor of 2 with the displaced oscillator strength appearing at higher energies, perhaps as plasmon bumps. It has been argued, e.g. by Beeferman and Ehrenreich (1970) and by Hermanson (1972) that the GG approximation is seriously unbalanced and that dressing of the quasi-particles should always be considered together with vertex corrections since there are very large cancellations between the two effects. It may also be noted that the correct plasmon behaviour of the electron gas dielectric function is violated by the GG approximation.

We conclude this section by a few remarks on photoemission. In X-ray photoemission the transition probability is proportional to (Hedin and Lundqvist (1969))

$$w \sim \sum_s |\langle N-1,s| \sum_j \langle k|\hat{e}\bar{p}|j\rangle c_j |N\rangle|^2 \delta(\varepsilon_k - \varepsilon_s - \omega), \qquad (83)$$

which result follows if we regard the emitted (fast) photoelectron as decoupled from the rest of the electrons so that the final state has the form $c_k{}^\dagger|N-1,s\rangle$ and the energy $\varepsilon_k + E_{N-1,s}$. Here k stands for a Bloch state and ε_k for a band energy. The approximation in equation (83) is equivalent to writing $P = GG_0$. Since for high energies $G \to G_0$ we here have a case where the GG approximation is quite appropriate. We note that knowledge of the one-electron spectral function A (cf equation (36)) is sufficient for the calculation of w, and thus of the result from an idealized XPS experiment. In practice there are, however, many experimental complications; the excitation radiation has a width of 0.5-1 eV and the photoelectron suffers energy losses on its way out of the solid. An indication of what type of resolution that can be obtained today is given in figure 7 from Baer and Busch (1973).

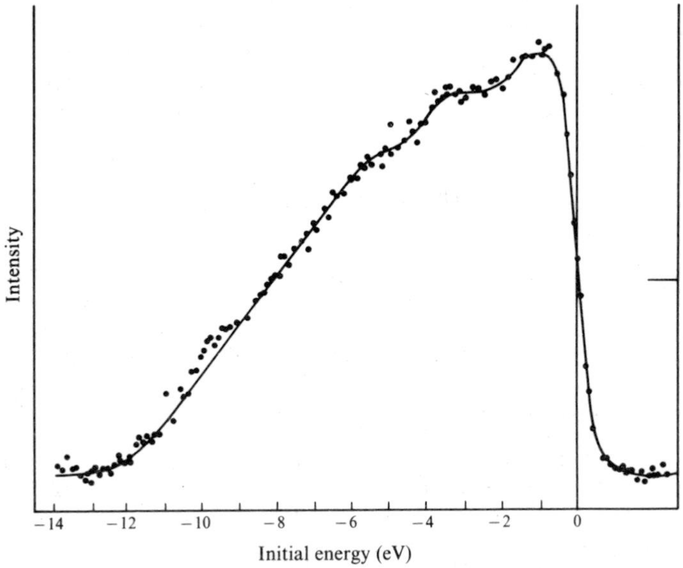

Figure 7 - X-ray photoemission result from the conduction band in aluminium. After Baer and Busch (1973).

If the excitation radiation is in the energy range of, say, 4-40 eV instead of the keV range, we speak of UV photoemission spectroscopy (UPS). As a first approximation we may perhaps still consider the outgoing photoelectron as decoupled from the remaining electrons and use equation (83). The lower the energy becomes, however, the stronger will be the necessity to include vertex corrections together with the dressing effect. As an indication of the critical point we may take the energy when the z-value of the electron gas quasi-particle has gone up to, say, 0.90. This occurs at a few times the plasmon energy according to estimates by B.I. Lundqvist (1969).

For further discussions of the many experimental methods for determination of band structures we refer to the general references mentioned in the beginning of this section.

B. EXPLICIT APPROXIMATIONS FOR THE CRYSTAL POTENTIAL

1 A Dynamic Generalization of the Hartree-Fock Exchange Potential

In part A we discussed qualitative features and the formal machinery, in part B we will go into the details of constructing explicit crystal potentials. The basic part of any crystal potential is the Hartree or Coulomb potential

$$V_{\mathrm{H}}(\bar{r}) = \int v(\bar{r} - \bar{r}')\rho(\bar{r}')d^3r'. \tag{84}$$

In the Hartree-Fock approximation one has in addition the non-local exchange potential

$$V_{\mathrm{ex}}(\bar{r},\bar{r}') = - v(\bar{r} - \bar{r}')\rho(\bar{r},\bar{r}'), \tag{85}$$

where $\rho(\bar{r},\bar{r}')$ is the density matrix

$$\rho(\bar{r},\bar{r}') = \sum_{i}^{occ} \phi_i(\bar{r})\phi_i^*(\bar{r}'). \tag{86}$$

We remind that the effect of a non-local potential on a wave-function is the following

$$(V_{\mathrm{ex}}\phi)_{\bar{r}} = \int V_{\mathrm{ex}}(\bar{r},\bar{r}')\phi(\bar{r}')d^3r'. \tag{87}$$

When we go beyond the Hartree-Fock approximation the choice of crystal potential is no longer obvious, but depends on the questions we want to have answered. One possible choice is the local potential defined by Kohn and Sham (1965) which in principle gives the exact charge density and the total energy but has no connection with, e.g., optical spectra. We will return to this potential in section 3. Another possible choice is the self-energy Σ, which appears in the equation of motion (equation

(28)), for the one-electron Green's function. This potential gives results for the excitation spectra of $N \pm 1$ electron systems as we may guess from the spectral resolution in equation (29) of the Green's function. Still other choices of crystal potential may be contemplated depending on what particular properties we want to study. In sections 1 and 2 we will, however, only discuss the self-energy, while in section 3 we will also take up the Kohn-Sham ground state potential.

We Fourier transform equation (28) with respect to time

$$(\omega - T - V_H)G(\bar{r},\bar{r}';\omega) - \int \Sigma(\bar{r},\bar{r}'';\omega)G(\bar{r}'',\bar{r}';\omega)d^3r'' = \delta(\bar{r} - \bar{r}'). \quad (88)$$

Suppose that one of the energies in equation (29), say ω_s, is discrete. We now study the behaviour of $G(\omega)$ when ω approaches ω_s. Then the 's' term grows without bounds while all others tend to constants. We realize that equation (88) only can be satisfied if f_s obeys the equation

$$(\varepsilon_s - T - V_H)f_s(\bar{r}) - \int \Sigma(\bar{r},\bar{r}';\varepsilon_s)f_s(\bar{r}')d^3r' = 0. \quad (89)$$

This is a straightforward generalization of the Hartree-Fock one-electron equation where Σ has replaced V_{ex}. We call it a *dynamic* generalization since $\Sigma(\bar{r},\bar{r}';\omega)$ depends on the energy ω. The eigenvalues of equation (89) are excitation energies of the $N \pm 1$ particle system, but the f_s have no simple direct interpretation.

For large systems the energy spectra are continuous and we cannot immediately interpret equation (89). In the electron gas case, however, the situation is still very simple. We can then make a Fourier transform of equation (88) with respect to $\bar{r}$

$$[\omega - \varepsilon_k - \Sigma(k,\omega)]G(k,\omega) = 1. \quad (90)$$

From equation (34) we have for the spectral function $A(k,\omega)$

$$A(k\omega) = \pm \frac{1}{\pi} \mathrm{Im}G(k\omega)$$

$$= \pm \frac{1}{\pi} \frac{\mathrm{Im}\Sigma(k\omega)}{[\omega - \varepsilon_k - \mathrm{Re}\Sigma(k\omega)]^2 + [\mathrm{Im}\Sigma(k\omega)]^2} . \quad (91)$$

$A(k\omega)$ will have a quasi-particle peak if there is a solution to Dyson's equation

$$E = \varepsilon_k + \mathrm{Re}\Sigma(k\omega), \quad (92)$$

which gives an E_k such that $\Gamma_k = \mathrm{Im}\Sigma(k,E_k)$ is small. For the electron gas there seems to be a reasonably well-defined quasi-particle peak (reasonably small Γ_k) for all k, as discussed in connection with figure 1. A discussion of quasi-particles in crystals is more complicated but it can be shown (Layzer (1963)) that equation (89) then gives a complex quasi-particle energy like Dyson's equation does for the electron gas.

What is now the use of working out a good approximation for Σ and then solving equation (89)? One answer is immediate, we can describe the clearcut one-particle excitations that are observed in X-ray photoemission, and to some extent in UV photoemission. Knowing Σ we may also study elastic scattering of electrons off surfaces (LEED). This problem is, however, more complicated than a usual band structure calculation since the translational symmetry is broken at the surface. The solution of equation (89) is also of relevance for the study of optical properties. Thus we will obtain precise values of the bulk chemical potential and the band gap in an insulator, since electrons at the Fermi surface as well as at the top of the valence band and the bottom of the conduction band have no damping from electron-electron collisions. It also seems quite meaningful to solve the band structure problem over an extended energy region, calculating real energy eigenvalues with $\mathrm{Re}\Sigma$ and then computing the polarizability $P_0 \sim G_0G_0$. By such a procedure the energy gaps will be placed precisely, which, as we have discussed earlier, is crucial for a mapping of the main features of an optical spectrum, while the oscillator strengths though they have no very precise meaning, may still be good enough on this level of approximation. A very precise calculation would on the other hand require a rather involved calculation using both dressed Green's functions and vertex corrections.

We now turn to explicit approximations for Σ. First we rewrite the Hartree-Fock exchange potential in equation (85) as

$$V_{ex}(\bar{r},\bar{r}') = \frac{i}{2\pi}\int e^{i\omega'\delta} v(\bar{r} - \bar{r}')G_0(\bar{r},\bar{r}';\omega + \omega')d\omega', \qquad (93)$$

which result can easily be checked by use of the analytical properties of G_0 shown in equation (29). A physically appealing generalization of V_{ex} is the following

$$\Sigma(\bar{r},\bar{r}';\omega) = \frac{i}{2\pi}\int e^{i\omega'\delta} W(\bar{r},\bar{r}';\omega')G_0(\bar{r},\bar{r}';\omega + \omega')d\omega', \qquad (94)$$

where W is a dynamically screened interaction

$$W(\bar{r},\bar{r}';\omega) = \int v(\bar{r} - \bar{r}'')\varepsilon^{-1}(\bar{r}'',\bar{r}';\omega)d^3r''. \qquad (95)$$

We will essentially keep only to this approximation for Σ and show that it includes a wide body of effects that do not appear in Hartree-Fock theory, and we will for short call it the GW approximation. For an electron gas equation (94) can be Fourier transformed

$$\Sigma(k\omega) = \frac{i}{(2\pi)^4}\int e^{i\omega'\delta}\,\frac{v(q)}{\varepsilon(q\omega)}\,\frac{d^3q\,d\omega'}{\omega + \omega' - \varepsilon_{k+q}}\,. \tag{96}$$

We take out Hartree-Fock exchange explicitly, which leaves a part that may be termed a polarization contribution since it vanishes in the case of zero polarization. We thus have

$$\Sigma(k\omega) = V_{ex}(k) + \Sigma_{pol}(k\omega), \tag{97}$$

where

$$V_{ex}(k) = -\frac{e^2 k_F}{\pi}\left[1 + \frac{(k_F^2 - k^2)}{(2k_F k)}\right]\log\left|\frac{(k_F + k)}{(k_F - k)}\right|,$$

$$\Sigma_{pol}(k\omega) = \frac{i}{(2\pi)^4}\int v(q)\left(\frac{1}{\varepsilon(q\omega)} - 1\right)\frac{d^3q\,d\omega'}{\omega + \omega' - \varepsilon_{k+q}}\,. \tag{98}$$

We have earlier (cf equation (59)) argued for a simplified expression for $\varepsilon(q\omega)$

$$\frac{1}{\varepsilon(q\omega)} = 1 + \frac{\omega_p^2}{\omega^2 - \omega(q)^2}, \tag{99}$$

which gives for Σ_{pol}

$$\Sigma_{pol}(k\omega) = \frac{i}{(2\pi)^4}\int v(q)\,\frac{\omega_p^2}{2\omega(q)}\,\frac{2\omega(q)}{(\omega')^2 - \omega(q)^2}\,\frac{d^3q\,d\omega'}{\omega + \omega' - \varepsilon_{k+q}}\,. \tag{100}$$

Equation (100) is the well-known lowest order contribution to the self-energy of an electron coupled to a boson field as was discussed by Hedin et al. (1967), and by Lundqvist (1967a,b; 1968). The boson field is in our case the density-density fluctuations of $\varepsilon^{-1}(q\omega)$ as approximated by the simplified 'plasmon' propagator $2\omega(q)/(\omega^2 - \omega(q)^2)$. The coupling coefficient is according to equation (100)

$$g(q)^2 = v(q)\,\frac{\omega_p^2}{2\omega(q)}\,. \tag{101}$$

The GW approximation thus describes a quasi-electron that has ordinary exchange interactions with all other electrons and which in addition couples to the collective motion of the whole electron cloud.

The integral over ω' in equation (100) can be done analytically ($\omega(q)$ has an infinitesimal negative imaginary part, cf equations (23,43))

$$\Sigma_{\text{pol}}(k\omega) = \frac{1}{\Omega}\sum_q g(q)^2 \frac{n_{\text{k+q}}}{\omega - \varepsilon_{\text{k+q}} + \omega(q)} - \frac{1}{\Omega}\sum_q g(q)^2 \frac{1 - n_{\text{k+q}}}{\varepsilon_{\text{k+q}} - \omega + \omega(q)} . \tag{102}$$

The quasi-particle energy is obtained from equation (92) and we have

$$E_{\text{k}} = \varepsilon_{\text{k}} + V_{\text{ex}}(k) + \Sigma_{\text{pol}}(k,\varepsilon_{\text{k}}), \tag{103}$$

where the replacement of E_{k} by $\varepsilon_{\text{k}} = \hbar^2 k^2/2m$ in the energy argument of Σ_{pol} has been discussed by Hedin (1965a). This result for the quasi-particle energy has been derived by Overhauser (1971) by quite a different approach. He first simplifies the interaction between a test charge and the other electrons by putting all the oscillator strength of each density-density fluctuation into a sharp plasmon mode like we did in equation (59). By modifying the coupling by a factor $1 - G(q)$ he then hopes to change from the case of a test particle coupling to the plasmons to the case of an electron in the system coupling to plasmons. The effect of the electron-plasmon coupling is finally evaluated in second order perturbation theory, which results in equation (40) in Overhauser's (1971) paper.

Overhauser chooses his G-factor from a study of the Kohn-Sham theory applied to the dielectric function. This is of course not very relevant for excitation spectra and quasi-particles but should still give a hint on the importance of the effect, which actually turns out to be fairly minor. This is consistent with results obtained by Hedin (1965a) where the GW approximation was obtained as a first term in a systematic expansion of Σ and the next order term was numerically estimated and found to be small. Overhauser's results is also consistent with restuls obtained by Rice (1965) in a study of the Landau parameters, where a $G(q)$-factor was included. We like to emphasize, however, that while the GW approximation seems to be surprisingly good for calculation of quasi-particle properties, it cannot give but a very rough description of the spectral properties of $A(k\omega)$ (cf Hedin et al. (1970)).

2 DIFFERENT LIMITING CASES OF THE DYNAMIC POTENTIAL

In the last section we introduced the GW approximation for the self-energy Σ (cf equation (94)) and discussed it for an electron gas. In this section we will discuss it for a solid in general, be it an insulator or a metal, and derive several important limiting cases. Some of the results we obtain are also of interest for atoms and molecules. We will show that the GW approximation can be interpreted in a physically appealing way as giving a screening of the Hartree-Fock exchange potential plus a Coulomb hole term. We will then derive the contribution of the GW term to core electron energies and show that the essential effect comes form the relaxation of the outer electrons. We will show that for valence or conduction electrons the GW term can be split into contributions from free correlated ion cores plus a GW contribution from the valence electrons only. We will finally show how the GW approximation is related to the electronic polaron.

We start by rewriting the GW expression in equation (94) on a few different explicit forms which are convenient for the further discussion. We first write out the explicit form for G_0

$$G_0(\bar{r},\bar{r}';\omega) = \sum_k \frac{\phi_k(\bar{r})\phi_k^*(\bar{r}')}{\omega - \varepsilon_k} , \qquad (104)$$

where $\{\phi_k\}$ is a complete orthonormal set of functions, which are solutions of a convenient zero order problem, and $\{\varepsilon_k\}$ are the associated one-electron energies. Inserting this expression for G_0 in equation (94) gives

$$\Sigma(\bar{r},\bar{r}';\omega) = \frac{i}{2\pi} \sum_k \int e^{i\omega'\delta} \frac{W(\bar{r},\bar{r}';\omega')\phi_k(\bar{r})\phi_k^*(\bar{r}')}{\omega + \omega' - \varepsilon_k} d\omega' . \qquad (105)$$

We assume that we have already solved the one-electron equation containing Σ and we want only to discuss the contribution from Σ to the energy eigenvalues. We thus only need the expectation value

$$\Sigma_k = \int \phi_k^*(r)\Sigma(r,r';\varepsilon_k)\phi_k(r')d^3r d^3r'$$

$$= \frac{i}{2\pi} \sum_k \int e^{i\omega\delta} \frac{\langle kk'|W(\omega)|k'k\rangle}{\varepsilon_k - \varepsilon_{k'} + \omega} d\omega. \qquad (106)$$

We have here introduced the conventional notation for a matrix element of a two-electron operator

$$\langle k_1 k_2|W(\omega)|k_3 k_4\rangle$$

$$= \int \phi_{k_1}^*(r)\phi_{k_2}^*(r')W(r,r';\omega)\phi_{k_3}(r)\phi_{k_4}(r')d^3r d^3r' . \qquad (107)$$

We may split Σ_k into a Hartree-Fock exchange contribution plus a polarization part

$$\Sigma_k = V_{ex}(k) + \Sigma_k^{pol}, \tag{108}$$

where

$$V_{ex}(k) = - \sum_{k'}^{occ} \langle kk'|v|k'k\rangle,$$

$$\Sigma_k^{pol} = \frac{i}{2\pi} \sum_{k'} \int \frac{\langle kk'|W(\omega) - v|k'k\rangle}{\varepsilon_k - \varepsilon_{k'} + \omega} d\omega. \tag{109}$$

From equations (25,43) we obtain (ε^{-1} is here defined with the time-ordered function F rather than the commutator expression H)

$$W(\bar{r},\bar{r}';\omega) - v(\bar{r} - \bar{r}') = \sum_n \frac{2\omega_n}{\omega^2 - \omega_n^2} C_n(\bar{r},\bar{r}'), \qquad \text{Im}\omega_n < 0, \tag{110}$$

where

$$C_n(\bar{r},\bar{r}') = \int \rho_{0n}'(\bar{r})\rho_{n0}'(\bar{r}'')v(\bar{r}'' - \bar{r}')d^3r''.$$

We may look at the function C_n as a more or less positive quantity. Thus $W(0) - v$ should be essentially negative as we may expect also from the expression $W(0) - v = (\varepsilon^{-1}(0) - 1)v$; $\varepsilon^{-1}(0)$ being smaller than unity in the long wavelength limit. Inserting equation (110) in equation (109), and performing the ω-integration we have

$$\Sigma_k^{pol} = \sum_{k'} \sum_n \langle kk'|C_n|k'k\rangle \left[\frac{n_{k'}}{\varepsilon_k - \varepsilon_{k'} + \omega_n} - \frac{1 - n_{k'}}{\varepsilon_{k'} - \varepsilon_k + \omega_n}\right], \tag{111}$$

which corresponds to equation (102) for the electron gas. Equation (111) will be the starting point for the derivation of a number of limiting cases. We note that for a metal most ω_n are of the order of the plasmon energy, while for an insulator all ω_n necessarily are larger than the band gap.

We will now rearrange Σ_k into screened exchange plus a Coulomb hole term, having in mind the conduction electrons in a metal. This is easily done by noting that the two $n_{k'}$ terms in equation (111) together give the full sum in equation (110) with an ω value of $\varepsilon_k - \varepsilon_{k'}$

$$\Sigma_k = - \sum_{k'} \langle kk'|W(\varepsilon_k - \varepsilon_{k'})|k'k\rangle - \qquad \text{(Contd)}$$

(Contd) $$- \sum_{k'} \sum_{n} \langle kk' | C_n | k'k \rangle \frac{1}{\varepsilon_{k'} - \varepsilon_k + \omega_n} . \tag{112}$$

To obtain a physical interpretation of the last term we make the rather crude approximation of neglecting $\varepsilon_k - \varepsilon_{k'}$ compared to ω_n in the energy denominator. We then have the $\omega = 0$ case of equation (110) and obtain

$$\Sigma_k^{CH} \cong \tfrac{1}{2} \sum_{k'} \langle kk' | W(0) - v | k'k \rangle . \tag{113}$$

Since the sum over all k' gives a delta function, Σ_k^{CH} is the expectation value of a local potential

$$V^{CH}(\bar{r}) \cong \tfrac{1}{2} \int [\varepsilon^{-1}(\bar{r},\bar{r}';0) - \delta(\bar{r} - \bar{r}')] v(\bar{r}' - \bar{r}) d^3r' . \tag{114}$$

The physical interpretation of this expression is quite clear. A test particle at point $\bar{r}$ gives rise to an induced charge of $[\varepsilon^{-1}(\bar{r},\bar{r}';0) - \delta(\bar{r} - \bar{r}')]$ at the point $\bar{r}'$ and the potential from this induced charge at the position of the test particle is, as we see from equation (114), $2V^{CH}(\bar{r})$. The result $2V^{CH}$ of our classical calculation should be reduced by a factor of two by accounting for adiabatic switching-on. The last term in equation (112) can thus be associated with the classical interaction between an electron and its own polarization cloud, and we call it a 'Coulomb hole' contribution. For an electron gas at metallic densities the crude approximation in equation (113) overestimates Σ_k^{CH} by some 25% when k is smaller than, say, 1.5 k_F. The Coulomb hole term is numerically larger than the screened exchange term by a factor of five or more at metallic densities and $k < 1.5\ k_F$, and thus is the dominant contribution in band structure calculations for the optical region.

For a fast electron the screening of the exchange term becomes unimportant and the unscreened exchange itself becomes small

$$V_{ex}(k) \cong \frac{2}{3} V_{ex}(k_F) \left(\frac{k_F}{k}\right)^2 , \qquad k >> k_F . \tag{115}$$

The Coulomb hole term (= last term in equation (112)) can be estimated by using plane waves and free electron energies

$$\Sigma_k^{CH} \cong - \frac{1}{\Omega} \sum_{q} \sum_{n} \frac{C_n(q,q)}{(h^2/2m)(2\bar{k}\bar{q} + q^2) + \omega_n} . \tag{116}$$

With k large we can neglect q^2 in the denominator and since $C_n(\bar{q},\bar{q})$ does not depend on the sign of $\bar{q}$ we have the result

$$\Sigma_k^{CH} \cong \frac{1}{2}\frac{1}{\Omega}\sum_q [W(\bar{q},\bar{v}\bar{q}) - v(q)]. \tag{117}$$

This result is just half (adiabatic switching!) of the Coulomb interaction between a test charge moving with velocity $\bar{v}$ and its induced polarization cloud (Hedin et al. (1967)). Further simplifications based on the large k limit give after a short calculation the well-known result (see, e.g., Lindhard (1954))

$$\Sigma_k^{CH} \cong -\frac{\pi}{4}\frac{e^2\omega_p}{v} = -\frac{\pi}{4}\frac{h\omega_p}{k_F a_0}\frac{k_F}{k}. \tag{118}$$

The Coulomb hole term thus tends to zero only very slowly as compared to the exchange term in equation (115). For, say, a 1 keV electron in aluminium, $k \approx 9\ k_F$ and $\Sigma_k^{CH} = -1.4$ eV.

We next study the correlation contribution to core electron energies. The largest contribution in equation (111) clearly will come from k'-states in the same band B_c as the core level we study

$$\Sigma_c^{pol} \cong \sum_{k' \in B_c}\sum_n \langle ck'|C_n|k'c\rangle \frac{1}{\varepsilon_c - \varepsilon_{k'} + \omega_n}. \tag{119}$$

If the core band is narrow we can neglect $\varepsilon_c - \varepsilon_{k'}$ and we can further use localized (atomic) wave functions ϕ_c obtaining

$$\Sigma_c^{pol} \cong -\tfrac{1}{2}\langle cc|W(0) - v|cc\rangle. \tag{120}$$

This result may be compared to the Coulomb hole result in equation (113). Again the polarization contribution is purely classical, being just one half (adiabatic switch off!) of the Coulomb interaction between the core electron charge density and the induced charge in the valence electron charge density (the polarization) caused by the presence of the core hole. In the core electron case we, however, have the reverse sign of the term as compared to equation (113), corresponding to the fact that we study a 'hole' state rather than an 'electron' state. It is true that equation (113) also may be used for 'hole' states ($k < k_F$) but then correlation effects enter the problem also through a very large screening (reduction) of the exchange term and a comparison with the core electron case cannot be made. Irrespective of the way we divide the correlation effect on different terms we have the general rule that 'hole' states are raised in energy whilst 'electron' states are lowered, as illustrated for an electron gas in figure 8. The result we have obtained for the polarization or correlation contribution to core levels also follows if we make two self-consistent Hartree-Fock calculations and subtract the total energies (Hedin

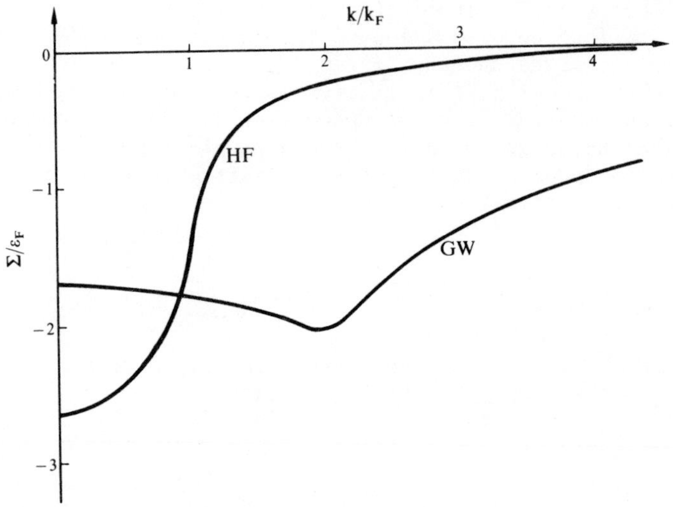

Figure 8 - Comparison between the exchange (HF) and the exchange-correlation (GW) potential for an electron gas at the density of Sodium (r_s = 4).

and Johansson (1969)), but the reason for the good agreement with experiment is then less clear.

We return to a more detailed study of the valence and conduction electron energies. The terms in equation (111) where k' runs over core states now are very small both due to large energy denominators and to small overlap between the wavefunctions in the matrix element. Neglecting these terms and putting valence-valence exchange back together with the Σ^{pol} contribution we have (cf equation (106))

$$\Sigma_k = V_{ex}{}^c(k) + \frac{i}{2\pi}\int e^{i\omega\delta} \sum_{k'}^{val} \frac{\langle kk'|W(\omega)|k'k\rangle}{\varepsilon_k - \varepsilon_{k'} + \omega}\, d\omega. \tag{121}$$

It is possible (Hedin (1965b)) to approximately split the polarization propagator P into a core part and a valence part, $P = P^c + P^v$ which gives a corresponding split of the screened potential, $W = W^v + W^v P^c W^v + \ldots$. The second term gives roughly the same contribution as the correlation potentials from free ion cores, and we thus have the simple result

$$\Sigma_k \cong V_{ex}{}^c + V_{corr} + \Sigma_k{}^v. \tag{122}$$

The effective potential acting on a valence electron thus consists of the correlated potentials from free ion cores plus a self-energy contribution from interacting valence electrons moving in a static field from free ion cores.

The concept of an electronic polaron was introduced by Toyozawa (1954) for the discussion of insulators and has recently attracted a renewed interest (see, e.g., Kunz (1972)). We will now show how the electronic polaron is related to the GW approximation. We start as usual from equation (111). In an insulator ω_n must be larger than the band gap. We make the approximation that ω_n equals some effective ε_0 for all n. Looking at equation (110) the sum $\sum C_n$ may in the same spirit of approximation be replaced by $-(\varepsilon_0/2)(W(0) - v)$, and thus we have

$$\Sigma_k^{pol} = -\frac{\varepsilon_0}{2}\sum_{k'} \langle kk'|W(0) - v|k'k\rangle \times\left[\frac{n_{k'}}{\varepsilon_k - \varepsilon_{k'} + \varepsilon_0} - \frac{1 - n_{k'}}{\varepsilon_{k'} - \varepsilon_k + \varepsilon_0}\right] . \quad (123)$$

If the wavefunctions are extended we may (perhaps) neglect the dispersion in the dielectric function and replace $\varepsilon(q,0)$ by the static (optical) dielectric function ε_s, giving

$$\Sigma_k^{pol} = \frac{\varepsilon_0}{2}\left(1 - \frac{1}{\varepsilon_s}\right)\sum_{k'} \langle kk'|v|k'k\rangle \times\left[\frac{n_{k'}}{\varepsilon_k - \varepsilon_{k'} + \varepsilon_0} - \frac{1 - n_{k'}}{\varepsilon_{k'} - \varepsilon_k + \varepsilon_0}\right]$$

$$= \frac{\varepsilon_0}{2}\left(1 - \frac{1}{\varepsilon_s}\right)\frac{1}{\Omega}\sum_{k'q} \left|\int \phi_k^*(\bar{r})e^{i\bar{q}\bar{r}}\phi_{k'}(\bar{r})d^3r\right|^2 v(q) \times\left[\frac{n_{k'}}{\varepsilon_k - \varepsilon_{k'} + \varepsilon_0} - \frac{1 - n_{k'}}{\varepsilon_{k'} - \varepsilon_k + \varepsilon_0}\right], \quad (124)$$

which is the usual expression in the theory of the electronic polaron. The matrix element vanishes unless $k' = k + q + K$, where K is a reciprocal lattice vector. Neglecting interband coupling gives

$$\Sigma_k^{pol} = \frac{\varepsilon_0}{2}\left(1 - \frac{1}{\varepsilon_s}\right)\frac{1}{\Omega}\sum_{q}^{BZ} \left|\int_{\Omega_0} u_k^*(\bar{r})u_{k+q}(\bar{r})d\bar{r}\right|^2 v(q) \times\left[\frac{n_{k+q}}{\varepsilon_k - \varepsilon_{k+q} + \varepsilon_0} - \frac{1 - n_{k+q}}{\varepsilon_{k+q} - \varepsilon_k + \varepsilon_0}\right] . \quad (125)$$

Only one of the terms in the square bracket appears at a time depending on whether k belongs to the valence band or the conduction band. For, say, the state in the bottom of the conduction band we have, putting the matrix element equal to unity and using an effective mass

$$\Sigma_k^{pol} = -\frac{\varepsilon_0}{2}\left(1 - \frac{1}{\varepsilon_s}\right)\frac{1}{(2\pi)^3}\int_{q<k_{BZ}} \frac{d^3q\,v(q)}{(\hbar^2q^2/2m^*) + \varepsilon_0} . \qquad (126)$$

an expression derived by Haken and Schottky (1958), and extensively discussed by Fowler (1966).

Another approach to the electronic polaron has been taken by Hermanson (1972a) who approximated the dielectric function by the plasmon pole formula of equation (99). By inserting this simple expression into equation (109) and performing the ω-integration one obtains

$$\Sigma_k^{pol} = \frac{1}{\Omega}\sum_{k'q} \left|\int \phi_k^*(\bar{r})e^{i\bar{q}\bar{r}}\phi_{k'}(\bar{r})d^3r\right|^2 \frac{\omega_p^2}{2\omega(q)}$$

$$\times v(q)\left[\frac{n_{k'}}{\varepsilon_k - \varepsilon_{k'} + \omega(q)} - \frac{1 - n_{k'}}{\varepsilon_{k'} - \varepsilon_k + \omega(q)}\right] . \qquad (127)$$

For small q-values equations (127,124) agree provided we identify ε_0 with ω_p. For larger q-values Hermanson's coupling coefficient becomes smaller than Toyozawa's, and Hermanson's approach thus gives rise to a smaller correlation correction.

The theory of the electronic polaron predicts that the band gap obtained in Hartree-Fock theory should be decreased, which indeed seems to be the case experimentally. This result follows immediately also from equation (111) by noting that only one of the terms in the square brackets dominates, depending on whether k belongs to the conduction band or the valence band, and is simply a particular case of our general rule that correlation raises occupied states and lowers empty states.

The electronic polaron theory is clearly a rather crude approximation of the GW expression. In particular it seems unjustified for the valence states where an approximation like that used for the core electrons may perhaps come closer to the correct result.

We conclude in this section with a few critical remarks about the GW approximation. This first criticism concerns the unknown accuracy of the approximation. In the electron gas case we can estimate the error to about 0.02 Ry, but no one really knows what the error will be for a solid. The second criticism concerns the uncertainty in the definition of the GW approximation, precisely what G and what dielectric function should we use? One would perhaps like to do a self-consistent calculation with the same G in $\Sigma = GW$ as in the solution of the equation

$$(\omega - V_H - \Sigma)G = 1.$$

But then still the question remains of what P to use in $\varepsilon = 1 - Pv$. Certainly we would not like to use $P = GG$ which violates sum rules and is generally ill-behaved. It is also doubtful if the replacement of G_0 by G in Σ will mean any improvement. By G_0 we then understand a function of the form given in equation (104) while G is a function where the quasi-particle strength is considerably reduced from unity.

We believe, however, that despite these two difficulties the GW approximation is such a large improvement over Hartree-Fock that it is very well worth further detailed study.

3 LOCAL DENSITY APPROXIMATIONS

Some ten years ago Hohenberg and Kohn (1964) proved a very interesting theorem: "If we know the charge density of an electron system in its ground state, then in principle we also know the one-body potential necessary to produce this charge density". A consequence of the theorem is that we cannot have two different one-body potentials which give rise to precisely the same charge density (unless the difference in potential is only a constant).

The theorem may be formulated mathematically as follows. Let Ψ_0 be the exact ground state solution to the Schrödinger equation

$$\left[\sum_{i=1}^{N} - \frac{\hbar^2}{2m}\nabla_i^2 + \frac{1}{2}\sum_{ij}{}' \frac{e^2}{r_{ij}} + \sum_{i=1}^{N} w(r_i)\right]\Psi_0 = E_0\Psi_0. \tag{128}$$

Further let the number of electrons N and the one-body potential $w(\bar{r})$ be such that this solution is non-degenerate. Let $\rho(\bar{r})$ be the charge density

$$\rho(\bar{r}) = \langle\Psi_0|\psi^\dagger(\bar{r})\psi(\bar{r})|\Psi_0\rangle = \int|\Psi_0(\bar{r},\bar{r}_2,\bar{r}_3..\bar{r}_N)|^2 d^3r_2 d^3r_3 .. d^3r_N. \tag{129}$$

Then according to the Hohenberg and Kohn theorem $w(\bar{r})$ is (within a constant) a unique, universal functional of $\rho(\bar{r})$.

The Hohenberg and Kohn theorem may be expressed symbolically as

$$w(\bar{r}) = F(\bar{r};[\rho]). \tag{130}$$

This simple-looking result entails a number of interesting consequences. Thus the ground state wavefunction Ψ_0 must be a functional of ρ, since if we know ρ, we know w and can solve equation (128). If we know Ψ_0 we can calculate the ground state energy E_0 which thus also is a functional of ρ. This functional can be written in the form

$$E_0 = T[\rho] + V[\rho] + \int \rho(\bar{r}) w(\bar{r}) d^3r, \quad (131)$$

where T and V are the expectation values of the kinetic and potential energies. The variational principle implies

$$\frac{\delta}{\delta\rho(\bar{r})} (T[\rho] + V[\rho]) + w(\bar{r}) = \lambda, \quad (132)$$

where λ is a Lagrangian parameter from the condition of particle conservation.

The variational principle expressed in equation (132) has been utilized by Kohn and Sham (1965) to formulate the following one-body problem. Solve the one-body equation

$$\left[-\frac{\hbar^2}{2m} \nabla^2 + V_{\mathrm{H}}(\bar{r}) + v_{\mathrm{xc}}(\bar{r}) \right] \phi_i(\bar{r}) = \varepsilon_i \phi_i(\bar{r}), \quad (133)$$

and evaluate

$$\rho(\bar{r}) = \sum_i^{occ} |\phi_i(\bar{r})|^2, \quad (134)$$

where the sum 'i' runs over the N states that have the lowest energies ε_i, Then if v_{xc} equals the functional derivative

$$v_{\mathrm{xc}}(\bar{r}) = \frac{\delta E_{\mathrm{xc}}[\rho]}{\delta\rho(\bar{r})}, \quad (135)$$

we have obtained with equation (134) the correct ground state charge density. We have also obtained the correct total energy through the expression

$$E_0 = \sum_i^{occ} \varepsilon_i - \frac{e^2}{2} \int \frac{\rho(\bar{r})\rho(\bar{r}')}{|\bar{r} - \bar{r}'|} d^3r d^3r'$$

$$- \int \rho(\bar{r}) v_{\mathrm{xc}}(\bar{r}) d^3r + E_{\mathrm{xc}}[\rho]. \quad (136)$$

Thus we only have to find out the universal function $E_{\mathrm{xc}}[\rho]$ which, however, is not a trivial task.

Equation (133) has been taken as a basis for a number of energy band calculations. We like to emphasize a few point regarding the interpretation of such calculations.

(i) There is no theoretical justification for expecting the energy eigenvalues ε_i to show correlation with experimentally observed excitation energies. Thus if correlation should occur with a certain approximation for v_{xc}, this has no significance except as a semi-empirical rule of thumb ("If we do so and so we get such and such result").

(ii) The wave functions $\phi_i(\bar{r})$ have no meaning individually, only sums like that in equation (134) have. Thus there is no point putting them in a Slater determinant and evaluating expectation values, except if we want to look for semi-empirical rules of thumb.

(iii) A corollary of point (ii) is that the virial theorem can not be applied to expectation values calculated from a Slater determinant of ϕ_i-functions. Consider first an electron gas. By symmetry the functions $\phi_i(\bar{r})$ must be plane waves and the expectation value of the kinetic energy with respect to a Slater determinant of them gives $(3\varepsilon_F/5)$ as for non-interacting electrons. The true expectation value $\langle\Psi_0|T|\Psi_0\rangle$ is quite different. Thus at the density of sodium ($r_s = 4$), $\langle\Psi_0|T|\Psi_0\rangle$ is some 40% larger than $(3\varepsilon_F/5)$ according to estimates by Hedin (1965a). In the Hartree-Fock approximation for an electron gas the two expectation values happen to agree but when the form of the wavefunctions no longer is determined by symmetry there is no longer any reason that the expectation values should agree. We may also express this by saying that there is no reason two density matrices, $\sum_i\phi_i(\bar{r})\phi_i^*(\bar{r}')$ ($\phi_i = \phi_i^{KS}$ respectively $\phi_i = \phi_i^{HF}$) should be equal just because their diagonal elements (the charge density) are. We have elaborated on the question of the virial theorem because the literature in the journals on this point is somewhat confusing.

The Hohenberg and Kohn theorem can also be applied to the calculation of excitation energies. Since Ψ_0 is a functional of ρ any expectation value like the one-electron Green's function $G = -i\langle\Psi_0|T\psi\psi^\dagger|\Psi_0\rangle$ must also be a functional of ρ. Since G uniquely defines a self-energy Σ, also $\Sigma(\bar{r},\bar{r}';\omega)$ must be a functional of ρ. Explicit approximate forms for Σ have been proposed by Hedin and Lundqvist (1971) for the paramagnetic case and by von Barth and Hedin (1972) for the spin-polarized case. Numerically the suggested approximate potentials for calculation of excitation spectra and for charge densities happen to be fairly equal in the case of slowly varying densities and for energies below the threshold for plasmon creation. This numerical coincidence may perhaps motivate the use of ground state energies ε_i in some approximate discussions of optical properties and other excitations.

It is not yet known what accuracy the available approximation schemes for the local density theories may have. The approximations are only well founded for a much slower variation of the densities than we actually meet in practice and only the much larger computational effort required to evaluate approximations like the GW scheme can motivate the pursuit of local density approximations in their present form

REFERENCES

Adler, S.L. (1962). Quantum Theory of the Dielectric Constant in real solids, *Phys. Rev.*, **126**, 413.

Azaroff, L.V. (ed.) (1973). *X-ray Spectroscopy*, (McGraw Hill, New York).

Baer, Y. and Busch, G. (1973). X-ray Photoemission from Aluminium, *Phys. Rev. Lett.*, **30**, 280.

von Barth, U. and Hedin, L. (1972). A Local Exchange-Correlation Potential for the Spin-Polarized Case: I, *J. Phys.*, **C5**, 1629.

Beeferman, L.W. and Ehrenreich, H. (1970). Optical Properties of Metals: Many-Electron Effects, *Phys. Rev.*, **B2**, 364.

Bennett,L.H. (ed.) (1971). *Electronic Density of States*, (Special Publication No. 323, National Bureau of Standards).

Bergstresser, T.K. and Rubloff, G.W. (1973). Local Field Effects in the Optical Properties of Solids: The Far-Ultraviolet Spectra of Ionic Crystals, *Phys. Rev. Lett.*, **30**, 795.

Ching, W.Y. and Callaway, J. (1973). Interband Optical Conductivity of Potassium, *Phys. Rev. Lett.*, **30**, 441.

Fowler, W.B. (1966). Influence of Electronic Polarization on the Optical Properties of Insulators, *Phys. Rev.*, **151**, 657.

Haken, H. and Schottky, W. (1958). Die Behandlung des Exzitons nach der Vielelektronentheorie, *Z. Phys. Chem.*, **16**, 218.

Hedin, L. (1965a). New Method for Calculating the One-Particle Green's Function with Application to the Electron-Gas Problem, *Phys. Rev.*, **139**, A796.

Hedin, L. (1965b). Effect of Electron Correlation on Band Structure of Solids, *Ark. Fys.*, **30**, 231.

Hedin, L., Lundqvist, B.I. and Lundqvist, S. (1967). On the Single-Particle Spectrum of an Electron Gas, *Int. J. Quant. Chem.*, **1S**, 791.

Hedin, L. and Johansson, A. (1969). Polarization Corrections to Core Levels, *J. Phys.*, **B2**, 1336.

Hedin, L. and Lundqvist, S. (1969). Effects of Electron-Electron and Electron-Phonon Interactions on the One-Electron States of Solids, in *Solid State Physics, Vol. 23*, (eds. Seitz, Turnbull and Ehrenreich), (Academic Press, New York), pp. 1-181.

Hedin, L., Lundqvist, B.I. and Lundqvist, S. (1970). Beyond One-Electron Approximation: Density of States for Interacting Electrons, *J. Res. Nat. Bur. Stand.*, **74A**, 417.

Hedin, L. and Lundqvist, B.I. (1971). Explicit Local Exchange-Correlation Potentials, *J. Phys.*, **C4**, 2064.

Hedin, L. (1973). In *Electrons in Crystalline Solids, Proceedings of the 1972 Trieste Winter College*.

Hermanson, J. (1972). Plasmon Sidebands in Alkali Metals, *Phys. Rev.*, **B6**, 400.

Hermanson, J. (1972a). Simple Model of Electronic Correlation in Insulators, *Phys. Rev.*, **B6**, 2427.

Hohenberg, P. and Kohn, W. (1964). Inhomogeneous Electron Gas, *Phys. Rev.*, **136**, B864.

Kohn, W. and Sham, L.J. (1965). Self-Consistent Equations Including Exchange and Correlation Effects, *Phys. Rev.*, **140**, A1133.

Kunz, A.B. (1972). Electronic Polarons in Non-Metals, *Phys. Rev.*, **B6**, 606.

Layzer, A.J. (1963). Properties of the One-Particle Green's Function for Non-Uniform Many-Fermion Systems, *Phys. Rev.*, **129**, 897.

Lindhard, J. (1954). *Dan. Math. Phys. Medd.*, **28**, No. 8.

Lundqvist, B.I. (1967a). Single-Particle Spectrum of the Degenerate Electron Gas. I The Structure of the Spectral Weight Function, *Phys. Kondens. Mater.*, **6**, 193.

Lundqvist, B.I. (1967b). Single-Particle Spectrum of the Degenerate Electron Gas. II Numerical Results for Electrons Coupled to Plasmons, *Phys. Kondens. Mater.*, **6**, 206.

Lundqvist, B.I. (1968). Single-Particle Spectrum of the Degenerate Electron Gas. III Numerical Results in the Random Phase Approximation, *Phys. Kondens. Mater.*, **7**, 117.

Lundqvist, B.I. (1969). Some Numerical Results on Quasi-Particle Properties in the Electron Gas, *Phys. Stat. Sol.*, **32**, 273.

Overhauser, A.W. (1971). Simplified Theory of Electron Correlations in Metals, *Phys. Rev.*, **B3**, 1888.

Rice, T.M. (1965). The Effects of Electron-Electron Interaction on the Properties of Metals, *Ann. Phys.*, **31**, 100.

De Reggi, A.S. and Rea, R.S. (1973). Optical Transitions Between Conduction Bands in Copper and Copper-Based Alloys, *Phys. Rev. Lett.*, **30**, 549.

Toyozawa, Y. (1954). Theory of the Electronic Polaron and Ionization of a Trapped Electron by an Exciton, *Prog. Theor. Phys.*, **12**, 421.

Williams, A.R., Janak, J.F. and Moruzzi, V.L. (1972). One-Electron Analysis of Optical Data in Copper, *Phys. Rev. Lett.*, **28**, 671.

Wiser, N. (1963). Dielectric Constant with Local Fields Included, *Phys. Rev.*, **129**, 62.

MANY BODY PERTURBATION THEORY
FOR NON-UNIFORM ELECTRONIC SYSTEMS

D.F. SCOFIELD†

Aerospace Research Laboratories,
Wright-Patterson AFB, Ohio 45433

1. INTRODUCTION

In the now famous papers by P. Hohenberg and W. Kohn [*1*] and in a sequel by W. Kohn and L.J. Sham [*2*] use was made of the fact that the total energy of an interacting electron system is a functional of the electronic charge density:

$$E[n(r)] = \int \boldsymbol{v}(r)n(r)\mathrm{d}r + F[n(r)], \tag{1}$$

where $v(r)$ is the external potential and F is a universal functional of the density. For the exact $n(r)$, $E[n(r)]$ is a minimum relative to variations of $n(r)$ consistent with a variation of $v(r)$. Unfortunately, for an approximate $F = F^{\mathrm{a}}$ the approximate energy functional E^{a} obtained from equation (1) by replacing $F[n]$ with $F^{\mathrm{a}}[n]$ does not have to have any relation in particular to E. That is, E^{a} may be greater or smaller or equal to E. An example is the Slater X_α scheme [*3*] in which one can vary α to obtain any E^{a} that one wants. Another fundamental objection is that given an F^{a} there may or may not exist a solution to the Kohn and Sham equations in a region of physical interest. Even when the approximate energy functional has the appropriate properties leading to a $n^{\mathrm{a}}(r)$ (there exists a point $\tilde{n}^{\mathrm{a}}(r)$ such that $E^{\mathrm{a}}[n] = \min$), the constraint used by Kohn and Sham that

$$\tilde{n}(r) = \sum_i |\tilde{\phi}_i(r)|^2, \qquad N = \int \tilde{n}(r)\mathrm{d}r, \tag{2}$$

† NAS-NRC Post-Doctoral Associate.

(using the Lagrange multiplier metyhod this constraint leads to the K-S equations upon minimizing the functional $[E^{a}[n(r)] - \lambda \times \int n(r)\mathrm{d}r]$ with respect to variations in n compatible with being produced by a variation in the external potential, supposing, of course, that the functional derivative exists) effectively raises the calculated energy even though it does not necessarily ensure N-representability. In principle other N-conserving constraints than (2) could be used.

Because of these drawbacks one has no *a priori* reason to believe that one approximation to $F[n]$ is better than another in a mathematical sense. One is free to choose any energy functional one wants upon the basis of experience, hope, etc.. The usefulness of the approximate scheme thereby obtained can be gauged only by the results one obtains. The approximate schemes can be regarded merely as density generating schemes. The eigenspectrum of the eigenvalue-eigenvector problems solved in obtaining $n(r)$ have no direct physical meaning. In practice, however, the approximation scheme in which the exchange part of the Coulomb energy is replaced by the function $\alpha n(r)^{4/3}$ has had remarkable success in obtaining approximate charge densities. Differences in the eigenvalues obtained by solving the K-S equations in atomic and solid state calculations have a remarkably close correspondency to excitation energies found by experiment. In the author's experience this 'Standard' approximation is not, however, good enough for accurate charge densities, cohesive energies or excitation energies even for such simple solid state materials as aluminum [4].

It is the author's feeling that one of the principle reasons for the failure of the usual approximation schemes for obtaining $F[n(r)]$ is that improper account has been taken of the inhomogeneous nature of the electron distribution. Expansions using gradients of the charge density [5,6] do little to improve the situation.

Of the various methods that could be employed to achieve more accurate results we will discuss only two. (1) The Brueckner Goldstone many body perturbation theory [7] (BGMBPT) for inhomogeneous systems and (2) the self-consistent Green's function approach of Martin and Schwinger [8] and Dyson [9]. In section 2 the review of some aspects of the (BGMBPT) will be given. The method of energy denominator insertions to sum certain classes of diagrams will be discussed. The use of a frequency independent K-matrix will be used as an example of the integral equation method of summing classes of diagrams. The consideration of a full time dependent theory will be discussed in section 4 following some mathematical preliminaries in section 3. The mathematical ideas provide a framework for discussing both the Hohenberg-Kohn-Sham and the Dyson-Martin-Schwinger approaches.

2. BRUECKNER GOLDSTONE MANY BODY PERTURBATION THEORY FOR NON-UNIFORM SYSTEMS. ATOMS AND SOLIDS

Background

By now there are numerous tutorials on MBPT for both uniform and non-uniform systems. The first are not of much interest to us here. In the second category one can have formulations of the theory based upon Feynman propagators or what one might call Brueckner-Goldstone propagators where the specific time ordering of a diagram is indicated and diagrams with the same shape (topology) but different time orders are counted as distinct. In the Brueckner-Goldstone approach one includes propagation of hole states and particle states separately. This leads to the Brueckner Goldstone Many Body Perturbation Theory [*10*,*11*]. This theory has turned out to be the most convenient method to calculate properties of atoms and nuclei. Applications of BGMBPT to atoms has been extensively reviewed by H.P. Kelly [*12*], a pioneer in the field. There is no corresponding work for solids. The reason for this is that very little computational experience in real solids has been gained. The reason for this is that the solid state calculations are much more complicated. It has been only with the advent of very large scale computers that the solid state computations have become feasible.

The reason for the greater complexity for crystals is well known: (i) The number of electrons is infinite so that the sums over one electron states become integrals and (ii) the direct part of the bare Coulomb interaction leads to divergences in the second and higher orders of perturbation theory. For example, the second order direct correlation (ring) diagram in the linked cluster expansion of the energy:

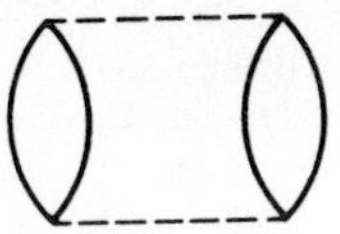

{ finite number for an atom
{ ∞ for a crystal

This situation gets worse. The third order ring diagram:

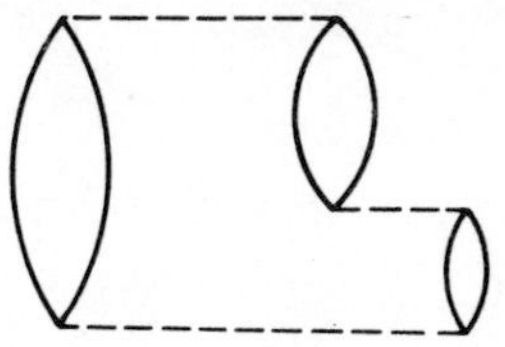

which is finite and smaller in magnitude than the second order diagram for atoms is even more divergent than the second order diagram for solids. For crystals one must sum the ring diagrams to infinite order before one gets a finite result. Obviously one does not merely add up these infinities to get this finite result. One must in effect not do perturbation theory. The divergent partial sums were obtained under the false assumption that a perturbation expansion was possible. In solids such an expansion is not possible because of the long range of the direct part of the Coulomb interaction in the many body theory. The way out of this difficulty is to use other methods than straight-forward perturbation theory. In the case of crystals one must calculate the screened interactions by solving a linear integral equation of the Fredholm type in x- and ω-space:

$$V^{S}(x,x',\omega) = V_0{}^{D}(x,x') + \iint dx''dx'''V_0{}^{D}(x,x'')\,\frac{d\omega'}{2\pi}$$

$$\times K(x'',x''',\omega + \omega')V^{S}(x''',x',\omega'),$$

or symbolically

$$V^{S} = V_0{}^{D} + V_0{}^{D}KV^{S},$$

so that

$$V^{S} = (1 - V_0{}^{D}K)^{-1}V_0{}^{S}.$$

The expansion of the operator $(1 - V_0{}^{D}K)^{-1}$ in a geometric series is not allowed. That is it diverges. It is just this expansion, allowed only when $V_0{}^{D}K$ is 'small', that leads to the perturbation expansion. The use of V^{S} in effect, then, sums an infinite class of diagrams. The necessity of this procedure obviously complicates the crystal problem. But as a dividend a much larger number of diagrams are included. Obviously, the same methods can be applied to the atomic problem. In fact, except for the fact that the symmetry orbital labels in atoms ($n\ell m$) and solids ($\vec{k}$) are different and the basis interaction matrix elements of $1/r_{12}$ are calculated differently, the only difference between the atomic and the crystal calculations is that sums over the one particle basis function indices that appear in the atomic calculations are replaced by integrals over the one particle indices. In the following we give a brief review of the Brueckner-Goldstone Theory for the total energy of a system of electrons.

Basis Sets for Perturbation Theory

The total Hamiltonian for the N-electron system interacting through the two body potential $1/r_{12}$ is assumed to be of the form

$$H = \sum_{i}^{N} T_{i} + \sum_{i<j}^{N} \frac{1}{r_{ij}} \,. \tag{1}$$

T_i is the sum of the kinetic-energy operator for the electron plus its interaction with the point nuclei of charge Z (we use atomic units throughout). Thus

$$T_i = -\frac{\nabla_i^2}{2} - \sum_n \frac{Z}{|r_i - R_n|} . \tag{2}$$

For an atom the sum in equation (2) obviously includes only one term. We now define an interaction potential and an unperturbed Hamiltonian H_0 by

$$H_0 = \sum_i^N (T_i + V_i), \tag{3}$$

such that

$$H = H_0 + H_I = H_0 + \left(\sum_{i<j}^N \frac{1}{r_{ij}} - \sum_i^N V_i \right) , \tag{4}$$

and V_i is a one body Hermitian potential.

V_i is chosen so that

$$H_0|\Phi_0\rangle = E_0|\Phi_0\rangle, \tag{5}$$

and $|\Phi_0\rangle$ is a single Slater determinant. We are not necessarily restricted to closed shell systems by this prescription [12]. Since H_0 is of the form $\sum_i^N F_i(r_i)$ we have

$$F_i(r_i)|\phi_i\rangle = \varepsilon_i|\phi_i\rangle, \qquad i = 1,2,\ldots,N,$$

$$\tilde{E}_0 = \sum_i^N \varepsilon_i, \tag{6}$$

and

$$|\Phi_0\rangle = \det \frac{\prod \phi_i}{\sqrt{N!}} .$$

The total energy is then given by the Brueckner-Goldstone Formula

$$E = \tilde{E}_0 + \sum_n \langle \Phi_0|H_I\left(\frac{1}{E_0 - H_0} H_I\right)^n |\Phi_0\rangle_L, \tag{7}$$

where the L indicates that the sum is over linked diagrams only. H_I includes the original single particle part $\sum_i -V_i$ as well as the $\sum_{i<j} \frac{1}{r_{ij}}$ term. The single particle part is denoted diagrammatically by

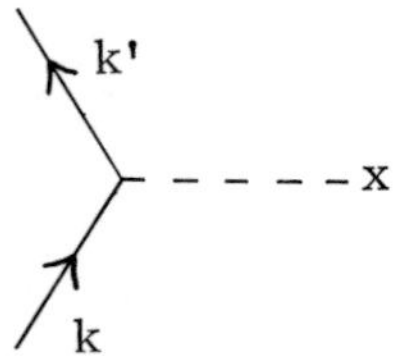

and the $1/r_{12}$ interaction by

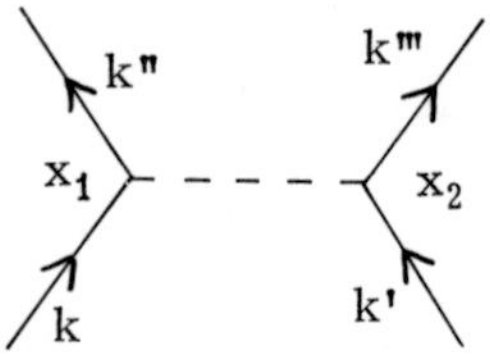

Upon inserting complete sets of N-particle Slater determinants to evaluate the product $(1/(E_0 - H_0)H_I)^n$ etc. in equation (7) one obtains the terms indicated by the diagrams given in the section on the total energy. A simple choice for H_0 is the Hartree-Fock operator in this case

$$F^{HF}(r_i)|\phi_i^{HF}\rangle = \varepsilon_i^{HF}|\phi_i^{HF}\rangle, \tag{8}$$

and the $|\phi_i^{HF}\rangle$ define a complete orthonormal basis of one particle functions. There is one problem with this basis, however, the virtual (or unoccupied) orbitals (called particle states) interact with the N occupied orbitals (called hole states) instead of $N - 1$ other occupied orbitals that an occupied orbital interacts with. This leads to slower convergence properties. To remedy this defect Kelly introduced the V^{N-1} potential for excited states [*13*]. The resulting basis set is then not exactly orthonormal but this effect is small and may be corrected for by perturbation theory. An alternative procedure is to note that the addition of any noncompact operator to F^{HF} which is zero when acting on the occupied states does not change the occupied spectrum. This additional operator can then be used to rearrange the virtual part of the spectrum to meet virtually any requirement [*14*]. In particular to make the virtual spectrum closer to the excitation spectrum of the system while leading to as much cancellation of the lower order diagrams as possible. In order to do this [*15*] define the projection operator

$$P = P^2 = \sum_{\substack{ij \\ occ}} |\phi_i^{HF}\rangle \Delta_{ij}^{-1} \langle \phi_j^{HF}|, \tag{9}$$

where

$$\Delta_{ij} = \langle \phi_i^{HF} | \phi_j^{HF} \rangle,$$

$$\theta = \theta^2 = 1 - P,$$

is then the projector onto the virtual space. That is if $|\phi_i^{HF}\rangle$ belongs to the occupied space

$$\theta |\phi_i\rangle \equiv 0.$$

Therefore for an arbitrary operator A,

$$F^{HF} + \theta A \theta \equiv F^{OAO} \tag{10}$$

is of the desired form. Various choices of A are possible. A simple one is

$$A(r) = -\frac{\beta}{r}.$$

which adds a charge of β to the nuclear charge Z making the Coulomb attraction of a virtual state toward the nucleus $(Z + \beta)/r$. This and other choices of A have been discussed and their effect on the correlation energies of atomic fluorine studied [*15*]. Effects of the use of an A which gives a V^{N-1} type potential for carbon has also been studied [*16*].

Whatever the A chosen, the one particle states which are exact (self-consistent) eigenstates of F^{OAO} are found using numerical techniques, in both the atomic [*17*] and the crystal case. In both the atom and the crystal $\varepsilon_i < 0$ and $\varepsilon_i \geqslant 0$ states must be calculated in order to obtain a complete set of one electron functions. If one is willing to use a basis which leaves H_0 nondiagonal, such as the complete basis formed from Slater functions then one may not need continuum functions. In principle such functions are not needed since the Hilbert space is separable and can be spanned by a denumerable set of functions. It has been found easier in practice to use continuum functions and exact eigenstates of an H_0. The basis sets for crystals that we use are the so-called boundary perturbation basis sets. These are Bloch sums of orbitals that vanish outside the Wigner-Seitz cell and exactly satisfy the cell boundary conditions, i.e.

$$\psi_k(r) = \sum_n \exp(ik \cdot R_n)\phi_k(r - R_n).$$

For the atom we use functions of the form

$$\phi_{n\ell m\sigma}(r) = R_{n\ell}(r)Y_{\ell m}(\theta,\phi)\chi_\sigma.$$

MATRIX ELEMENTS OF $1/r_{12}$

In order to evaluate any diagram we must not only have the one particle states but also the matrix elements of the perturbation. The most difficult matrix element to evaluate is, of course, the $1/r_{12}$ one.

In this case difference in the symmetry labels appropriate to crystals and atoms leads to a difference in the method of evaluating its matrix elements. In atoms it is natural to expand $1/r_{12}$ in terms of spherical harmonics:

$$\frac{1}{r_{12}} = \sum_{\ell=0}^{\infty} \frac{r_<^{\ell}}{r_>^{\ell+1}} \frac{(2\ell + 1)}{4\pi} \sum_{m=-\ell}^{m=+\ell} Y_{\ell m}^*(\theta_1,\phi_1)Y_{\ell m}(\theta_2,\phi_2). \qquad (11)$$

Since the basis functions $\phi_{n\ell m}(r)$ are taken as eigenfunctions of L^2, i.e.

$$\phi_{n\ell m}(r) = R_{n\ell}(r)Y_{\ell m}(r),$$

this leads to the expression for the direct Coulomb interaction:

$$\langle kk'|\frac{1}{r_{12}}|pq\rangle =$$

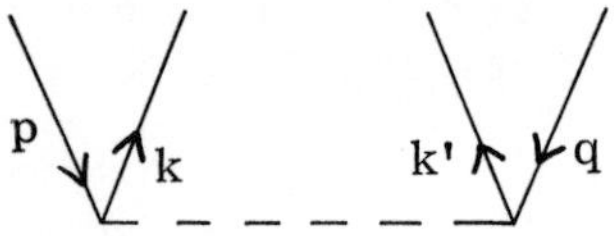

$$= \int \phi(n\ell m)_k^*(r_1)\phi(n\ell m)_{k'}^*(r_2)\, \frac{1}{r_{12}}\, \phi(n\ell m)_p(r_1)\phi(n\ell m)_q(r_2)dr_1dr_2$$

$$= \sum_{LM}^{\infty} \frac{2L + 1}{4} \int R(n\ell)_k(r_1)R(n\ell)_p(r_1)U_L(r_1,r_2)R(n\ell)_k(r_2)R(n\ell)_q(r_2)$$

$$\times\, r_1^2 dr_1 r_2^2 dr_2 \times \qquad \text{(Contd)}$$

$$\times \int Y_{LM}^*(\theta_1,\phi_1)Y_{(\ell m)_k}^*(\theta_1,\phi_1)Y_{(\ell m)_p}(\theta_1,\phi_1)d\Omega_1 \quad \text{(Contd)}$$

$$\times \int Y_{LM}(\theta_2,\phi_2)Y_{(\ell m)_{k'}}^*(\theta_2,\phi_2)Y_{(\ell m)_q}(\theta_2,\phi_2)d\Omega_2, \quad (12)$$

where

$$U_L(r_1,r_2) = \frac{r_<^{\,L}}{r_>^{\,L+1}} .$$

The angular integrations can be evaluated in terms of 3-J symbols using the relations

$$\int Y_{a\alpha}Y_{b\beta}Y_{c\gamma}d\Omega = \left(\frac{(2a+1)(2b+1)(2c+1)}{4\pi}\right)^{\frac{1}{2}} \times \begin{pmatrix} a b c \\ \alpha\beta\gamma \end{pmatrix}\begin{pmatrix} a b c \\ 000 \end{pmatrix} , \quad (13a)$$

$$Y_{kq}^* = (-)^q Y_{k-q}. \quad (13b)$$

The two-dimensional radial integral is evaluated by evaluating a potential function then finding the potential energy of the remaining part of the integrand in this potential as described by Hartree [*17*]. The sum over L and M above is actually finite because of the property of the 3-J symbols. This can be seen by noting that (1) while the system is non-uniform so that linear momentum is not conserved, the atomic system (H_0 really) is spherically symmetric so that L is a good quantum number and (2) L is conserved by the $1/r_{12}$ interaction.

For example, if we have $L = 0,1,2,3$ for excited states the following matrix elements are non-zero

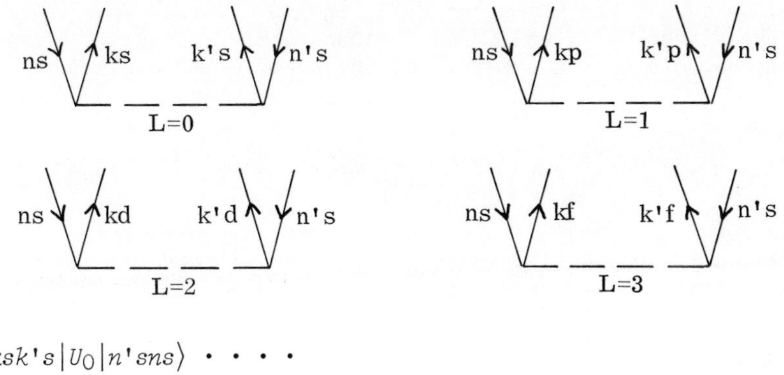

$\langle ksk's|U_0|n'sns\rangle$ . . . .

Similarly the following are non-zero

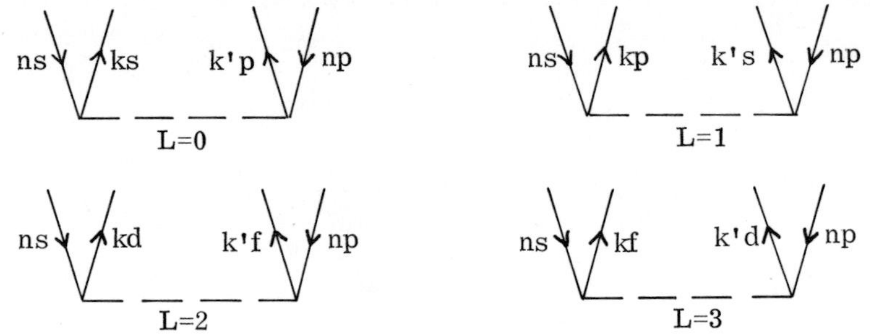

$\langle ksk'p|U_0|nsnp\rangle$ . . .

In general a matrix element then is of the form (labelling only the angular momentum)

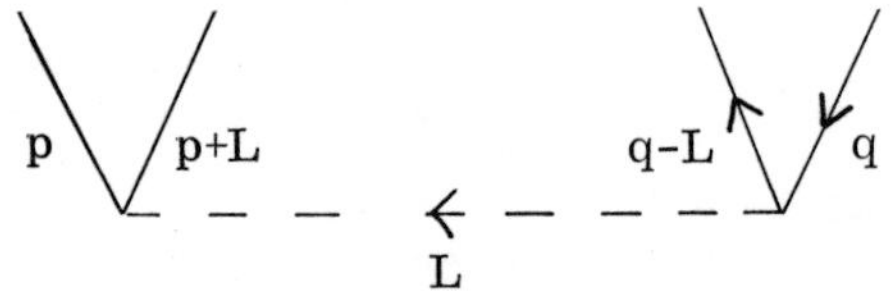

A similar situation arises in the solid state problem. There it is natural to expand $1/r_{12}$ in a Fourier sum

$$\frac{1}{r_{12}} = \frac{4\pi}{N\Omega_0}\sum_{Q'}{}' \frac{e^{iQ'\cdot(r_1-r_2)}}{|Q'|^2} . \tag{14}$$

where N equals the number of cells in the crystal, Ω_0 = volume of the Wigner-Seitz cell, $\{Q\}$ includes all wave vectors. However, because of the symmetry of the unperturbed basis functions

$$\psi_k(r) = \frac{1}{\sqrt{N}}\sum_n e^{ik\cdot R_n}\phi_k(r - R_n). \tag{15}$$

(The $\phi_k(r - R_n)$ that we use are non-zero only with the n-th WS cell) only reciprocal lattice vectors are selected in the matrix elements. For example the direct matrix element

$$\langle p+Q,q-Q|\frac{1}{r_{12}}|pq\rangle = \lim_{N\to\infty}\frac{4\pi}{N\Omega_0}\sum_{Q'}{}' \frac{1}{|Q'|^2}\,\Phi_{p+Q,p}(Q')\Phi_{q-Q,q}(-Q'), \tag{16a}$$

where

$$\Phi_{p+Q,p}(Q') = \int_{\substack{WS \\ cell}} \phi_{p+Q}^{*}(r)e^{iQ'\cdot r}\phi_p(r)dr$$

$$= \begin{cases} 0 \text{ unless } Q=Q' \text{ modulo} \\ \quad \text{a reciprocal lattice vector.} \end{cases}$$

We can ignore the divergent $Q' = 0$ term here and elsewhere merely by systematically renormalizing the interaction. The $Q' = 0$ electron-electron interaction and the interaction of the electrons with the lattice for zero momentum transfer exactly cancel. Similarly the exchange matrix element

$$\langle p + Q, q - Q|\frac{1}{r_{12}}|qp\rangle = \lim_{N\to\infty} \frac{4\pi}{N\Omega_0} \sum_{Q'}' \frac{\Phi_{p+Q,q}(Q')\Phi_{q-Q,p}(-Q')}{|Q' - p + q|^2} .$$

involves only the Φ's.

Harris, Monkhorst and Kumar [19] have presented a more general form of these integrals when the basis orbitals $\phi_k(r - R_n)$ overlap from cell to cell and the unit cell has more than one atom, i.e. the Bravais lattice has a basis with more than one atom. The Fourier transforms, $\Phi_{p+q,q}(Q)$ are found by performing an integration over the Wigner-Seitz sphere and then adding corrections to recover the integral over the Wigner-Seitz cell. The sphere integration requires the evaluation of a one-dimensional radial integral but the correction integral is a fully fledged three-dimensional integral. This integral is evaluated using Diophantine integration techniques [20].

THE TOTAL ENERGY

Let us now consider what diagrams should be evaluated. When F^{HF} is used one need calculate only the second and higher order diagrams to obtain the correlation energy. In atoms since one usually uses the restricted Hartree-Fock operator F^{RHF}, for open shell atoms one has to include some diagrams which account for the polarization of the core that would automatically be accounted for if the unrestricted F^{HF} operator were used.

In diagrams equation (7) yields

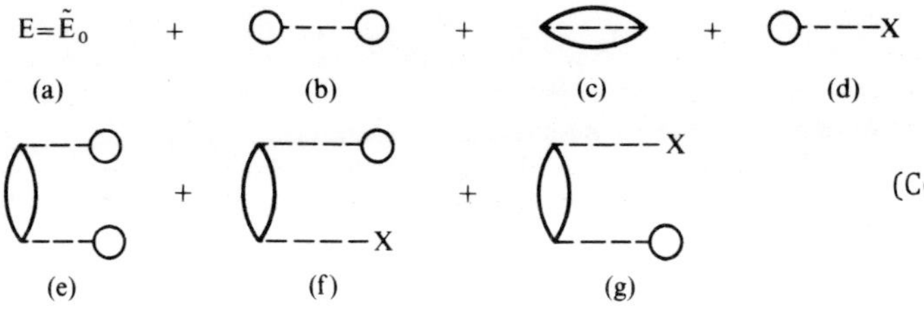

(Contd)

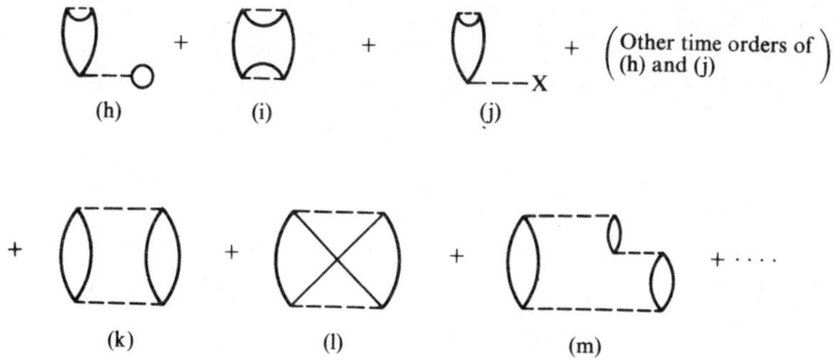

Diagrams (a-d) give the Hartree-Fock energy when F^{HF} is used. When F^{HF} is used single excitation diagrams (d-j) sum to zero because of Brillouin's Theorem [*21*] — the Hartree-Fock wave function is that single determinantal wave function which is stable under the H_0 interaction to particle hole excitations, i.e. the variation

$$\delta\langle\Phi_0|H_0 + V|\Phi_0\rangle = \langle\delta\Phi_0|H_0 + V|\Phi_0\rangle = 0, \tag{18}$$

so that if $|\Phi_{\mathrm{ph}}\rangle$ is a particle hole excitation of $|\Phi_0\rangle$,

$$\langle\phi_{\mathrm{ph}}|H_{\mathrm{I}}|\phi_0\rangle = 0. \tag{19}$$

In crystals we do not make the restricted Hartree-Fock approximation but can use the exact F^{HF} so that only diagrams (k, ℓ,etc.) must be calculated. One could use Hartree-Fock-Slater basis functions which are somewhat easier to compute but then diagrams (d-j) have to be calculated. In atoms the sum of diagrams (d-j) can be minimized by choosing α in Slater's scheme so that diagrams (a-c) add to give the Hartree-Fock energy. Unfortunately one usually doesn't know this energy for solids so that if one uses HFS one must calculate all these diagrams.

DIAGONAL ENERGY DENOMINATOR INSERTIONS

In addition to the previous diagrams one usually adds third order particle-hole, hole-hole correction and other diagrams involving a pair of hole electrons p and q shown below [*16*] (exchanges not shown). The hole particle interaction in (e) also is to be included between k' and q. The interaction of the passive unexcited state in (f) also is to be included on the line k'. The intermediate interaction in (g) also is to be included on line k'. In (b) for atoms only included for bound virtual states (for unbound, continuum states a separate calculation should be performed). All these diagrams may be included through infinite order by changing the energy denominator in the second order direct diagram by an amount $\Delta(pqkk')$ given below.

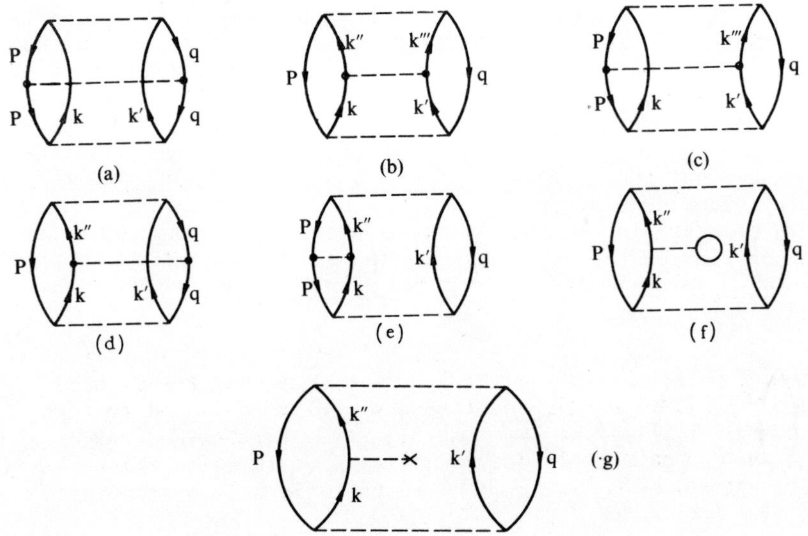

Thus using the rules for BG graphs all these diagrams may be summed to give the direct 'second order' energy

$$E_2 = \sum_{\substack{pq\ occ \\ kk'\ unocc}} \frac{|\langle kk'|\frac{1}{r_{12}}|pq\rangle|^2}{\varepsilon_p + \varepsilon_q - \varepsilon_k - \varepsilon_{k'} + \Delta(pqkk')} \tag{20}$$

where [*16*]

$$\begin{aligned}\Delta(pqkk') = &- \langle pq|\frac{1}{r_{12}}|\widetilde{pq}\rangle - \langle kk'|\frac{1}{r_{12}}|\widetilde{kk'}\rangle \\ &+ \langle pk'|\frac{1}{r_{12}}|\widetilde{pk'}\rangle + \langle qk|\frac{1}{r_{12}}|\widetilde{qk}\rangle \\ &- \left[\sum_{n\neq p} \langle nk|\frac{1}{r_{12}}|\widetilde{nk}\rangle - \langle k|V|k\rangle\right] \\ &- \left[\sum_{n\neq q} \langle nk'|\frac{1}{r_{12}}|\widetilde{nk'}\rangle - \langle k'|V|k'\rangle\right] \\ &+ \left[\sum_{n} \langle np|\frac{1}{r_{12}}|\widetilde{np}\rangle - \langle p|V|p\rangle\right] \\ &+ \left[\sum_{n} \langle nq|\frac{1}{r_{12}}|\widetilde{nq}\rangle - \langle q|V|q\rangle\right].\end{aligned}$$

The notation $|\widetilde{pq}\rangle$ means to include both direct and (-) exchange matrix elements involving the pair pq. These energy denominator insertions can be viewed as resulting from a renormalization of the N body determental propagators $1/(E_0 - H_0)$ appearing in equation (7) by including the diagonal part of the N-body perturbation $(\sum 1/r_{ij} - \sum V_i - \theta A\theta)$ into H_0. Such a renormalization does not change the wave functions. It only changes the energy denominators.

This change in the energy denominator can drastically alter the behavior of the divergence of the second order diagram in a crystal. For example consider the plane wave case (for a fictional Fermi gas), then

$\frac{1}{N}$ 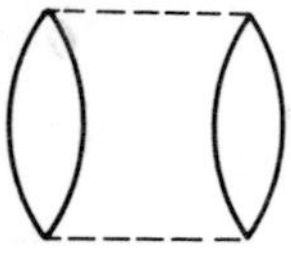

$$= -\frac{3}{8\pi^5}\int\frac{d^3q}{q^4}\int_{\substack{|k|<1\\|k+q|>1}} d^3k\int_{\substack{|\ell|<1\\|q-\ell|>1}}\frac{d^3\ell}{q^2 + q\cdot(k-\ell) + \Delta},$$

where all momenta are expressed in terms of the Fermi momentum. The largest contribution of the integrand is for small q. Because of the limits on the integration, for small q, k and ℓ lie within a differential shell of thickness proportional to $|q|$. Now for small $|q|$ the integrand is proportional to Δ^{-1} (in the absence of Δ it would be proportional to $|q|^{-1}$). The integration volume over k and q is then proportional to $|q|^2$ so it follows that

$\frac{1}{N}$ 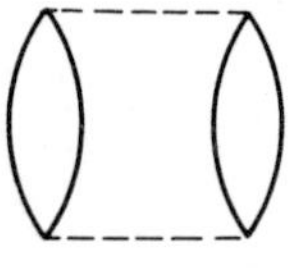

$$\approx \frac{1}{N}\int\frac{d^3q}{q^2\Delta} \approx C\Delta^{-1} < \infty,$$

(unless $\Delta \leqslant 0$) without Δ the result is infinite. This renormalization of the propagators will be met again. It is an entirely different kind of effect than the screening of the interaction which is called renormalization of the interaction. When the interaction is screened the direct matrix elements of the Coulomb interaction for small q are replaced by $1/(q^2 + k^2(q))$, where q is the momentum transfer and $k(q)$ is a screening function. Even though we can get rid of the divergence in second order by renormalizing the propagator we still haven't taken care of the long range behavior of the direct part of the Coulomb interaction. The screening in fact might be much more important, though there are not any calculations in real solids yet that allow one to make a comparison.

INTEGRAL EQUATION METHOD

Let us now consider the calculation of the sum of all diagrams of type (b) above. These are called particle ladder diagrams. If we consider the series of diagrams

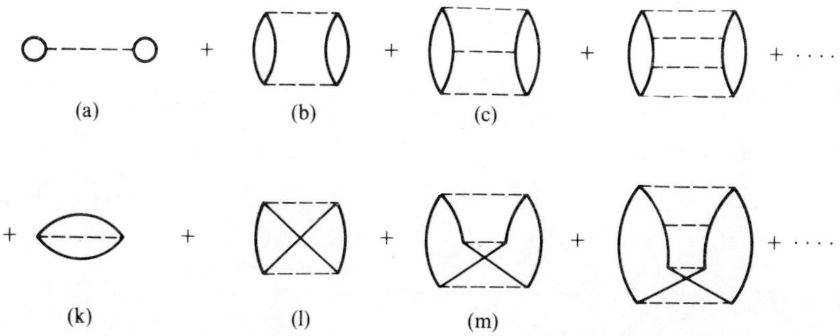

we can deduce a method of summing them. Let us write down the expressions for these diagrams for the pair ij

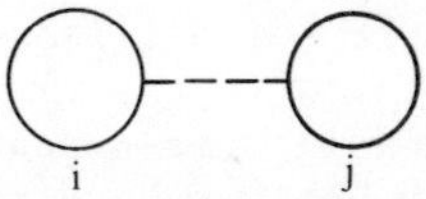

$$= \frac{1}{2} \sum_{\substack{ij \\ occ}} \langle ij|\frac{1}{r_{12}}|ij\rangle ,$$

(a)

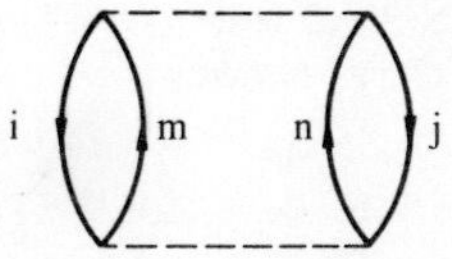

$$= \frac{1}{2} \sum_{\substack{ij\ occ \\ mn\ unocc}} \frac{\langle ij|\frac{1}{r_{12}}|mn\rangle\langle mn|\frac{1}{r_{12}}|ij\rangle}{\varepsilon_i + \varepsilon_j - \varepsilon_m - \varepsilon_n} ,$$

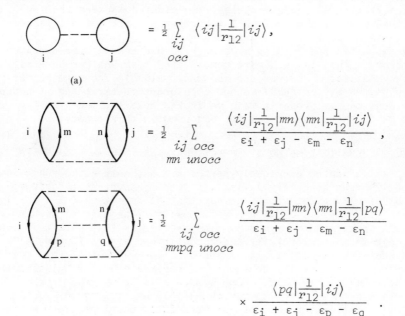

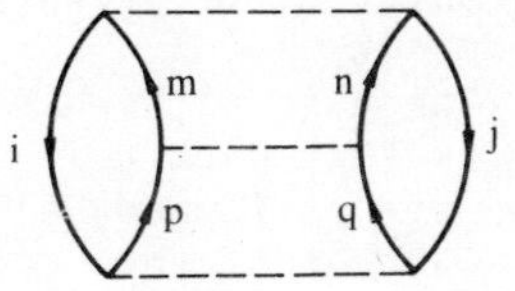

$$= \frac{1}{2} \sum_{\substack{ij\ occ \\ mnpq\ unocc}} \frac{\langle ij|\frac{1}{r_{12}}|mn\rangle\langle mn|\frac{1}{r_{12}}|pq\rangle}{\varepsilon_i + \varepsilon_j - \varepsilon_m - \varepsilon_n}$$

$$\times \frac{\langle pq|\frac{1}{r_{12}}|ij\rangle}{\varepsilon_i + \varepsilon_j - \varepsilon_p - \varepsilon_q} .$$

The states ij appear in each diagram as well as mn. Other states along with matrix elements and energy denominators appear for each addition of an internal interaction line as can be seen comparing (b) and (c). Therefore we may sum these diagrams by solving the integral equation:

$$\langle mn|K|ij\rangle = \langle mn|\frac{1}{r_{12}}|ij\rangle + \sum_{\substack{p\ell \\ unocc}} \frac{\langle mn|\frac{1}{r_{12}}|p\ell\rangle\langle p\ell|K|ij\rangle}{\varepsilon_i + \varepsilon_j - \varepsilon_p - \varepsilon_\ell} . \qquad (21)$$

A careful comparison of equation (21) with the series of graphs (a,b,c,...) shows that

$$\frac{1}{2}\sum_{\substack{ij\\ occ}} \langle ij|K|ij\rangle =$$

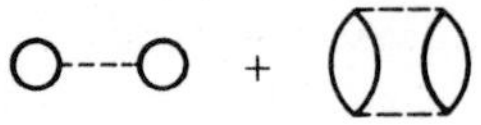

+ ....

and also shows that

$$-\frac{1}{2}\sum_{\substack{ij\\ occ}} \langle ij|K|ji\rangle =$$

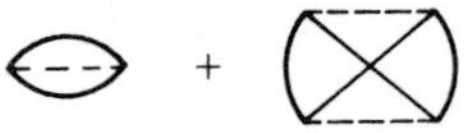

+ ....

K thus provides us with an effective interaction for summing all these ladder energy diagrams. Unfortunately the screening it provides is mostly short-range. In fact the method of summing these diagrams was invented to screen the short-range pairing interaction in nuclear theory. In order to screen the long-range part of the Coulomb interaction we must sum not ladder diagrams but ring diagrams. In order to deal with the ring diagrams effectively it is necessary to use a full time-dependent theory. This is much more complicated than what we have discussed thus far. However, it has many valuable consequences only suggested by the BGMPT method. In order to treat the time-dependent theory properly we give in the next section a review of some of the mathematical methods used in this theory.

3. MATHEMATICAL METHODS. FUNCTIONAL ANALYSIS IN BANACH SPACES

As indicated previously perturbation theory without any embellishment generally fails. It either converges too slowly or not at all. That is not to say that PT is not useful. Far from it. PT provides us with a very powerful language with which to talk about quantum systems. The problem is that when this language is taken literally and computations are performed, all kinds of nasty difficulties appear. The way out of this difficulty is to directly consider the non-perturbative solution to the equations at hand. Some progress is being made in this direction by using Padé approximates [*23*] and continued fractions [*24*]. The latter has not yet been fully developed into a tractable many particle theory so we defer its treatment here. Our approach here will be more conventional in that we will solve the Dyson's equation in an abstract setting using the ideas presented immediately below [*25*]. In the next section an example is worked out in complete detail.

We shall be concerned mainly with the theory of the solution of non-linear operator equations in a Banach space. (S. Banach, 1892-1945, Polish Mathematician). The Dyson equation is an example of such an equation.

BANACH SPACES

A complete normed linear space is called a *Banach space*. A *normed space* is a vector space X in which a function $\|\cdot\|$ is defined which satisfies the conditions of a norm: let u and v be elements of the space and let α be an element of a scalar field Λ, then the function $\|\cdot\|$ must satisfy

$$\|u\| \geqslant 0 \qquad \|u\| = 0 \Leftrightarrow u = 0,$$
$$\|\alpha u\| = |\alpha|\,\|u\|, \tag{1}$$
$$\|u + v\| \leqslant \|u\| + \|v\|.$$

By a *vector space* we mean a collection of elements $u, v, \ldots$, called vectors, for which linear operations (addition $u + v$ of two vectors u, v and multiplication $\alpha \cdot u$ of a vector u by a scalar $\alpha \in \Lambda$) are defined and obey the usual rules of such operations. If $\|u\| = 1$ u is said to be *normalized*.

CONVERGENCE IN BANACH SPACES

In the space X the *convergence* of a sequence of vectors to a $u \in X$ can be defined by $\|u_n - u\| \to 0$. The limit u, whenever it exists, is uniquely determined by the sequence $\{u_n\}$.

The condition that $\|u_n - u\| \to 0$ implies that $\|u_n - u_m\| \to 0$. When the space X is of infinite dimension, however, the converse is not necessarily true. That is even though $\|u_n - u_m\| \to 0$, u, the limit of the sequence $\{u_n\}$ does not necessarily belong to X. A normed space in which the convergence of every such sequence ($\{u_n\} \ni \|u_n - u_m\| \to 0$, called a *Cauchy sequence* or *fundamental sequence*) implies that u is an element of the space of the u_m, X, is called *complete*. Completeness is indispensible in most applications of normed linear space. Hence it is natural to deal with complete spaces. It might be recalled that a finite-dimensional normed space is automatically complete. As an example of these notions consider the normed linear space X of rational fractions p/q which we can write as $v = (p,q)$ where p and q are integers. The norm $\|\cdot\|$ is the usual absolute value $|p/q|$. Two vectors (p,q) and (p',q') are said to be *equivalent* if $p/q = p'/q'$. The scalar field Λ is the natural numbers. In this case multiplication by a scalar is defined by $\alpha(p,q) = (\alpha p,q)$. Addition of two elements (p,q) and (p',q') is given by $(p,q) + (p',q') = (pq' + qp',qq')$. Multiplication of two elements can also be defined but is not needed for the space to be a normed linear space. This defines a normed linear space. Now let us consider a sequence of vectors $\{u_n\} = \{(p_n,q_n)\}$ which has the numerical value p_n/q_n. Now as everyone knows $\lim p_n/q_n$ can exist ($\|u - u_n\| \to 0$) and yet not be a rational number. In fact such limits of rationals are a classical way of obtaining irrational numbers. For example, let $u_0 = (1,1)$ and $(p_m,q_m) = (p_{m-1},2q_{m-1}) + (3q_{m-1},2p_{m-1})$ whose norm $x_m = \|(p_m,q_m)\|$ is

$\frac{1}{2}(x_{m-1} + 3/x_{m-1})$. A little computation will show the reader that x_m converges to number $x^* = \sqrt{3}$. The rational numbers with a scalar field of the natural numbers does not form a complete normed linear space. This space can be completed, however. This means that X can be identified with a linear subspace in a complete normed space $\tilde{X}$. Now $\tilde{X}$ can be so chosen that X is *dense* in $\tilde{X}$ (i.e., the limit of every convergent sequence in X is in $\tilde{X}$). X is constructed as the set of all equivalence classes of Cauchy sequences $\{u_n\}$ in X. Two Cauchy sequences $\{u_n\}$ and $\{v_n\}$ are said to be *equivalent* if $\lim \|u_n - v_n\| = 0$. In our example, two sequences $\{(p_n,q_n)\}$ and $\{(p_n',q_n')\}$, are equivalent if they converge to the same real number. Thus by embedding the rationals in the Banach space of the real numbers R^1 we have completed the original space and this completion is with the same norm.

Banach spaces of higher dimensions, infinite dimensions are of course possible, have many of the properties of n-dimensional real manifolds R^n. They are vector spaces, they have the notion of *distance* provided by the definition of a norm (the distance d between u and $v \in X$ is $\|u - v\|$; recall the definition of norm above satisfies the intuitive notion of distance) and every Cauchy sequence converges. On the other hand, the norm does not have to arise from the definition of an inner product as in Hilbert spaces. Thus all Hilbert spaces are infinite-dimensional Banach spaces but not all infinite-dimensional Banach spaces need be Hilbert spaces. Banach spaces are abstractions of real (and complex) number systems in which it is possible to pose and solve numerous problems in mathematical physics from a unified standpoint. Typical examples in addition to those mentioned above are:

(i) The n-dimensional real vector spaces R_n of elements

$$\chi = (\xi_1,\xi_2,\ldots,\xi_n),$$

with norms

$$\|\chi\|_p = \left(\sum_{i=1}^{n} |\xi_i|^p\right)^{1/p} \qquad 1 \leqslant p \leqslant \infty, \tag{2a}$$

or

$$\|\chi\|_\infty = \max_i |\xi_i|; \tag{2b}$$

and

(ii) The space $C[0,1]$ of continuous functions $x = x(s)$, with finite support $0 \leqslant s \leqslant 1$ with

$$\|x\| = \max_{[0,1]} |x(s)|, \tag{3}$$

(if $\|x\| = \int_0^1 x(s)ds$, then under this norm $C[0,1]$ is not complete);

(iii) The space of n-dimensional square matrices with norm

$$\|M\| = \text{least upper bound}_{\|x\|=1} \|Mx\| \equiv \sup_{\|x\|=1} \|Mx\|$$

where x is a n-component vector whose norm is defined, for example, by

$$\|x\| = \left(\sum_i^n |\xi_i|^2\right)^{\frac{1}{2}}.$$

In fact we can define three different norms for x in (iii)

$$\text{I.} \qquad \|x\|_I = \max_i |\xi_i|, \tag{4a}$$

$$\text{II.} \qquad \|x\|_{II} = \sum_i^n |\xi_i|, \tag{4b}$$

$$\text{III.} \qquad \|x\|_{III} = \left(\sum_i^n |\xi_i|^p\right)^{1/p} \qquad 1 < p < \infty, \tag{4c}$$

each of which induces a different norm on the Banach space of matrices. The matrix norm is said to be *subordinate* to the vector norm chosen. An important relation between the norms I, II and III ($p = 2$) is

$$\|x\|_I \leqslant \|x\|_{II} \leqslant n\|x\|_I, \tag{5a}$$

and

$$\|x\|_I \leqslant \|x\|_{III} \leqslant \sqrt{n}\|x\|_I. \tag{5b}$$

When such relations exist between two norms, the norms are said to be *topologically equivalent*. In finite-dimensional normed linear spaces, in fact, any two norms are topologically equivalent. If $\|\cdot\|$ and $\|\cdot\|$ are topologically equivalent then we have

$$\lim_{m\to\infty} \|x_m - x^*\|_I = 0 \Leftrightarrow \lim_{m\to\infty} \|x_m - x^*\|_{II} = 0.$$

Thus for topologically equivalent Banach spaces we can choose the norm that is most convenient for computation to investigate convergence.

OPERATORS ON BANACH SPACES

As is well known, matrices transform a vector into another vector via the definition of matrix products. Matrices are the prototype of the notion of operators. An *operator* T on a space X into a space Y is a set of ordered pairs (x,y) of elements $x \in X$, $y \in Y$ such that there exists one and only one $(x,y) \in T$ for each $x \in X$. The fact that $(x,y) \in T$ will be denoted by

$$T(x) = y.$$

If $Y = X$, T is said to be an *operator in* X. An example of an operator is an $m{\times}n$ matrix $\mathbf{M}$ which transforms the n-dimensional vector $x \in X$ into a vector $y \in Y$ of dimension m, i.e.

$$\mathbf{M}x = y.$$

Matrices are special cases of *linear operators* which obey

$$L(x_1 + x_2) = Lx_1 + Lx_2,$$

$$L\lambda x = \lambda Lx.$$

Abstracting from the familiar ideas from matrix theory one can define several more concepts. If S and T are operators from a space X into a space Y, their *sum*, denoted by $(S + T)$, is the operator defined by

$$(S + T)x = Sx + Tx \tag{6}$$

for each $x \in X$. If S is an operator from X to Y and R is an operator from Y to Z their *product* (or *composition*) is the operator from X to Z denoted by (RS) and defined by

$$(RS)x = R(Sx). \tag{7}$$

for all $x \in X$. For an operator P in a space X, the n-th *power* P^n of P can be defined by (I is the *identity operator* $Ix = x$):

$$P^0 = I, \qquad P^n = PP^{n-1}, \qquad n \quad 1,2,\ldots .$$

If $P^n = R$, P is called the n-th *root* of R. If for a given operator S on X to Y, there exists an operator R on Y to X such that

$$RSx = x.$$

Then R is called a left inverse of S. Similarly if L is an operator from Y to X and

$$SLy = y.$$

Then L is a right inverse of S. If the spaces X and Y are the same then $L = R$ and we write $L = R = S^{-1}$. Thereby defining the inverse operator of the operator S

$$SS^{-1} = I = S^{-1}S.$$

An operator S from X to Y is said to be *continuous* if it transforms every convergent (in the norm of X) sequence of elements in X into a convergent (in the norm of Y) sequence of elements in Y. That is, S is continuous if

$$\lim_{n\to\infty} x_n = x \Rightarrow \lim_{n\to\infty} Sx_n = Sx.$$

If S is such that for every *bounded sequence* of elements $\{x_n\}$ in a subset M of X ($\{x_n\}$ is *bounded* $\Leftrightarrow$ $\|x_n\| < \infty$ for all n) one can select a subsequence $\{x_{n_i}\}$ such that $\{Sx_{n_i}\}$ converges in Y then S is *compact*. In other words S is compact if it transforms every bounded set into a compact set. An operator S is said to be *completely continuous* if it is continuous and compact. An operator S is said to be bounded in the set $M \subset X$ if and only if there exists a non-negative number μ such that

$$\|Sx_0 - Sx_1\|_Y \leqslant \mu \|x_0 - x_1\|_X \tag{8}$$

for all $x_0, x_1 \in X$. The greatest lower bound of the numbers satisfying this condition is called the norm of S and is denoted by $\|S\|$. A bounded operator is continuous. If both R and S are bounded then from equation (6) it follows that

$$\|R + S\| \leqslant \|R\| + \|S\| \tag{9a}$$

and

$$\|RS\| \leqslant \|R\| \cdot \|S\|. \tag{9b}$$

If L is a continuous linear operator from a Banach space X into a Banach space Y one can show, because $L + L = 2L$ etc., that if $\alpha \in \Lambda$, the scalar field of X, that

$$L\alpha x = \alpha L x,$$

consequently

$$\|L\alpha X\| = |\alpha| \, \|Lx\|. \tag{10}$$

In addition, one can show continuity implies boundedness. That is a continuous linear operation L from a Banach space X into a Banach space Y is bounded on X. The bound of L is denoted by $||L||$. It can be proven that definition (8) together with the continuity property implies that

$$||L|| = \sup_{||x||=1} ||Lx||, \tag{11}$$

and that consequently

$$||Lx|| \leqslant ||L|| \cdot ||x|| \tag{12}$$

Equation (10) together with equation (11) imply that

$$||\alpha L|| = |\alpha| \, ||L||.$$

Because of the facts that

$$||L|| > 0, \qquad ||L|| = 0 <=> L = 0,$$

and equations (9), the space $L(X,Y)$ of continuous linear operators on the Banach space X to the Banach space Y is a normed linear space. A fundamental theorem is that it is also complete under the norm given by equation (11) [*26*]. Thus $L(X,Y)$ is itself a Banach space. For some $L(X,Y)$ the product of two operators can be defined. If $X = Y$ then $A,B \in L(X,Y)$ are operators from X to X which can be used to define an operator (AB) from X to X. Note AB not necessarily equals BA. Since

$$||AB|| \leqslant ||A|| \, ||B||.$$

one then has what is called a *Banach algebra*. The Banach space of square matrices forms a Banach algebra.

Because $L(X,Y)$ is a Banach space one can define linear operators to $L(X,Y)$. A linear operator from a space X to the space $L(X,Y)$ is called a *bilinear operator* from X to Y [*26*]. If B is a bilinear operator from X to Y then for all $x_1 \in X$, Bx_1 is a linear operator from X to Y. Therefore for all $x_2 \in X$ we have [*26*]

$$Bx_1x_2 = (Bx_1)x_2 = y \in Y, \tag{13a}$$

$$Bx_1(x_2 + x_3) = Bx_1x_2 + Bx_1x_3, \tag{13b}$$

$$B(x_1 + x_2)x_3 = Bx_1x_3 + Bx_2x_3, \tag{13c}$$

$$B(\lambda x_1)(\mu x_2) = (\lambda\mu)Bx_1x_2. \tag{13d}$$

Furthermore, the operators B and Bx are bounded and

$$||Bx_1x_2|| \leqslant ||B|| \cdot ||x_1|| \cdot ||x_2|| \tag{14}$$

A bilinear operator satisfying $Bxy = Byx$ is called *symmetric*.

DIFFERENTIATION IN BANACH SPACES

Having defined the algebraic properties of linear and bilinear operators on Banach spaces and the concept of the norm along with that of limit we are in a position to develop a differential and integral calculus of operators [*26*].

Let $F(x)$ be an operator from a Banach space X into a Banach space Y. $F(x)$ is said to be *Frechet differentiable* at the point $x \in X$ if a bounded linear operator L from X to Y exists such that for all Δx near x

$$\lim_{||\Delta x|| \to 0} \frac{||F(x + \Delta x) - F(x) - L\Delta x||}{||\Delta x||}, \tag{15}$$

and we write

$$\left.\frac{\delta F(x)}{\delta x}\right|_{x} = L(x) = F'(x). \tag{16}$$

The norm of $F'(x)$ is that in the Banach space of linear operators from X to Y (denoted by $L(X,Y)$). F is said to be *differentiable on a subset* S_1 of S if it is differentiable at each point of S_1 and to *continuously differentiable* on S_1 if $F'(x)$ is continuous for each $x \in S_1$.

If Q is an operator from a Banach space X into a Banach space Z and P is an operator for Z into a Banach space Y then if their product, define above, has a Frechet derivative at x_0 we have the following *chain rule*

$$\frac{\delta}{\delta x}\,[P(Q(x))]_{x_0} = \left.\frac{\delta P(Q(x))}{\delta Q(x)}\right|_{Q(x_0)} \cdot \left.\frac{\delta Q(x)}{\delta x}\right|_{x_0}. \tag{17}$$

Some operators may be directly differentiated by application of equation (15). If L is a bounded linear operator from X into Y then

$$\frac{\delta L(x)}{\delta x} = L \tag{18}$$

and

$$\frac{\delta^2 L(x)}{\delta x^2} = 0.$$

Combining equations (18,17) gives

$$\frac{\delta}{\delta x}\,[(LQ)(x)]_{x_0} = L\,\frac{\delta Q(x)}{\delta x}\bigg|_{x_0},$$

since

$$\frac{\delta}{\delta x}\,LQ = \frac{\delta L}{\delta x}\,Q + L\,\frac{\delta Q}{\delta x} = L\,\frac{\delta Q}{\delta x}\,.$$

Thus in the differential calculus of Banach spaces, bounded linear operators play similar roles to constant multipliers in ordinary calculus [*26*].

The second Frechet derivative at x_0 of an operator $F(x)$ which is once Frechet differentiable in a neighbourhood $||x - x_0|| < \delta$, $\delta > 0$ about the point x_0 is defined as the bilinear operator B from X to Y (if it exists) such that

$$\lim_{||\Delta x||\to 0} \frac{||F'(x_0 + \Delta x) - F'(x_0) - B\Delta x||}{||\Delta x||} = 0.$$

B is in fact symmetric. We write

$$\frac{\delta^2 F(x)}{\delta x^2}\bigg|_{x_0} = B(x_0),$$

and have $B(x_0)xy = B(x_0)yx$ by the symmetry of B. The norm of $F''(x\ )$ is that in $L(X,L(X,Y))$. That is

$$||F''(x_0)x_1|| \leqslant ||F''(x_0)||\cdot||x_1||,$$

and thus (see equation (14))

$$||F''(x_0)x_1x_2|| \leqslant ||F''(x_0)||\cdot||x_1||\cdot||x_2||.$$

One can consider F'' as an element of the space of bounded bilinear operators from $X\times X$ into Y.

In addition to the chain rule, above, we have the obvious property that

$$\frac{\delta}{\delta x}\,(F(x) + G(x))|_{x_0} = \frac{\delta F(x)}{\delta x}\bigg|_{x_0} + \frac{\delta G(x)}{\delta x}\bigg|_{x_0}.$$

Also if F is continuously differentiable on $\{x|x = x_0 + th, 0 \leqslant t \leqslant 1\}$, then

$$||F(x_0 + h) - F(x_0)|| \leqslant \sup_{0\leqslant t\leqslant 1} ||F'(x_0 + th)||\cdot||h||,$$

and if twice continuously on the same set, then

$$\|F(x_0 + h) - F(x_0) - F'(x_0)h\| \leqslant \sup_{0 \leqslant t \leqslant 1} \tfrac{1}{2}\|F''(x + th)\| \cdot \|h\|^2.$$

These should be compared with the mean value theorem for ordinary differentiable real functions $f = f(x)$

$$f(b) - f(a) = f'(\theta)(b - a),$$

where θ is some number in (a,b).

Integration in Banach Spaces

If F is continuously differentiable on $\{x | x = x_0 + th, 0 \leqslant t \leqslant 1\}$ then

$$F(x + h) - F(x) = \int_0^1 F'(x + th)h dt,$$

where the integral is in the Riemann sense. Thus there is defined the inverse of the Frechet derivative.

Functionals

A *functional* is an operator which maps a Banach space X into its scalar field Λ. The Frechet derivative of a functional at $x_0 \in X$ is then a bounded linear functional on X. The set of all bounded linear functionals on X forms a Banach space denoted by X^* called the conjugate space to X. A Banach space is *reflexive* if $X^* = X$. If $f \in X^*$ we write

$$f(x) = (f,x).$$

If X is a Hilbert space the notation (f,X) will denote the inner product of the elements $x \in H$ and $f \in H^* = H$. Thus if $f(x)$ is a functional that is differentiable at x_0 we have

$$\frac{\delta}{\delta x} f(x)\big|_{x_0} = y_0,$$

and we write

$$f'(x_0)h = (y_0,h)$$

which is a linear functional of h.

Potentials and Gradients

If $f(x)$ is a Frechet differentiable functional with domain D in X, the operator

$$F(x) = \delta f(x)/\delta x \tag{19}$$

is also called the *gradient* of $f(x)$. The functional $f(x)$ is called the *potential* of the operator $F(x)$. $F(x)$ is an operator from X to X^*. It is not necessarily a linear operator since it is the derivative of a functional. Its derivative, however, will be a linear operator if it exists, i.e., $f''(x_0)$if it exists is a linear operator. $f''(x_0)$ allows us to define the *bilinear functional*

$$f''(x_0)hk = (f''(x_0)h,k).$$

If we were given an operator $F(x)$ from X to X^* the question might arise whether or not it is the gradient of some functional $f(x)$. An example of this question would arise in the situation where we had obtained an equation for the charge density of a system via approximating certain terms. We then would ask could we have gotten this by minimizing some energy functional of reasonable form with respect to the density imposing the constraint that the number of particles remain constant, etc..

The following theorem gives necessary and sufficient conditions for $F(x)$ to be a gradient [27]:

THEOREM: *If $F(x)$ is continuous in the simply connected region R of X it is the gradient of some functional if and only if the curvilinear Riemann-Stieltjes integral*

$$I = \int_L (F(x), \mathrm{d}x)$$

is independent of the path in R.

In this connection, one should recall the similar requirement for a vector to be the gradient of a scalar potential. The following theorem gives a way of computing potentials [27]:

THEOREM: *If $F(x)$ is a gradient it has the unique potential*

$$f(x) = f(x_0) + \int_0^1 (F\{x_0 + t(x - x_0)\}, x - x_0)\mathrm{d}t$$

or (20)

$$(f,x) = (f,x_0) + \int_0^1 (F\{x_0 + t(x - x_0)\}, x - x_0)\mathrm{d}t$$

which takes on the value $f(x_0)$ at the point x_0.

If $f(x)$ is twice differentiable this relation may be integrated by parts to yield

$$f(x) = f(x_0) + \int_0^1 (F(tx), tx)\, \frac{\mathrm{d}t}{t}\,. \qquad (21)$$

A point x_0 at which

$$\frac{\delta f(x)}{\delta x} = 0$$

is called a *critical point* of the functional $f(x)$. Such a point is a solution to the functional equation

$$F(x) = 0.$$

Dyson's equation is such an equation when a definite approximation to the self-energy is made. Its solution gives the single particle Green's function associated with the approximate self-energy.

An interior point x_0 of a set S such that $f(x) \geqslant f(x_0)$ is called a *relative minimum* of the functional $f(x_0)$. If $f(x_0)$ is an energy functional and has such a relative minimum at x_0 then x_0 is the x which gives the minimum energy. A point $x_0 \in S$ such that $f(x) \leqslant f(x_0)$ is called a relative *maximum point*. Relative maximum and minimum points of $F(x)$ are called relative *extremal points* of $f(x)$.

A relative extremal point x_0 of a Frechet differentiable functional $f(x)$ is a critical point of $f(x)$. Thus at a relative minimum x_0 of a correct energy functional we have that $F(x) = 0$. This theorem provides a variational theorem for the functional $f(x)$. It shows that at x_0 the first variation of $f(x)$ — the first Frechet derivative of $f(x)$ at x_0 — is zero. It also shows that the procedure of approximation $F(x)$ and then using the exact $f(x)$ to find the energy is incorrect. The converse of the theorem: that $F(x_0) = 0$ implies an extremal point of $f(x)$ — is obviously false, x_0 may be an inflection point. In addition, by the properties of such variations we see that if we constrain the variation say by imposing an external condition of conservation of particle number using a single determinantal expression for it, that we will get a larger $f(x)$ (at best equal to the unconstrained or unrestricted variational minimum). Thus one finds a relative minimum rather than an absolute minimum of $f(x)$. Thus the Dyson equation method with the Feynman formula for the energy functional and the Hohenberg-Kohn-Sham methods with a given functional are from this standpoint correct approaches. The Dyson equation approach, however, is an unrestricted formulation with particle conservation built in, not added as an external constraint that is difficult to specify. Moreover, it allows for the systematic extension of the energy functional by simply adding more terms to the self-energy operator Σ^*. It should also be obvious here that non-self-consistent theories can easily give any answer one chooses — including the right one — but should be suspect.

To procede further we must review some definitions.

A sequence $\{x_m\}$ in X is said to *converge weakly* to $x \in X$ if

$$\lim_{m\to\infty} (y,x_m) = (y,x)$$

for all $y \in X^*$. *Strong convergence* is the convergence in the norm $\|\cdot\|$ discussed previously. Strong convergence implies weak convergence but the reverse is not necessarily the case. But in infinite-dimensional spaces the two are equivalent. A sphere of radius R at x_0 is the set of all x such that $\|x - x_0\| < R$. A weakly compact sphere is a sphere S in which every infinite sequence of elements of S has a weakly convergent subsequence, the limit of which does not have to belong to S. All reflexive Banach spaces and therefore all Hilbert spaces have weakly compact spheres. The functional $f(x)$ is said to be *weakly lower semi-continuous* at x_0 if for all sequences $\{x_n\}$ which converge weakly to x_0:

$$f(x_n) \leqslant \lim_{n\to\infty} \text{greatest lower bound } f(x_n) \equiv \lim_{n\to\infty} \inf f(x_n).$$

A very important theorem concerning the bounds of weakly lower semi-continuous functions is:

THEOREM: *A weakly lower semi-continuous functional is bounded below and attains its lower bound on each bounded and weakly closed subset of the Banach space X.*

This theorem together with a maximum principle, similar to that used in the theory of elliptic partial differential equations, that within a bounded weakly closed and hence closed subset of a weakly compact sphere in X

$$f(x) > f(x_0)$$

for all x on the boundary of S. Then the Frechet derivative of the functional has at least one critical point. It thus attains its minimum at x_0 which is interior to S. If f has a second Frechet derivative and furthermore

$$f''(x)hh \geqslant \|h\|\xi(\|h\|),$$

where $\xi(t)$ is continuous and non-negative for $t \geqslant 0$ and $\lim_{t\to\infty} \xi(t) = \infty$ then f has at least one critical point.

SOLUTIONS OF NON-LINEAR OPERATOR EQUATIONS

The theorems above give us some information about the existence of solutions to functional equations. However, we still have not given any methods for obtaining the solution of such functional equations. One such theorem is the following:

THEOREM: Contraction Mapping Principle: *Let S be a closed sphere in a Banach space X and let F be an operator defined on X to X satisfying the Lipschitz condition*

$$\|Fx_1 - Fx_2\| \leqslant \alpha\|x_1 - x_2\|, \qquad x_1, x_2 \in S,$$

where $\alpha < 1$. *If the operator F transforms S into itself then the equation*

$$x = F(x)$$

has a unique solution x^* *in* S.

This solution can be computed by the successive approximations method:

$$x_0 \in S,$$
$$x_n = F(x_{n-1}), \qquad n = 1,\ldots, \tag{22}$$

and x_0 is an arbitrary element in x_0. The rate of convergence is given by

$$\|x_n - x^*\| \leqslant \frac{\alpha^n}{1-\alpha}\|x_1 - x_o\|, \qquad n = 1,\ldots\,. \tag{23}$$

The principle of contraction mapping is related to Schauder's principle that states that if a continuous operator F transforms a closed convex set S of a Banach space into a compact subset of S then there exists a point $x \in S$ such that $x = F(x)$.

In the above theorem $F(x)$ can be for instance the functional derivative of a functional $f(x)$. If the Banach space is reflexive then $F(x)$ is a mapping of X to X. $F(x)$ is thus an operator on X to X. It is not necessarily linear. An example of such an operator is the Hartree-Fock operator when considered as acting on a function ϕ. Squares and complex conjugates appear.

NEWTON'S METHOD

The non-linear operator equation

$$F(x) = 0 \tag{24}$$

may also be solved using *Newton's Method* which [*26*,*27*] may be expressed as

$$x_0 = y,$$
$$x_{n+1} = x_n - F'(x_n)^{-1}F(x_n), \qquad n = 1,\ldots\,, \tag{25}$$

supposing that $F'(x_n)$ exists and the linear operator $F'(x_n)$ has an inverse at x_n.

Supposing that the sequence $\{x_m\}$ is infinite (if it terminates at some integer K because $F(x_n) = 0$ we say the sequence converges at the n-th step) it will either converge or diverge. Denote x^* by (the convergent limit) i.e.,

$$x^* = \lim_{m\to\infty} x_m.$$

The following three theorems [26] then give sufficient conditions for x^* to be a solution to equation (24).

THEOREM 1: *If* $F'(x)$ *is continuous at* $x = x^*$, *then* $F(x^*) = 0$.

THEOREM 2: *If* $\|F'(x)\| \leqslant \mu$ *in some closed sphere which contains* $\{x_m\}$ *then* x^* *is a solution to equation* (24).

THEOREM 3: *If* $\|F''(x_0)\| \leqslant k$ *in some closed sphere* $\bar{S}(x_0,r)$ $0 < r < \infty$ *which contains* $\{x_m\}$ *then* x^* *is a solution of equation* (24).

A theorem of Kantorovic gives a computational theory of the process.

THEOREM 4: (Kantorovic): *If*

$$\|F''(x)\| \leqslant k \tag{26a}$$

in some closed sphere $\bar{S}(x_0,r)$ *and*

$$h_0 = B_0 \eta_0 k \leqslant \tfrac{1}{2}, \tag{26b}$$

then the Newton sequence (25) *starting with* x_0 *will converge to a solution* x^* *of equation* (24) *which exists in* $\bar{S}(x_0,r)$ *provided that*

$$r \geqslant r_0 = \frac{1 - (1 - 2h_0)^{\frac{1}{2}}}{h_0}\,\eta_0, \tag{26c}$$

where

$$\|x_1 - x_0\| \leqslant \eta_0 \tag{26d}$$

and

$$\|F'(x_0)^{-1}\| \leqslant B_0. \tag{26e}$$

Providing that $F'(x_0)$ exists, merely by making η_0 as small as needed one can satisfy (26b,c). If x_1 and x_2 are close enough then convergence is assured. The convergence of Newton's method is thus entirely different from that in the contraction mapping principle method where a very strong Lipschitz condition must be verified. In addition the convergence rates of the two methods are very different. The Contraction Mapping Principle converges as

$$||x_{n+1} - x_n|| \leqslant \alpha ||x_n - x_{n-1}||,$$

whereas Newton's method converges according to

$$||x_{n+1} - x_n|| \leqslant \beta ||x_n - x_{n-1}||^2.$$

The contraction mapping method converges only geometrically. Whereas Newton's method converges quadratically.

EXAMPLES

To give some further examples we show how one may (1) derive the potential of the Dyson operator, (2) derive various solution schemes to Dyson's equation using the notions of the Frechet derivative, and (3) derive a generalized Hellman-Feynman theorem.

We write Dyson's equation in the form:

$$G^{-1} = G_0{}^{-1} - \Sigma^*[G], \tag{27}$$

familiar from self-consistent Green's function theory. $\Sigma^*[G]$ contains only skeleton graphs. $G(x,x',\omega)$ is a function in the space $L^2[R^7,dx]$ of square Lebesgue integrable functions. $\Sigma^*[G]$ is a non-linear operator with argument G. From equation (20) calling the functional E we have

$$- \Delta E = \int_0^1 (F\{G + t(G_0 - G)\}, G_0 - G)dt,$$

where

$$F[G] = - G^{-1} + G_0{}^{-1} - \Sigma^*[G].$$

Thus

$$- \Delta E = - (\int_0^1 \{G + t(G_0 - G)\}^{-1}, dt(G_0 - G)$$

$$+ (G_0{}^{-1}, \int_0^1 (G_0 - G)dt) - (\int_0^1 \Sigma^*[G + t(G_0 - G)], (G_0 - G))dt.$$

So performing the integrations:

$$\Delta E = (\ln[G + t(G_0 - G)]\Big|_0^1) - (G_0^{-1},(G_0 - G))$$

$$+ (\int_0^1 \Sigma^*[G + t(G_0 - G)],(G_0 - G)dt).$$

Now $\Sigma^*[G]$ is of the general form

$$\Sigma^*[G] = \sum_{n=1}^{\infty} V^n G^{2n-1}.$$

Thus the last term in the expression for ΔE may be written as

$$\Phi[G] - \Phi[G_0] = \sum_n (\int_0^1 V^n(G + t(G_0 - G))^{2n-1},(G_0 - G)dt)$$

$$= \sum_n \frac{1}{2n} (V^n(G + t(G_0 - G))^{2n}\Big|_0^1)$$

$$= - \tfrac{1}{2} \sum_n \frac{1}{n} V^n G^{2n} + \tfrac{1}{2} \sum_n \frac{1}{n} V^n G_0^{2n},$$

thus

$$\Delta E = - (\ln G_0^{-1}G) + (G_0^{-1}G - 1) + \Phi[G] - \Phi[G_0].$$

The operation (...) is a trace operation, for instance the operation

$$\frac{1}{2\pi} \sum_{kk'} \delta(k - k')\int_C d\omega,$$

and G, G_0 and V are expressed in terms of their Fourier transforms. When $V = 0$, $\Delta E = 0$. Also when $G = G_0$ and $V \neq 0$, $\Delta E = 0$. That is the system can propagate with $G = G_0$ in the presence of a peculiar perturbation without changing energy. By the theorems above we have

$$\frac{\delta \Delta E}{\delta G} = 0. \tag{29}$$

Equation (29) among other things shows that ΔE is stable with respect to variations in the unperturbed spectrum.

Another interesting exercise is to derive a generalized Hellman-Feynman Theorem (Feynman, R.P. (1939). *Phys. Rev.*, **56**, 340). Consider

$$F(\Psi) = (H + \lambda\theta(\Psi) - E(\lambda))\Psi_\lambda$$

$$= (H(\lambda) - E(\lambda))\Psi_\lambda,$$

where $\theta(\Psi)$ is a many particle operator (not necessarily linear). We suppose that it is continuously differentiable with respect to Ψ.

Then the potential of F (which implicitly depends upon the parameter λ) may be obtained via

$$\Delta = f(\Psi) - f(\Psi_0) = \int_0^1 (F(t\Psi), t\Psi)\,\frac{dt}{t}.$$

Now $(H - E(\lambda))$ is a linear operator. Hence

$$(H - E(\lambda))(t\Psi) = t(H - E(\lambda))\Psi.$$

Thus

$$\Delta = \int_0^1 ([H - E(\lambda)]\Psi,\Psi)t\,dt + \lambda\int_0^1 (\theta[t\Psi],t\Psi)\,\frac{dt}{t}.$$

Now at $\Psi = \Psi_0$, $\Delta = 0$ so that we have

$$\tfrac{1}{2}([H - E(\lambda)]\Psi,\Psi) = -\lambda\int_0^1 (\theta(t\Psi),\Psi)dt.$$

Now we write $\theta(t\Psi) = P'(t\Psi)$. We therefore obtain

$$\tfrac{1}{2}([H - E(\lambda)]\Psi,\Psi) = -\lambda\int_0^1 (P'(t\Psi),\Psi dt),$$

$$= -\lambda[(P(\Psi)) - (P(0))].$$

Suppose $P(0) \equiv 0$.

Then we obtain for the expectation value of P the relation

$$\frac{(P(\Psi_{\lambda=0}))}{(\Psi_0,\Psi_0)} = \lim_{\lambda\to 0}\frac{1}{2}\,\frac{((E(\lambda) - H)\Psi_\lambda,\Psi_\lambda)}{(\Psi_\lambda,\lambda\Psi_\lambda)}.$$

If θ were a constant Hermitian linear operator then $P(\Psi) = \tfrac{1}{2}(\theta\Psi,\Psi)$ and we regain

$$\frac{1}{2}\frac{(\theta\Psi_0,\Psi_0)}{(\Psi_0,\Psi_0)} = \frac{1}{2}\lim_{\lambda\to 0}\frac{((E(\lambda) - H)\Psi_\lambda,\Psi_\lambda)}{(\Psi_\lambda,\lambda\Psi_\lambda)} ,$$

so that

$$\langle\theta\rangle = \lim_{\lambda\to 0}\frac{(\Psi_\lambda,(E(\lambda) - H)\Psi_\lambda)}{\lambda(\Psi_\lambda,\Psi_\lambda)} .$$

Let us now consider various methods of solving Dyson's equation. One would, of course, be the contraction mapping principle:

$$G^{n+1} = (G_0{}^{-1} - \Sigma^*[G^n])^{-1}. \tag{30}$$

This defines a sequence which we can call a generalized continued fraction. We see that equation (30) continually makes energy denominator insertions. We now procede to use a matrix representation method in the solution of equation (30).

The matrix representations of G and Σ^* are elements of a Banach algebra. The Frechet derivative of Σ^* is a linear operator $\tilde{\Sigma}^*$ whose indices are labelled by matrix indices (the indices of the vector space of the G's). Thus

$$\frac{\delta\Sigma^*[G]}{\delta G} \to \left(\frac{\delta\Sigma_{ij}{}^*[G]}{\delta G_{\ell k}}\right) ,$$

$$\frac{\delta\Sigma_{ij}}{\delta G_{\ell k}} = \tilde{\Sigma}(ik)(j\ell)[G].$$

We can use this fact to expand Σ^* in a Taylor series about G_0 obtaining:

$$1 = G_0{}^{-1}G - \Sigma^*[G]G$$

$$= G_0{}^{-1}G - (\Sigma^*[G_0] + \tilde{\Sigma}[G_0](G - G_0) + \dots)G.$$

This series would terminate with only the terms shown, if for instance

$\Sigma^*[G] =$ 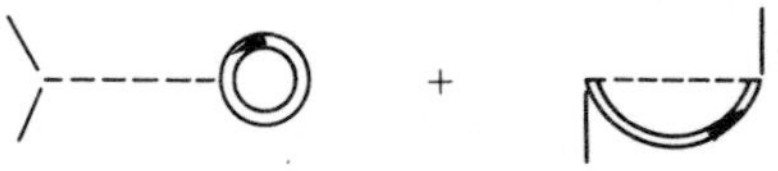 (32)

(The double lines indicate that the exact propagator G is to be used). Equation (31) would then be a quadratic equation in a Banach space. Its solution by the contraction mapping principle leads to a continued fraction for G whose convergence properties are well established. We are usually interested, however, in cases where the interaction is renormalized also. This would lead to an infinite series since all derivatives of $\Sigma^*[G]$ would exist. This may be circumvented, however, by a two step process. For example suppose the screening is RPA. Then

$\Sigma^{\star}[G] \quad =$ 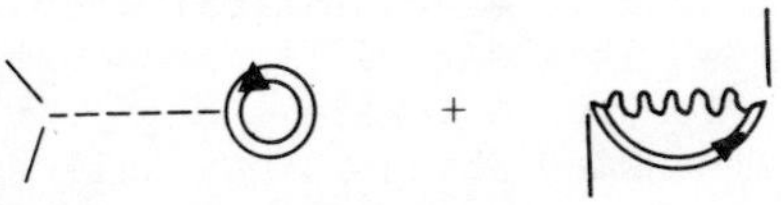 (33)

where "〰" $= (1 - V\Pi^0[G])^{-1}V = \varepsilon_{RPA}{}^{-1}[G]V = \tilde{V}$. Then $\delta\Sigma^*/\delta G$, using the chain rule, is

$$\frac{\delta\Sigma^*}{\delta G} = \tilde{I}V + \frac{\delta}{\delta G}\,[[1 - V\Pi^0[G]]^{-1}VG]$$

$$= \tilde{I}V - (1 - V\Pi^0[G])^{-1}\,\frac{\delta}{\delta G}\,[-V\Pi^0[G]]$$

$$\times(1 - V\Pi^0[G])^{-1}VG - (1 - V\Pi^0[G])^{-1}VI$$

$$= \tilde{I}V - \varepsilon_{RPA}{}^{-1}VI - \varepsilon_{RPA}{}^{-1}[G]V\,\frac{\delta\Pi^0[G]}{\delta G}\,\varepsilon_{RPA}{}^{-1}VG$$

$$= \tilde{I}V - \tilde{V}I - \tilde{V}\,\frac{\delta\Pi^0}{\delta G}\,\tilde{V}G. \tag{34}$$

The tilde on the $\tilde{I}$ is used to differentiate the difference between the linear operator $I = \delta G(x,x')/\delta G(x''',x'')$ and the linear operator $\tilde{I} = \delta G(x,x)/\delta G(x''',x'')$. $\delta\Pi^0[G]/\delta G$ is a linear operator also. We can therefore substitute equation (34) into the approximation given by equation (31)

However, we note that Newton's method also requires only one derivative. Using

$$F[G] = G^{-1} - G_0{}^{-1} + \Sigma^*[G] \tag{35}$$

in the Newton's method, we need $F'[G]$ according to equation (25):

$$F'[G] = -\,G^{-1}IG^{-1} + \frac{\delta\Sigma^*[G]}{\delta G}\,,$$

which by equation (34) yields

$$F'[G] = - G^{-1}IG^{-1} + V\tilde{I} - \tilde{V}I - \tilde{V}\,\frac{\delta\Pi^0}{\delta G}\,\tilde{V}G.$$

Using equation (25) we obtain after some rearrangement

$$F'[G_n]G_{n+1} = - 2G_n^{-1} + G_0^{-1} + \frac{\delta\Sigma^*}{\delta G_n}\,G_n - \Sigma^*[G_n], \tag{36}$$

which we recall is a quadratically convergent process for G. Equation (36) is a set of linear equations in the matrix representation of G for G_{n+1} since $F'[G_n]$ is a linear operator.

Another scheme, which is basically a contraction mapping method, but which enables one to screen the interaction is described in the next section. We outline it here since in that section the indices and algebra can obscure the general line of reasoning. We note that this kind of approach is probably better suited to inhomogeneous systems than the HKS approach. This is because here the variations of the charge density (Green's function really but $iG(r,r,0^+) = n(r)$) with respect to variations in the external potential are explicitly taken into account. Thus one is not restricted to systems with slowly varying density. We write Dyson's equation in the form given by Martin and Schwinger [8]

$$G_0^{-1}G + iV\left(G - \frac{\delta}{\delta U}\right)G[U]\bigg|_{U=0} = 1, \tag{37}$$

(this is a functional differential equation) where

$$\Sigma^* = i\left[VG - V\,\frac{\delta G[U]}{\delta U}\right]_{U=0} G^{-1},$$

and obtain $\delta G/\delta U$ from the Dyson's equation for G in the presence of the external potential U:

$$G^{-1}[U] = G_0^{-1}[U] - \Sigma^*[G[U]].$$

Then, using the chain rule

$$- G^{-1}\,\frac{\delta G}{\delta U}\,G^{-1} = I - \frac{\delta\Sigma^*}{\delta U} = I - \frac{\delta\Sigma^*}{\delta G}\,\frac{\delta G}{\delta U}\,.$$

This is a self-consistent equation for $\delta G/\delta U|_{V=0} \equiv \tilde{G}$, i.e.

$$\tilde{G} = - GIG + GJ\tilde{G}G, \tag{38}$$

where J is the linear operator $\delta\Sigma^*/\delta G$. If we take for instance

$\Sigma^{*}[G] =$ 

then J = a constant linear operator. Substituting equation (38) into equation (37) then yields.

$$G_0^{-1}G + iV(G - \tilde{G}) = 1, \tag{39}$$

which must be solved for G. It is a linear equation because $\tilde{G}$ is a linear operator. Equations (38,39) are then iterated until self-consistency is achieved. The solution obtained is the GRPA one.

We now show how G can be used in a Newton's iteration. Our equations are of the form

$$G_0^{-1}G + iVGG - iV\tilde{G}[G] = 1, \tag{40}$$

$$\tilde{G} = - GIG + GJ[G]\tilde{G}[G]G. \tag{41}$$

Define

$$F[G] = 1 - G_0^{-1}G - iVGG + iV\tilde{G}[G].$$

For Newton's method:

$$F'[G_n][G_n - G_{n+1}] = F[G_n], \tag{42}$$

We need $F'[G]$:

$$F'[G] = - G_0^{-1}I - iVIG - iVGI + iV\,\frac{\delta\tilde{G}}{\delta G}\,. \tag{43}$$

Now

$$\frac{\delta\tilde{G}}{\delta G} = - IIG - GII + \left(\frac{\delta GJ}{\delta G}\right)\tilde{G}G + GJ\,\frac{\delta\tilde{G}}{\delta G}\,G + GJ\tilde{G}I, \tag{44}$$

which must be solved for $\delta\tilde{G}/\delta G$. This equation is entirely determined since J is assumed given. If $J = \delta[$  $+$  $]/\delta G$ then $\delta J/\delta G \equiv 0$ and we obtain a simplification of our equations for GRPA. Substituting the solution of equation (44) into equation (43) and then using equation (43) in equation (42) would yield a quadratically convergent GRPA iteration.

4. FRECHET DERIVATIVE AND THE DIELECTRIC FUNCTION. SELF-CONSISTENT PERTURBATION THEORY

We consider the case of an inhomogeneous medium in 3-space with time. We therefore have for instantaneous interactions in the unperturbed system, translational invariance with respect to the time parameter. We therefore may use a 4-vector notation without the constraint that time and space coordinates are connected by being in a pseudo-metric space. Within this direct product space $R^3 \times \mathcal{T}$ we define the Frechet derivative (or functional derivative) $D_{\alpha'\beta'}(x',y')$ with respect to the external potential $U_{\alpha'\beta'}(x',y')$ where x and y are 4-vectors and α, β are spin indices ($U_{\alpha'\beta'}(x',y')$ is an inhomogeneous spin dependent external potential) of the functional $F_{\alpha\beta}(xy|U)$ by [28]

$$D_{\alpha'\beta'}(x',y')F_{\alpha\beta}(xy|U) = \frac{\delta F_{\alpha\beta}(xy|U)}{\delta U_{\beta'\alpha'}(y',x')} \tag{1}$$

$$\equiv F_{\alpha\alpha'\beta\beta'}(xx'yy'|U),$$

where $\delta F_{\alpha\beta}(xy|U)/\delta U_{\beta'\alpha'}(y',x')$ is the Frechet derivative as described previously. We require that

$$D_{\alpha\beta}(x,y)U_{\alpha'\beta'}(x',y') = \frac{\delta U_{\alpha'\beta'}(x',y')}{\delta U_{\beta\alpha}(y,x)} \tag{2}$$

$$= \delta_{\alpha\beta'}\delta_{\alpha'\beta}\delta(x - y')\delta(x' - y).$$

In terms of this derivative and the one particle Green's function, the two particle Green's function may be expressed as follows [8]

$$G_{\alpha\alpha'\beta\beta'}(x_1x_1'x_2x_2') = G_{\alpha\beta}(x_1x_2)G_{\alpha'\beta'}(x_1'x_2')$$

$$+ D_{\alpha'\beta'}(x_1',x_2')G_{\alpha\beta}(x_1x_2|U)\big|_{U=0}, \tag{3}$$

where $G_{\alpha\beta}(x_1x_2|U)$ is the single particle Green's function in the presence of the external potential U. We write $G_{\alpha\beta}(x_1x_2|U)\big|_{U=0} \equiv G_{\alpha\beta}(x_1x_2)$. $G_{\alpha\beta}(x_1x_2|U)$ is thus a functional of the external potential which in the limit $U = 0$ equals the exact Green's function of the interacting system in the absence of the external potential. The two particle Green's function is of interest in itself since the poles and residues of its particle-hole parts yield the excitation energies of the N-particle system and can be used to compute the generalized oscillator strengths. This is all the information necessary for a calculation of all one particle transition matrix elements.

By the use of Dyson's equation in the presence of the external potential:

$$G_{\alpha\beta}{}^{-1}(x_1x_2|U) = G_{0\alpha\beta}{}^{-1}(x_1x_2|U) - \Sigma_{\alpha\beta}{}^{*}(x_1x_2|U), \tag{4}$$

one can find a self-consistent equation for

$$D_{\alpha'\beta'}(x_1',x_2')G_{\alpha\beta}(x_1x_2|U)\Big|_{U=0} = \frac{\delta G_{\alpha\beta}(x_1x_2|U)}{\delta U_{\beta'\alpha'}(x_2'x_1')}\Bigg|_{U=0}$$

$$\equiv \tilde{G}_{\alpha\alpha'\beta\beta'}(x_1x_1'x_2x_2'), \tag{5}$$

the last term in equation (3).

The reason for doing this is that one can find the full two particle Green's function in this manner. Another more important reason for our purposes is that a certain contraction of the full two particle Green's function is related to the dielectric function as we have indicated in the previous section.

In Dyson's equation (equation (4)) Σ^* is called the proper self-energy. In the formulation we shall use, Σ^* is considered to be a non-linear operator of argument equal to the exact Green's function. In order to avoid double counting terms Σ^* includes no self-energy insertions. Σ^* is then expressed in terms of skeleton graphs only, i.e., in diagrams

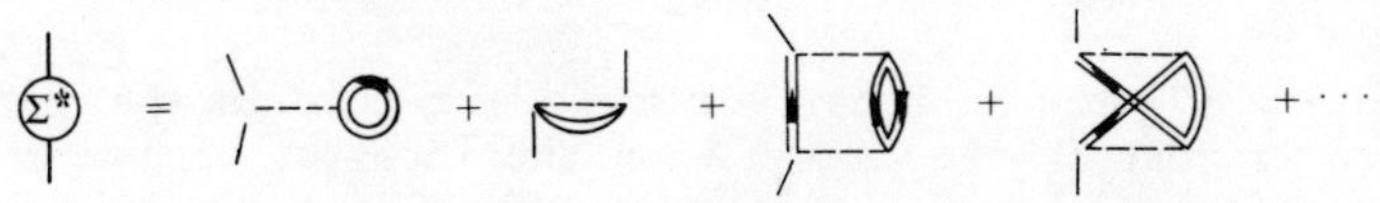

(short stubs do not indicate propagators)

The internal propagation is given by the full single particle Green's function. $G_0{}^{-1}(x_1x_2|U)$ in equation (4) is given by

$$G_{0\alpha\beta}{}^{-1}(x_1x_2|U) = (i\delta(t_1 - t_2)\frac{\partial}{\partial t_2} - H_0(x_1)\delta(x_1 - x_2))_{\alpha\beta} + U_{\alpha\beta}(x_1,x_2) \tag{6}$$

(this is slightly ambiguous). What is meant is that $G_0{}^{-1}$ together with Dyson's equation may be expressed in terms of the integro-differential equation (suppressing spin):

$$(i\frac{\partial}{\partial t_1} - H_0(x_1) + U(x_1,x_2))G(x_1x_2|U) - \int dx'\Sigma^*(x_1x'|U)G(x'x_2|U)$$

$$= \delta(x_1 - x_2).$$

One obtains the equation for $\tilde{G}$ by first evaluating the Frechet derivative of the identity

$$\sum_{\gamma} \int dx_2 G_{\alpha\gamma}(x_1x_2) G_{\gamma\beta}^{-1}(x_2x_3) = \delta_{\alpha\beta}\delta(x_1 - x_3) \tag{7}$$

with respect to $U_{\alpha'\beta'}(x_1',x_2')$ to obtain

$$\sum_{\gamma} \int dx_2 \left\{ \frac{\delta G_{\alpha\beta}(x_1x_2|U)}{\delta U_{\beta'\alpha'}(x_2',x_1')} \; G_{\gamma\beta}^{-1}(x_2x_3|U) \right.$$

$$\left. + \; G_{\alpha\gamma}(x_1x_2) \frac{\delta G_{\gamma\beta}^{-1}(x_2x_3|U)}{\delta U_{\beta'\alpha'}(x_2',x_1')} \right\} \Bigg|_{U=0} = 0.$$

To simplify this expression we multiply by $G_{\beta\delta}(x_3x_4)$ and 'integrate' $\sum_{\beta} \int dx_3$ obtaining

$$\sum_{\beta\gamma} \int dx_2 dx_3 \frac{\delta G_{\alpha\gamma}(x_1x_2|U)}{\delta U_{\beta'\alpha'}(x_2',x_1')} G_{\gamma\beta}^{-1}(x_2x_3) G_{\beta\delta}(x_3x_4)$$

$$+ \sum_{\beta\gamma} \int dx_2 dx_3 G_{\alpha\gamma}(x_1x_2) \frac{\delta G_{\gamma\beta}^{-1}(x_2x_3|U)}{\delta U_{\beta'\alpha'}(x_2',x_1')} G_{\beta\delta}(x_3x_4) = 0.$$

By equation (7) the last part of the first term in the above equation is just $\delta_{\gamma\delta}\delta(x_2 - x_4)$ when the integration is performed. Therefore we have

$$- \frac{\delta G_{\alpha\beta}(x_1x_4|U)}{\delta U_{\beta'\alpha'}(x_2',x_1')}$$

$$= \sum_{\beta\gamma} \int dx_2 dx_3 G_{\alpha\beta}(x_1x_3) \frac{\delta G_{\gamma\beta}^{-1}(x_2x_3|U)}{\delta U_{\beta'\alpha'}(x_2',x_1')} G_{\beta\gamma}(x_3x_4). \tag{8}$$

From Dyson's equation we have

$$G_{\gamma\beta}^{-1}(x_2x_3|U) = G_{0\gamma\beta}^{-1}(x_2x_3) + U_{\gamma\beta}(x_2x_3) - \Sigma^*_{\gamma\beta}(x_2x_3|U).$$

Therefore using equation (2) we obtain

$$\frac{\delta G_{\gamma\beta}^{-1}(x_2x_3|U)}{\delta U_{\beta'\alpha'}(x_2',x_1')} = \delta_{\alpha'\beta'}\delta_{\beta'\gamma}\delta(x_1' - x_3)\delta(x_2' - x_2) - \frac{\delta \Sigma_{\gamma\beta}^*(x_2x_3|U)}{\delta U_{\beta'\alpha'}(x_2',x_1')}.$$

Introducing an obvious notation, analogous to that of equation (5), we may write the above equation as

$$\frac{\delta G_{\gamma\beta}{}^{-1}(x_2x_3|U)}{\delta U_{\beta'\alpha'}(x_2',x_1')}$$

$$= \delta_{\alpha'\beta}\delta_{\beta'\gamma}\delta(x_2' - x_2)\delta(x_1' - x_3) - \tilde{\Sigma}_{\gamma\alpha'\beta\beta'}(x_2x_1';x_3x_2'),$$

substituting this into equation (8) one obtains

$$-\frac{\delta G_{\alpha\delta}(x_1x_4|U)}{\delta U_{\beta'\alpha'}(x_2',x_1')}$$

$$= \sum_{\beta\gamma}\int dx_2dx_3G_{\alpha\gamma}(x_1x_2)\times G_{\beta\delta}(x_3x_4)$$

$$\times(\delta_{\alpha'\beta}\delta_{\beta'\gamma}\delta(x_1' - x_3)\delta(x_2' - x_2) - \tilde{\Sigma}_{\gamma\alpha'\beta\beta'}(x_2x_1';x_3x_2')),$$

so that

$$\tilde{G}_{\alpha\alpha'\beta\beta'}(x_1x_1';x_4x_2')$$

$$= - G_{\alpha\beta}(x_1x_2')G_{\alpha'\delta}(x_1'x_4)$$

$$+ \sum_{\beta\gamma}\int dx_2dx_3G_{\alpha\gamma}(x_1x_2)\tilde{\Sigma}_{\gamma\alpha'\beta\beta'}(x_2x_1';x_3x_2')G_{\beta\delta}(x_3x_4),$$

changing x_4 to x_2 and x_2 to x_4 and changing δ to β and β to γ, we obtain

$$\tilde{G}_{\alpha\alpha'\beta\beta'}(x_1x_1';x_2x_2')$$

$$= - G_{\alpha\beta'}(x_1x_2')G_{\alpha'\beta}(x_1'x_2)$$

$$+ \sum_{\delta\gamma}\int dx_3dx_4G(x_1x_4)\tilde{\Sigma}_{\gamma\alpha'\delta\beta'}(x_4x_1';x_3x_2')G_{\delta\beta}(x_3x_2).$$

Now $\tilde{\Sigma}$ may be expressed in terms of another functional derivative and $\tilde{G}$ making this equation into a self-consistent equation for determining $\tilde{G}$ by using the following chain rule

$$\tilde{\Sigma}_{\gamma\alpha'\delta\beta'}(x_4x_1';x_3x_2') = \frac{\delta\Sigma_{\gamma\delta}(x_4x_3|U)}{\delta U_{\beta'\alpha'}(x_2',x_1')} = \qquad \text{(Contd)}$$

$$(\text{Contd}) \quad = \sum_{\mu\nu} \int dx_5 dx_6 \, \frac{\delta\Sigma_{\gamma\delta}(x_4x_3|U)}{\delta G_{\mu\nu}(x_5x_6|U)} \, \frac{\delta G_{\mu\nu}(x_5x_6|U)}{\delta U_{\beta'\alpha'}(x_2',x_1')}\bigg|_{U=0} \tag{9a}$$

$$\equiv \sum_{\mu\nu} \int dx_5 dx_6 J_{\gamma\nu\delta\mu}(x_4x_6;x_3x_5)\tilde{G}_{\mu\alpha'\gamma\beta'}(x_5x_1';x_6x_2').$$

J is the so-called irreducible vertex which appears in the Bethe-Salpeter equation. Using the above equation in equation (9) we obtain (compare with equation (3.38))

$$\begin{aligned}\tilde{G}_{\alpha\alpha'\beta\beta'}(x_1x_1';x_2x_2') &= -\, G_{\alpha\beta'}(x_1x_2')G_{\alpha'\beta}(x_1'x_2) \\ &+ \sum_{\gamma\delta}\sum_{\mu\nu}\int dx_3dx_4dx_5dx_6 G_{\alpha\gamma}(x_1x_4)J_{\gamma\nu\delta\mu}(x_4x_6;x_3x_5) \\ &\times\tilde{G}_{\mu\alpha'\nu\beta'}(x_5x_1';x_6x_2')G_{\delta\beta}(x_3x_2).\end{aligned} \tag{10}$$

The full two particle Green's function may therefore be written as

$$\begin{aligned}G_{\alpha\alpha'\beta\beta'}(x_1x_1';x_2x_2') &= G_{\alpha\beta}(x_1x_2)G_{\alpha'\beta'}(x_1'x_2') - G_{\alpha\beta'}(x_1x_2')G_{\alpha'\beta}(x_1'x_2) \\ &+ \sum_{\delta\gamma}\sum_{\mu\nu}\int dx_3dx_4dx_5dx_6 G_{\alpha\gamma}(x_1x_4)J_{\gamma\nu\delta\mu}(x_4x_6;x_3x_5) \\ &\times\tilde{G}_{\mu\alpha'\nu\beta'}(x_5x_1';x_6x_2')G_{\delta\beta}(x_3x_2),\end{aligned} \tag{11}$$

which may be expressed diagramatically:

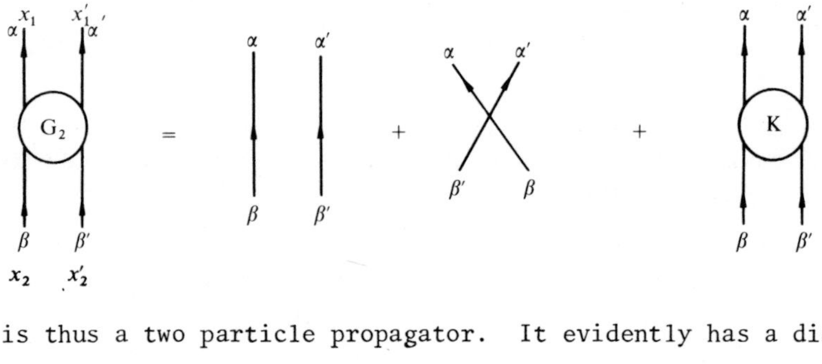

G_2 is thus a two particle propagator. It evidently has a diagrammatic expansion obtained by using equations (11,10). For example, the probability amplitude that if a particle is introduced at x_2 with spin β and a particle removed at x_2' (i.e. if a hole is introduced at x_2' then there will be a hole observed at x_2' and a particle at x_2) is given by the following part of G_2

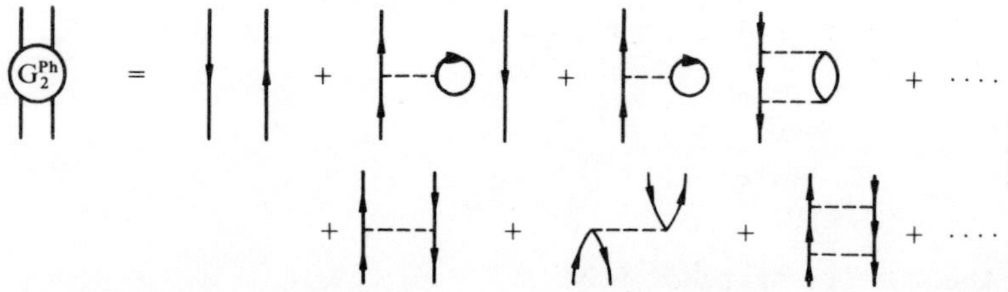

A similar contribution arises by reversing the direction of the lines. A useful approximation that we will make is to keep only the particle-hole parts of G_2 [*30*].

The polarization propagator is obtained from G by setting $x_1' = x_2'$ and $x_1 = x_2$ in equation (11). It should be noted that the internal propagation here is with the exact Green's function!

In electronic systems such as atoms with unpaired electrons, ferromagnetic crystals or systems for which spin dependent H_0's and spin dependent interactions are present, for instance due to the inclusion, perturbatively, of relativistic effects, the long range character of the Coulomb potential leads to divergencies in the straightforward perturbation expansion. Thus it is important to properly screen the interactions. This screening is related to the part of the two particle Green's function G_2 calculated above. The relation of G_2 to the dielectric response function and the screened potential are developed below. Our convention on the unscreened interaction V is chosen as follows to be consistent with our Green's function notation:

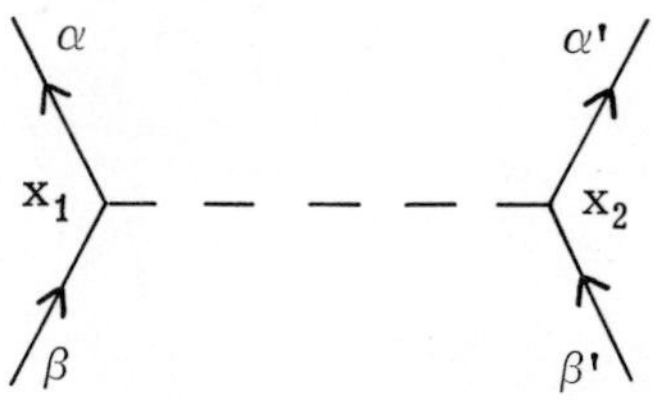

$= V_{\alpha\alpha'\beta\beta'}(x_1 x_2)$.

If the interaction is instantaneous:

$$V_{\alpha\alpha'\beta\beta}(x_1 x_2) = V_{\alpha\alpha'\beta\beta'}(\vec{x}_1\vec{x}_2)\delta(t_1 - t_2).$$

If the interaction is spin independent and instantaneous

$$V_{\alpha\alpha'\beta\beta'}(x_1x_2) = \delta_{\alpha\beta}\delta_{\alpha'\beta'}V(\vec{x}_1\vec{x}_2)\delta(t_1 - t_2).$$

Therefore

$$\sum_{\gamma} V_{\alpha\gamma\beta\gamma}(x_1x_2) = 2\delta_{\alpha\beta}V(\vec{x}_1\vec{x}_2)\delta(t_1 - t_2),$$

and

$$\sum_{\gamma'} V_{\alpha\alpha'\gamma'\gamma'}(x_1x_2) = \delta_{\alpha\alpha'}V(\vec{x}_1\vec{x}_2)\delta(t_1 - t_2)$$

in the most restricted case.

We can find the screened interaction by first finding the response function of the system for an external spin dependent probe $U_{\alpha\beta}(x_1,x_2)$. In the presence of $U_{\alpha\beta}$ the effective potential $V_{\alpha\beta}^{\mathrm{eff}}(x_1x_1')$ at the point x_1 depends not only upon $U_{\alpha\beta}$ at that point but also upon the response of the system through the Coulomb potential induced by the charge density $iG_{\alpha\beta}^{(-)}(x_3x_3^{\dagger})$:

$$V_{\alpha\beta}^{\mathrm{eff}}(x_1x_1) = U_{\alpha\beta}(x_1x_1) + i\int \sum_{\alpha'\beta'} d^3x_3 V_{\alpha\alpha'\beta\beta'}(x_1x_3)G_{\alpha'\beta'}^{(-)}(x_3x_3^{\dagger}|U). \tag{12}$$

The response function $\varepsilon_{\alpha\alpha'\beta\beta'}^{-1}(x_1x_2)$ is defined as the variation of the effective potential due to a variation in the applied field at another point $(U_{\alpha\beta}(x_2,x_2^{\dagger}) \equiv U_{\alpha\beta}(x_2))$

$$\varepsilon_{\alpha\alpha'\beta\beta'}^{-1}(x_1x_2) = \frac{\delta V_{\alpha\beta}^{\mathrm{eff}}(x_1x_1|U)}{\delta U_{\beta'\alpha'}(x_2^{\dagger}x_2)} . \tag{13}$$

In terms of the response function one has the screened potential [29]

$$V_{\alpha\alpha'\beta\beta'}^{\mathrm{scr}}(x_1x_2) = \sum_{\gamma\gamma'} \int d^4x_3 V_{\alpha\gamma\beta\gamma'}(x_1x_3)\varepsilon_{\gamma'\alpha'\gamma\beta'}^{-1}(x_2x_3). \tag{14}$$

Now from equations (2,12,5) we can find

$$\frac{\delta V_{\gamma'\gamma}^{\mathrm{eff}}(x_2x_2|U)}{\delta U_{\beta'\alpha'}(x_3^{\dagger}x_3)} = \delta_{\gamma'\beta'}\delta_{\gamma\alpha'}\delta(x_3 - x_2) + \qquad \text{(Contd)}$$

$$\text{(Contd)} \qquad + i \sum_{\mu\mu'} \int d^3x_4 V_{\gamma'\mu\gamma\mu'}(x_2x_4) \frac{\delta G_{\mu'\mu}(x_4x_4{}^\dagger)}{\delta U_{\beta'\alpha'}(x_3{}^\dagger x_3)}, \tag{15}$$

so

$$\varepsilon_{\gamma'\alpha'\gamma\beta'}{}^{-1}(x_2x_3) = \delta_{\gamma'\beta'}\delta_{\gamma\alpha'}\delta(x_3 - x_2)$$

$$+ i \sum_{\mu\mu'} \int d^3x_4 V_{\gamma'\mu\gamma\mu'}(x_2x_4)\tilde{G}_{\mu\alpha'\mu'\beta'}(x_4x_3;x_4{}^\dagger x_3{}^\dagger).$$

By using equation (15) in equation (14) we find the screened potential

$$V_{\alpha\alpha'\beta\beta'}{}^{\mathrm{scr}}(x_1x_2) = \sum_{\gamma\gamma'} \int d^4x_3 V_{\alpha\gamma\beta\gamma'}(x_1x_3)\delta_{\gamma'\beta'}\delta_{\gamma\alpha'}\delta(x_3 - x_2)$$

$$+ \sum_{\gamma\gamma'} \sum_{\mu\mu'} \int d^4x_3 d^3x_4 V_{\alpha\gamma\beta\gamma'}(x_1x_3)$$

$$\times\tilde{G}_{\mu\alpha'\mu'\beta'}(x_4x_3x_4{}^\dagger x_3{}^\dagger)V_{\gamma'\mu\gamma\mu'}(x_2x_4)$$

$$= V_{\alpha\alpha'\beta\beta'}(x_1x_2)$$

$$+ i \sum_{\gamma\gamma'} \sum_{\mu\mu'} \int d^4x_3 d^3x_4 V_{\alpha\gamma\beta\gamma'}(x_1x_3)$$

$$\times\tilde{G}_{\mu\alpha'\mu'\beta}(x_4x_3;x_4{}^\dagger x_3{}^\dagger)V_{\gamma'\mu\gamma\mu'}(x_2x_4).$$

Intuitively speaking the screened potential equals the bare potential plus a screening potential given by the last term in equation (15). We note that equation (10) gives $\tilde{G}$ so that once equation (10) is solved V^{scr} is entirely determined.

In a crystal where H_0 is the unrestricted Hartree-Fock operator F^{HF}, the first term of V^{scr} almost exactly cancels with the H_0 potential terms so that almost only the second term remains! In the GRPA approximation the effective potential then includes a screened direct interaction and a true screened exchange interaction. The plasmon frequencies are thus renormalized by the screened exchange. (The RPA screening in  does not screen the exchange since the RPA renormalization includes only direct diagrams).

SELF-CONSISTENT DETERMINATION OF THE GREEN'S FUNCTION

By making use of the expression for time derivative of the

field operator $\psi_\alpha(x)$ and the definition of the two particle Green's function [8] one can obtain the following coupled equations

$$\sum_\gamma \left(i\frac{\partial}{\partial t_1} - H_0(x_1)\right)_{\alpha\gamma} G_{\gamma\beta}(x_1x_2)$$

$$+ i \sum_{\alpha'\beta'\gamma} \int dx_3 V_{\alpha\alpha'\gamma\beta'}(x_1x_3) G_{\gamma\alpha'\beta\beta'}(x_1x_3x_2x_3{}^\dagger) = \delta(x_1 - x_2)\delta_{\alpha\beta}. \quad (16)$$

Comparing this with Dyson's equation

$$\sum_\gamma \left(i\frac{\partial}{\partial t_1} - H_0(x_1)\right)_{\alpha\gamma} G_{\gamma\beta}(x_1x_2)$$

$$- \sum_\gamma \int dx_3 \Sigma_{\alpha\gamma}{}^*(x_1x_3) G_{\gamma\beta}(x_3x_2) = \delta(x_1 - x_2)\delta_{\alpha\beta}, \quad (17)$$

we see that

$$\Sigma_{\alpha\beta}{}^*(x_1x_2) = -i \sum_{\alpha'\beta'\gamma\delta} \int dx_3 dx_4 V_{\alpha\alpha'\gamma\beta'}(x_1x_3)$$

$$\times G_{\gamma\alpha'\delta\beta'}(x_1x_3x_4x_3{}^\dagger) G_{\delta\beta}{}^{-1}(x_4x_2).$$

We thus see that the method of obtaining G through the solutions of the linear integral equation for $\tilde{G}$ is merely a method of summing an infinite number of diagrams. This is just the procedure used in section 2 to obtain the RPA screened potential. As one could guess we can make an intuitive derivation of the equation for G merely by looking at the set of skeleton diagrams for Σ^* that we want summed by G, remove one interaction line and one propagator from each Σ^* diagram and neglect the first GG term. The remaining terms then can be grouped into geometric series whose sum may be obtained by the solution of a linear integral equation.

Now equation (16) together with equations (11,10) for a given approximation to Σ^* in equation (9) comprise a set of self-consistent equations for determining the single particle Green's function. In this approximation procedure equation (10) provides a screened self-energy functional for equation (16).

In particular choosing Σ^* to contain only graph (a) below leads to the RPA screening in equation (11). The self-consistent procedure then produces self-consistent RPA. By including both (a) and (b) our self-consistent scheme produces self-consistent generalized (by the inclusion of exchange) RPA. The summation of all ring diagrams (diagrams (d,f etc.)) is accomplished by this technique. Since the diagrams are Feynman this summation includes all time orderings of the Brueckner-Goldstone diagrams. The actual expressions, in the matrix representation of eigenfunctions of H_0 is quite tedious to develop and will not be developed here because of lack of space. For the spin independent case the reader is referred to reference [*31*], where working equations for solid state problems are developed.

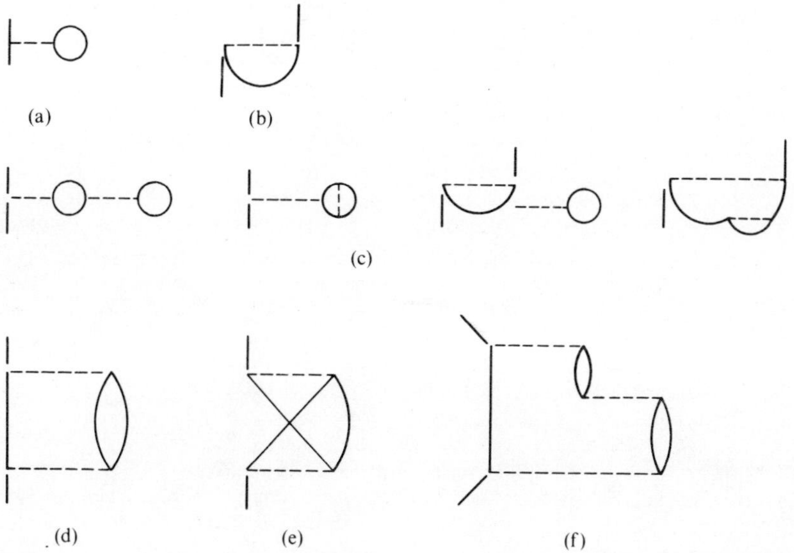

First and second order diagrams contributing to Σ^* (a) Hartree direct, (b) exchange, (c) Hartree-Fock with electron self-energy correction, (d) polarization (ring) and (e) exchange polarization, (f) third order ring. Diagrams (c) are not skeleton graphs.

In terms of diagrams our set of equations gives us a method of solving for example

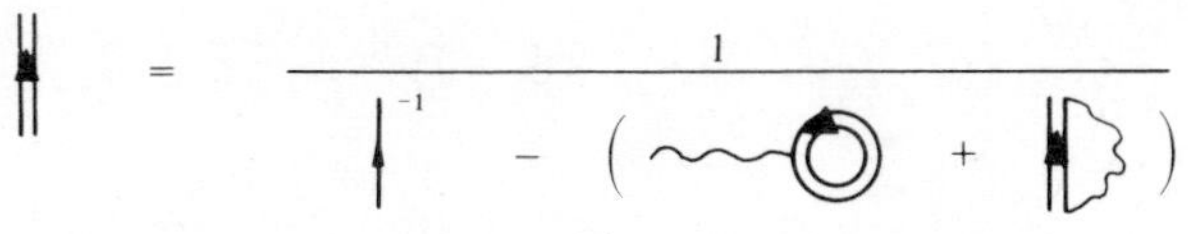

when  is the GRPA dielectric function screened interaction.

REFERENCES

1. Hohenberg, P. and Kohn, W. (1964). *Phys. Rev.*, **B136**, 864.
2. Kohn, W. and Sham, L.J. (1965). *Phys. Rev.*, **A140**, 1133.
3. Slater, J.C. (1951). *Phys. Rev.*, **81**, 385; Gaspar, R. (1954).

Acta Phys. Sci. Hung., **3**, 263.
4. Scofield, D.F. (1971). *Int. J. Quantum Chem.*, **5**, 489.
5. Herman, F., Van Dyke, J.P. and Ortenburger, I.B. (1969). *Phys. Rev. Lett.*, **22**, 807.
6. Ma, S. and Brueckner, K.A. (1968). *Phys. Rev.*, **165**, 18.
7. Some excellent references: Fetter, A.L. and Walecka, J.D. (1971). *Quantum Theory of Many-Particle Systems*, (McGraw-Hill Book Co., New York); March, N.H., Young, W.H. and Sampanthar, S. (1967). *The Many Body Problem in Quantum Mechanics*, (Cambridge University Press, London); Hedin, L. Lundquist S. (1969). In *Solid State Physics, Vol. 23*, (eds. Seitz, F. and Turnbull, D), (Academic Press, New York); Kumar, K. (1962). *Perturbation Theory and The Nuclear Many Body Problem*, (North Holland, Amsterdam); Pines, D. (1962). *The Many Body Problem*, (W.A. Benjamin, New York); Nozieres, P. (1964). *Theory of Interacting Fermi Systems*, (W.A. Benjamin, New York); Mattuck, R.D. (1967). *A Guide to Feynman Diagrams in The Many Body Problem*, (McGraw-Hill Book Co., New York).
8. Martin, P.C. and Schwinger, J.S. (1959). *Phys. Rev.*, **115**, 1342.
9. Dyson, F.J. (1949). *Phys. Rev.*, **75**, 486, 1736.
10. Brueckner, K.A. (1955). *Phys. Rev.*, **100**, 36.
11. Goldstone, J. (1957). *Proc. Roy. Soc.*, **A239**, 267.
12. Kelly, H.P. (1968). *Adv. Theor. Phys.*, **2**, 75.
13. Kelly, H.P. (1964). *Phys. Rev.*, **B136**, 896.
14. Huzinaga, S. and Arnau, C. (1970). *Phys. Rev.*, **A1**, 1285.
15. Scofield, D.F., Dutta, N.C. and Dutta, C.M. (1972). *Int. J. Quantum Chem.*, **6**, 9.
16. Miller, J.H. and Kelly, H.P. (1971). *Phys. Rev*, **A3**, 578.
17. Hartree, D.R. (1957). *The Calculation of Atomic Structures*, (John Wiley and Sons, New York).
18. Froese, C. (1963). *Can. J. Phys.*, **41**, 1895.
19. Harris, F.E., Kumar, L. and Monkhorst, H.J. (1973). *Phys. Rev.*, **B7**, 2850.
20. Hazelgrove, C.B. (1961). *Math. Comp.*, **15**, 323; Davis, P.J. and Rabinowitz, P. (1967). *Numerical Integration*, (Blaisdell Publishing Co., Waltham, Massachusetts).
21. Brillouin, L. (1933). *Actual, Sci. Ind.*, No. 71; (1934). *Actual. Sci. Ind.*, No. 159.
22. Slater, J.C. (1951). *Phys. Rev.*, **81**, 385; Gaspar, R. (1954). *Acta Phys. Sci. Hung.*, **3**, 263.
23. Graves-Morris, P.R. (ed.) (1973). *Padé Approximants*, (Lectures delivered at a Summer School Held at the University of Kent, July 1972, (The Institute of Physics, London).
24. Scofield, D.F. (1972). *Phys. Rev. Lett.*, **29**, 811.
25. The material presented here is standard. It may be found in any of the following: Kato, T. (1966). *Perturbation Theory for Linear Operators*, (Springer-Verlag, New York); Reisz, F. and Nagy, B. Sz. (1955). *Functional Analysis*, (Frederick Ungar, New York).

26. A very good source for what follows is: Rall, L.B. (1969). *Computational Solutions of Nonlinear Operator Equations*, (John Wiley and Sons, New York). Rall has also given a very nice survey of Vainberg's book in: Anselone, P.M. (1964). *Nonlinear Integral Equations*, (University of Wisconsin Press, Madison).

27. Vainberg, M.M. (1964). *Variational Methods for The Study of Nonlinear Operators*, (Holden-Day, San Francisco). This book also has a chapter on Newton's method in Banach spaces by L.V. Kantorovich and G.P. Akilov.

28. Zmuidzinas, J.S. (1970). *Phys. Rev.*, **B2**, 4445.

29. Kadanoff, L.P. and Baym, G. (1962). *Quantum Statistical Mechanics*, (W.A. Benjamin, New York).

30. Schneider, B., Taylor, H.S. and Yaris, R. (1970). *Phys. Rev.*, **A1**, 855.

31. Scofield, D.F. *Self-Consistent Green's Function Method for Energy Bands*, (to be published).

MANY-ELECTRON EFFECTS IN SPECTRA OF ELECTRONS BOUND IN ATOMS AND SOLIDS

S. LUNDQVIST

Chalmers University of Technology,
Göteborg, Sweden

1. INTRODUCTION

The study of elementary excitations and not so elementary excitations in solids forms the main theme of this Advanced Study Institute. It is the purpose of these lectures to give a brief introduction to the analog questions in atomic systems and with a particular point of view. Rather than just reviewing recent progress made in atomic theory I wish to emphasize that these problems can be discussed using essentially the same concepts and similar theoretical methods which are used for solids. Atomic theory has to a large extent followed its own different path, partly because many problems for light atoms can be solved with direct numerical methods but also because of a certain reluctance in applying the more powerful methods used in solid state theory. In removal of one electron, for example, one can in principle find the threshold energy by comparing two Hartree-Fock calculations which is possible for a system which is small enough but becomes impractical for a larger system. Similarly, autoionizing resonances can be handled by direct numerical methods for the lightest atoms. For heavier atoms and of course of extended systems the more powerful methods of many-body theory are needed to develop a theory from which the line-shape parameters can be calculated. The examples just given refer to cases where techniques used in many-body solid state theory could with advantage be applied to atomic systems, but where the simplest cases can be treated by brute force.

It is generally said that many-electron effects are small in atoms because of the strong attractive field from the nucleus. Calculations of correlation energies, atomic polarizabilities and other ground state properties indicate that the many-electron contribution in most cases are moderately small. However, we shall give evidence in these lectures that under certain conditions strong effects of interactions will occur in the high frequency spectra of an atomic system. These effects may even show up as giant resonances in the system and they have a character intermediate between plasmons in solids and vibrations in nuclei. They are described by application of appropriate extensions of the Random Phase Approximation (RPA).

A number of spectroscopic methods are now in use to study the high frequency spectra, but we shall concentrate on two essential methods. The one-electron spectrum is conveniently studied by the ESCA-method, the method of photo-electron spectroscopy and we shall restrict the discussion to the case of X-ray photo emission. The second method we shall discuss is the photoabsorption spectroscopy, and we shall mainly discuss the soft X-ray region. For energies high enough where one can neglect the final state interactions with the outgoing electron the two spectra will be closely related to each other.

These lectures will essentially be an introduction to the study of the spectrum by exciting strongly bound electrons and there the properties of the atomic shell are obviously essential. However, the atomic properties will obviously also show up in the corresponding spectra from solids. The theoretical techniques are essentially the same. The key differences will come about because the matrix element will now involve states of the outgoing electron in the solid rather than bound or continuum excited states of an isolated atom. An important aspect of the present theory is that in the spectral region of interest, many-electron resonances are always degenerate with single particle excitations and we are in most cases in a region of fairly strong damping and this is accounted for in the theoretical formulation we shall present.

2. REVIEW OF SOME EARLY THEORIES

Let us give an elementary introduction to the problem, taking the case of photoabsorption as an example. The photoabsorption cross section of the atom, i.e. the photoextinction coefficient per atom, can be written as

$$\sigma(\omega) = \frac{2\pi^2 e^2}{mc} g(\omega), \qquad (2.1)$$

where

$$\frac{2\pi^2 e^2}{mc} = 8.067\ 10^{-18}\ \text{Ry cm}^2,$$

with

$$1\ \text{Ry} = 13.6\ \text{eV},$$

and $g(\omega)$ is the *differential oscillator strength distribution*, which contains all the information about the absorption spectrum. For a survey of the dependence of $g(\omega)$ on frequency and atomic number Z it is convenient to separate the following three frequency ranges:

(1) *The low frequency range*, where $0 < (\omega/\text{Ry}) < 1$, we have essentially the line spectrum from optical spectroscopy;

(2) *The intermediate frequency range*, where $1 < \omega/\text{Ry} < Z^2$. Here the intermediate shells of the atom will give the main contribution and one would expect that these contributions overlap considerably so that $g(\omega)$ should depend smoothly on ω;

(3) *The high frequency range*, where $\omega/\text{Ry} > Z^2$. In this region we have the characteristic X-ray edges.

Whereas regions 1 and 3 have been studied experimentally and theoretically for a long time, 2 is the new region recently opened for research and where new physical effects can be found.

In the intermediate frequency region for a heavy atom one would expect a large number of electrons taking part in the absorption, and that the gross features of the spectrum could be described by a model treating the atom as an *electron fluid of non-uniform density*. Such a theory was formulated in the early thirties by Bloch [*1*]. He used the analogy with the hydrodynamical oscillations of a charged liquid drop and developed the theory of small oscillations around the equilibrium state of a Thomas-Fermi atom. The model was applied by Jensen [*2*] to a simplified model assuming a constant density up to the surface. This hydrodynamical model gave the following formulas for the eigenfrequencies and the oscillator strengths

$$\omega_n = k_n Z, \qquad f_n = q_n Z,$$

where Z is the number of electrons in the atom and k_n and q_n are numerical coefficients of the order of unity. The eigenmodes correspond to *plasma oscillations* of a charged liquid drop. The formulas show that the plasma oscillations fall in the intermediate frequency range between the optical and the X-ray spectra. The results were used to determine the energy loss of a fast particle and good agreement with experiments was found.

After the war there was a renewed interest in this problem and several physicists tried to apply the Bloch model to a more realistic model which corresponded to the electron density distribution in a real atom. These attempts were not successful and pointed to a serious deficiency of the model. One obtained a continuous rather than a discrete spectrum, and the wave-length of the oscillations went to zero in the outer low-density region when $r \to \infty$. This undesirable feature arises because the theory neglects the damping due to decay of the oscillations into single-particle excitations. Attempts to incorporate the effect of damping into the equations have not led to a successful theory.

In the frequency regime of interest damping due to single-particle excitations is always present and is often fairly strong. The conventional technique to first solve for approximate eigenmodes and then graft on the damping afterward does not seem to be the most efficient method. One would rather suggest some method to calculate directly the response of an atom to an external field and hence the absorption. Such a theory of atomic oscillations can be expressed in terms of the longitudinal dielectric constant of the atom $\varepsilon(\omega,r,r')$, which fulfils the equation:

$$\varepsilon(\omega,r,r') = \delta(r - r') + \int dr'' \frac{P(\omega,r,r')}{|r'' - r'|} ,$$

where P is the irreducible polarization propagator. The oscillator strength distribution and hence the absorption cross section is obtained from

$$g(\omega) = \frac{\omega}{2\pi^2} \text{Im} \int dr dr' \varepsilon^{-1}(\omega,r,r'), \tag{2.3}$$

and fulfils the usual sum rule

$$\int_0^\infty g(\omega)d\omega = Z.$$

In order to calculate the dielectric response, many workers have introduced a local density approximation by using the dielectric function for a uniform gas with the local density $\rho(r)$ as the density parameter. The simplest assumption neglects the spatial non-locality completely and describes the response by the classical formula

$$\varepsilon(r) = 1 - \frac{\omega_{p\ell}^{\,2}(r)}{\omega^2} ,$$

where $\omega_{p\ell}(r) = (4\pi e^2\rho(r))/m)^{\frac{1}{2}}$ is the classical plasma frequency. This ansatz gives

$$g(\omega) = \int dr\rho(r)\delta(\omega_{p\ell}(r) - \omega).$$

This result has a simple physical meaning: The single particle modes are screened out completely and the atom absorbs at each point by a collective mode given by the local plasma frequency. This gives a smooth absorption curve reflecting the charge distribution and shows no resonant behavior. Calculations based upon the full $\varepsilon(q,\omega)$ for an electron gas give very similar results.

The approach just mentioned is not a satisfactory procedure to investigate the possibility of many-electron resonances. The reason is that the non-locality length in the dielectric function is of the same order as the size of the atom itself and therefore the local approximation using the results for a uniform electron gas is not useful to develop a quantitative theory. So far no appreciable progress has been made treating the atomic oscillations as a drop of electron liquid of non-uniform density. An alternative approach is a many-body theory based upon a self-consistent theory, e.g. Hartree-Fock. In general then the results must be evaluated numerically from atomic wavefunctions. However, in the semi-classical limit of high quantum numbers a simple comprehensive discussion is possible, because in this limit the matrix elements of the Coulomb interaction can be expressed in terms of the dipole moments and the radius of the orbitals from which the transition take place. In the correspondence limit the Coulomb matrix elements between two excitations n and n' take the form

$$V_{nn'} = P_n P_{n'}(\lambda_n^2 + \lambda_{n'}^2). \tag{2.4}$$

The coupling constants λ_n are related to the mean radii R_n of the electron orbitals from which the excitations take place according to

$$\lambda_n^2 = R_n^{-3}.$$

With (2.4) it is straightforward to solve for the atomic polarizability in the RPA approximation. However, in this limit the system of dipolar excitations is equivalent to a system of coupled harmonic oscillators, described by the Hamiltonian

$$H = \frac{m}{2e^2}\sum_n \frac{1}{f_n}(\dot{P}_n^2 + \omega_n^2 P_n^2) + \frac{1}{2}\sum_{nn'}{}' P_n P_{n'}(\lambda_n^2 + \lambda_{n'}^2) + \sum P_n E_{ext}. \tag{2.5}$$

where $f_n = (2m/e^2)\omega_n P_n^2$ is the oscillator strength. One obtains the equations of motion

$$\ddot{P}_n + \omega_n^2 P_n = \frac{e^2}{m} f_n E_{ext} - \frac{e^2}{m} f_n(\lambda_n^2 \sum_{n'} P_{n'} + \sum_{n'} \lambda_n^2 P_n). \tag{2.6}$$

This can be solved readily to give the total dipole moment $P = \sum P_n$ in terms of the external field and we obtain for the polarizability $\alpha(\omega)$:

$$\alpha(\omega) = \left\{\frac{e^2}{m}\sum_n \frac{f_n}{\omega_n^2 - \omega^2}\right\} \times \left\{1 + \frac{2e^2}{m}\sum_n \frac{\lambda_n^2 f_n}{\omega_n^2 - \omega^2} - \frac{e^4}{2m^2}\sum_{nn'}(\lambda_n^2 - \lambda_{n'}^2)^2 \frac{f_n}{\omega_n^2 - \omega^2}\frac{f_{n'}}{\omega_{n'}^2 - \omega^2}\right\}^{-1}. \tag{2.7}$$

It follows from equation (2.7) that no resonances ever occur at the frequencies ω_n. Rather the resonances occur at new frequencies, obtained by putting the denominator equal to zero.

Let us for simplicity look at the spectra from an outer shell of equivalent electron in an atom. We introduce the notations

$$\varepsilon_{in} = 1 + \frac{2e^2}{m}\sum_n^{inner} \frac{\lambda_n^2 f_n}{\omega_n^2},$$

for the static effective screening constant of the inner shells,

$$- \chi(\omega) = \sum_n \frac{f_n}{\omega_n^2 - \omega^2}$$

for the zero order polarizability of the outer shell. We now only need one coupling constant and so we can write

$$\varepsilon_{in} - \lambda^2 \chi(\omega) = 0 \tag{2.8}$$

for the equation to determine the new modes.

The general nature of the solutions is shown graphically in figure 1. In the discrete part of the spectrum there is a shift in the excitation frequencies which increases with increasing coupling constant. There is also a solution in the continuous spectrum, where we have damping due to decay into single particle excitation. A closer study of the model shows that there is an appreciable *redistribution of oscillator strength.* The discrete line spectrum will be depleted in strength and the interaction shifts the oscillator strength in the continuum towards higher frequencies.

A simple application of the theory to the photoabsorption cross section of $4d^{10}$ shell in Xe was made by Brandt, Ederer and the author [*3*]. The coupling constant for the excitation from the $4d$ to the continuum levels was calculated from the formula

$$\lambda_n^2 = \frac{\langle n | r_{12}^{-1} | n \rangle}{P_n^2}$$

using Slater wave function. Instead of taking into account the variation with energy, they used a fixed value of λ calculated at the absorption maximum.

The single particle absorption spectrum for $4d$ transitions was calculated by Cooper [*4*]. It shows a strong peak at about 80 eV energy. This is because the continuum levels form a narrow band of virtual levels around that energy and there is a strong wave-function overlap with the bound $4d$ levels of Xe. Including the effect of the Coulomb interaction one finds an appreciable shift of the absorption peak from 80 eV to about 100 eV and a substantial broadening as is illustrated in figure 2.

Already this very simple model points out some features which are typical. The single particle excitation spectrum is of decisive importance and should be strongly peaked in order to obtain an appreciable effect. Only if this is fulfilled the interactions will give rise to a considerable shift in the spectrum. The result shown in figure 2 is often reproduced in the literature but is always said to be based on some speculative

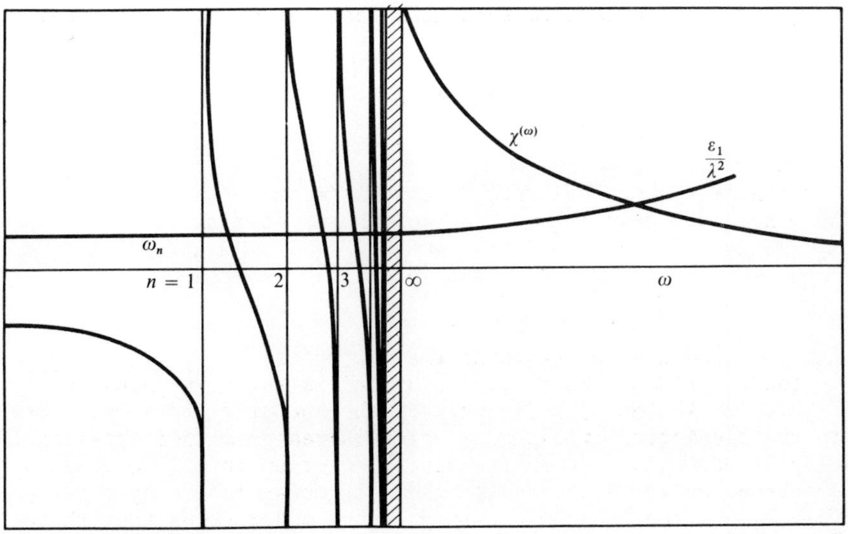

Figure 1 - Graphical solution of a dispersion relation typical of outer atomic shells.

intra-atomic plasmon theory. On the contrary as has been shown in this lecture the treatment is nothing but a simplified effective field treatment in which the Coulomb interactions have been approximated with a constant value. The treatment is equivalent to solve the RPA equations for a factorized kernel and is open to exactly the same criticism as in the other cases when factorization of the interaction has been used.

This review has led up to the recent developments in this field and before going into these developments it may be useful to summarize the experiences obtained by the earlier approaches.

It should then first be mentioned that the idea of an oscillating non-uniform electron gas has not yet been developed into a successful theory. The theory of free oscillations by Bloch does not seem to work and the extensions to a theory of a damped system have not given useful results. The method to calculate directly the response function using results from a uniform electron gas is open to many objections and could hardly lead to a quantitative theory. The more tractable way, which has led to some successful applications, is to start from the single-particle spectrum and calculate the effect of the interactions using the self-consistent field approach (RPA) and extensions thereof. This gives rise to modified excitations having one single particle component and one "collective" component caused by the interactions, and will under appropriate conditions show strong resonances in the continuous part of the spectrum.

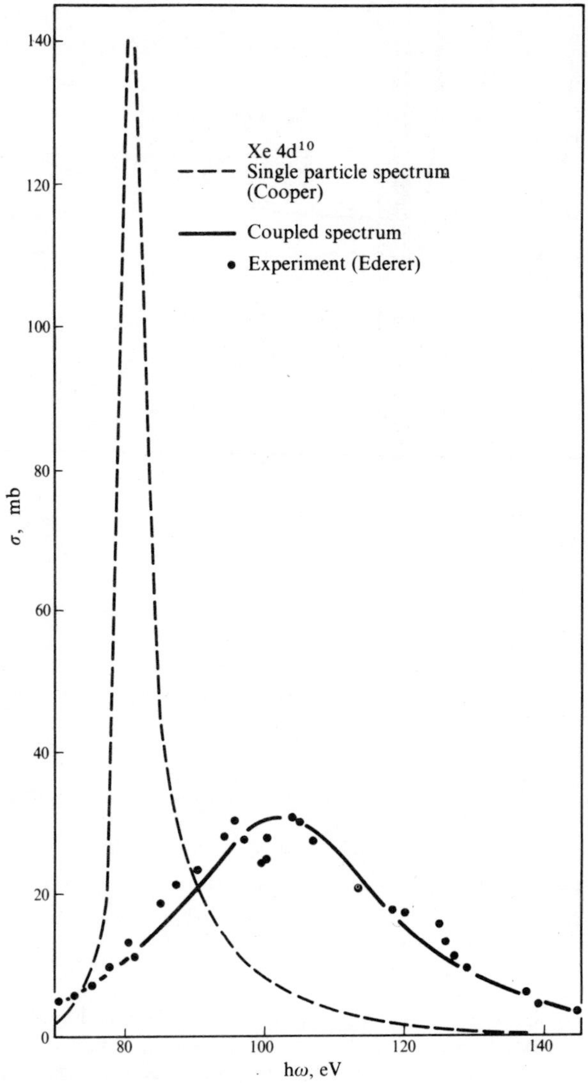

Figure 2 - Photoabsorption of the $4d^{10}$ shell of Xe.

3. THE ONE-ELECTRON SPECTRUM

We summarize a few well-known formulas for later references. The one-electron spectral weight function is defined as

$$A(\mathbf{x},\mathbf{x}',\omega) = \sum_s f_s(\mathbf{x}) f_s(\mathbf{x}')\delta(\omega - \varepsilon_s), \qquad (3.1)$$

where

$$f_s(\mathbf{x}) = \langle N - 1,s|\Psi(\mathbf{x})|N\rangle \tag{3.2}$$

and Ψ $(\mathbf{x})$ is the electron field operator. It gives the probability amplitude for reaching an excited state $|N - 1,s\rangle$ when an electron is suddenly removed from the ground state. The quantity ε_s is the excitation energy

$$\varepsilon_s = E(N) - E(N - 1,s). \tag{3.3}$$

These equations apply for ω smaller than the chemical potential μ. For $\omega > \mu$ we have instead the relations

$$\varepsilon_s = E(N + 1,s) - E(N) \tag{3.4}$$

and

$$f_s(\mathbf{x}) = \langle N|\Psi(\mathbf{x})|N + 1,s\rangle. \tag{3.5}$$

The spectral function obeys the sum rules

$$\begin{aligned} &\int_{-\infty}^{+\infty} A(\mathbf{x},\mathbf{x}',\omega)d\omega = \delta(\mathbf{x} - \mathbf{x}'), \\ &\int_{-\infty}^{\mu} A(\mathbf{x},\mathbf{x}',\omega)d\omega = \rho(\mathbf{x},\mathbf{x}') \end{aligned} \tag{3.6}$$

where $\rho(\mathbf{x},\mathbf{x}')$ is the density matrix.

In the applications it is often practical to introduce some suitable matrix representation, e.g. in terms of Hartree-Fock states, and we then have the relations

$$\int_{-\infty}^{+\infty} A_{kk'}(\omega)d\omega = \delta_{kk'}. \tag{3.7}$$

We can now define a one-electron density of states

$$N(\omega) = \mathrm{Tr}A(\omega) = \sum_k A_{kk}(\omega) \tag{3.8}$$

with the sum rule

$$\int_{-\infty}^{\mu} N(\omega)d\omega = N. \tag{3.9}$$

An example of the spectral weight matrix is given in figure 3.

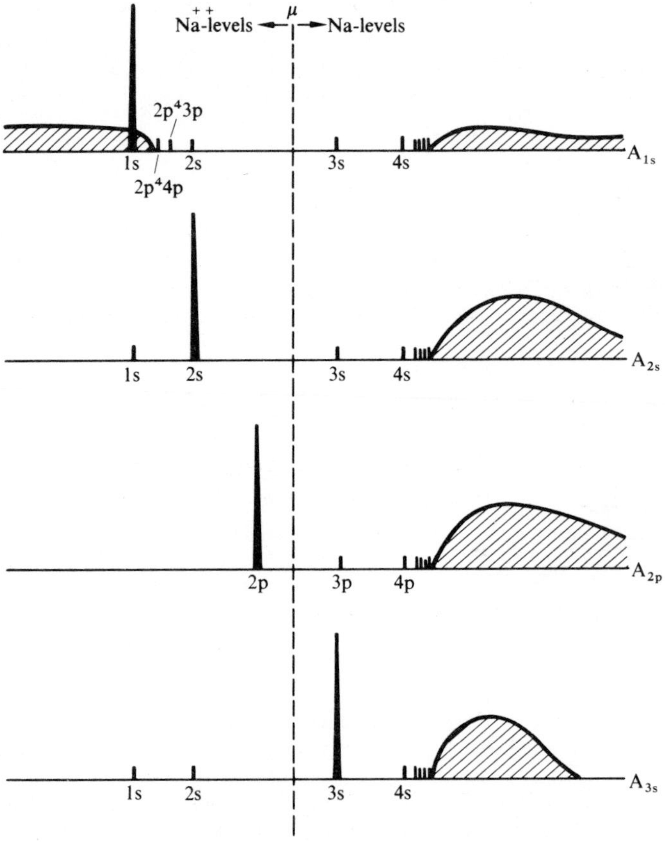

Figure 3 - Spectral weight function A for Na^{+}.

The sum rules give that the oscillator strength on each horizontal line adds up to unity and all strength to the left of the μ-line adds up to N.

In order to calculate the spectrum one has to make an approximation for the self-energy $\Sigma(\mathbf{x},\mathbf{x}'\omega)$ or, equivalently, for the Green function $G(x,x',\omega)$, related through the formulas

$$A(\omega) = \frac{1}{\pi} \operatorname{Im}G(\omega) = \frac{1}{\pi} \operatorname{Im}(\omega - h - \Sigma(\omega))^{-1}. \qquad (3.10)$$

h stands for the kinetic and Coulomb contributions. If we find a representation where the non-diagonal elements of Σ are small, we find for the diagonal elements of A the formula

$$A_k(\omega) = \frac{1}{\pi} \frac{|\mathrm{Im}\Sigma_k(\omega)|}{\omega - \varepsilon_k - \mathrm{Re}\Sigma_k(\omega)^2 + \mathrm{Im}\Sigma_k(\omega)^2} \; . \qquad (3.11)$$

This formula shows how structure in Σ implies structure in A. For Σ we shall at the moment not go beyond the lowest order diagram

Σ = 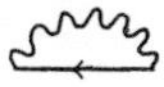

where W = ∿∿∿ is the screened Coulomb interaction and G_0 = ⟵ is the propagator in the independent particle approximation. We have that

$$W = \varepsilon^{-1} v = (1 - vP)^{-1} v \; , \qquad (3.12)$$

where v is the bare Coulomb interaction. The polarization kernel is often approximated by the RPA formula

$$P(x,x') = - iG_0(x,x')G_0(x'x) \; . \qquad (3.13)$$

In some applications one can use the expansion

$$W \simeq v + vPv + \ldots \; ,$$

and retain only the first two terms. The first term gives rise to the Hartree-Fock exchange potential and the second term is the second order dynamical polarization correction. In order to see qualitatively the effect of the dynamical polarization let us look at the second order term which takes the form

$$\Sigma_k^{(2)}(\omega) \simeq \sum_m \frac{V_m^2}{\varepsilon_m + \varepsilon_k - \omega - i\delta} \qquad (3.14)$$

According to equation (3.11) the spectrum takes the form illustrated in figure 4.

The schematic figure shows that one may have a considerable polarization shift of the independent particle level ε_k because of the interaction, with smaller shifts also occurring for the weaker components. This example illustrates how the self-energy operator can be used to find the structure of the spectrum of particles and holes without having to resort to a calculation of the difference between the total energies.

4. THE XPS SPECTRUM (ESCA)

In the X-ray photoemission experiment (ESCA) an energetic

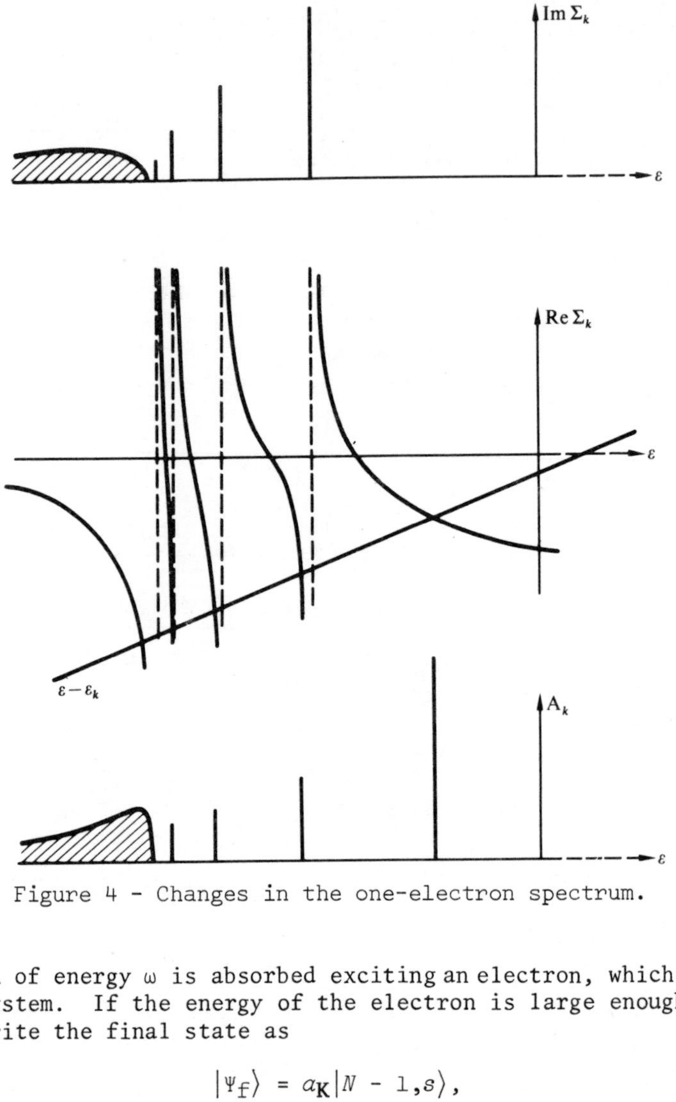

Figure 4 - Changes in the one-electron spectrum.

photon of energy ω is absorbed exciting an electron, which leaves the system. If the energy of the electron is large enough, we can write the final state as

$$|\Psi_f\rangle = a_{\mathbf{K}}|N - 1,s\rangle, \quad (4.1)$$

where K refers to an electron in the continuum with energy $\varepsilon_K = \hbar^2\mathbf{K}^2/2m$ which describes the photoelectron. The probability for this process is given by the Golden Rule,

$$w \sim \sum_f |\langle\Psi_f|P|\Psi_i\rangle|^2 \delta(\omega - E_f + E_i)$$

$$= \sum_{\mathbf{K},s} |\langle N - 1,s|a_{\mathbf{K}} \sum_{kk'} p_{kk'}a_k a_{k'}|N\rangle|^2 \delta(\omega - \varepsilon_{\mathbf{K}} + \varepsilon_s). \quad (4.2)$$

P is the total momentum and $p_{kk'}$ is the matrix element between two one-electron states. For a fast electron we have that

$$a_{\mathbf{k}}|N\rangle \simeq 0,$$

and hence we obtain

$$\omega \sim \sum_{\mathbf{K}} |\langle N - 1,s| \sum_{k} p_{\mathbf{K},k} a_k |N\rangle|^2 \delta(\omega - \varepsilon_{\mathbf{k}} + \varepsilon_s)$$

$$= \sum_{\mathbf{K}} \sum_{kk'} A_{kk'}(\varepsilon_{\mathbf{K}} - \omega) p_{\mathbf{K}k} p_{\mathbf{K}k'} \qquad \text{for } \varepsilon_{\mathbf{K}} - \omega < \mu. \qquad (4.3)$$

The energy distribution of the electrons will be given by the formula

$$I(\varepsilon) \sim \sqrt{\varepsilon} \sum_{kk'} A_{kk'}(\varepsilon - \omega) p_{\mathbf{K}k} p_{\mathbf{K}k'}$$

$$\simeq \sqrt{\varepsilon} \sum_{k} A_{kk}(\varepsilon - \omega) |p_{\mathbf{K}k}|^2 \qquad \text{for } \varepsilon - \omega < \mu. \qquad (4.4)$$

Here we have used that $\varepsilon = \varepsilon_{\mathbf{K}} \sim \mathbf{K}^2$, and in the last step we have dropped the non-diagonal matrix elements. If we finally introduce an average momentum matrix element, we obtain the simple approximate formula

$$I(\varepsilon) \sim \sqrt{\varepsilon} p_{\text{eff}}^{2} N(\varepsilon - \omega). \qquad (4.5)$$

The neglect of non-diagonal terms should be a good approximation if the one-electron basis functions are well chosen.

We shall only discuss the excitation of fairly strongly bound levels in the atomic core. If we neglect excitations of the core electron system we then have that the operator a_k in equation (4.3) must destroy a core electron ($a_k = a_c$).

If we neglect relaxation effects in the core-electron system, the core spectral function has the form

$$A_c(\omega) = \sum |\langle N - 1,s|N - 1\rangle|^2 \delta(\omega - \varepsilon_s). \qquad (4.6)$$

Where $|N - 1\rangle$ is the ground state of the valence electrons *before* the core electron was removed and $\langle N - 1,s|$ is an excited valence electron state *with* the core hole present.

We note that this assumption about no relaxation among the core electrons gives us the same formula as obtained if we use the *sudden approximation in perturbation theory*.

Going back to equation (4.4) or (4.5), we see that there are two quantities of importance for the photo-electron spectrum: the matrix element of the momentum operator and the spectral function A. To say something about the transition moment it is clear that high values of the photo-electron cross sections will occur only when the wave-length of the photo-electron is comparable with the dimensions of the core orbital. This implies for example that the contribution from the valence shells to the photo-electron current is small at high energies because of the wave function mismatch which makes the matrix element small. The matrix elements obviously depend of course also on the angular symmetry of the core levels. However, since these practically important questions relating to the matrix elements depend only on the one-electron states involved, we shall not discuss these problems further.

Going back to the other key feature, the spectral weight function A, we recall the brief discussion given about a typical example in figure 4. This shows how the one-particle core level ε_K shifts appreciably to the right whereas the other levels undergo smaller shifts. A beautiful example of an experimental result is given in figure 5, which shows a recent spectrum from Uppsala [5].

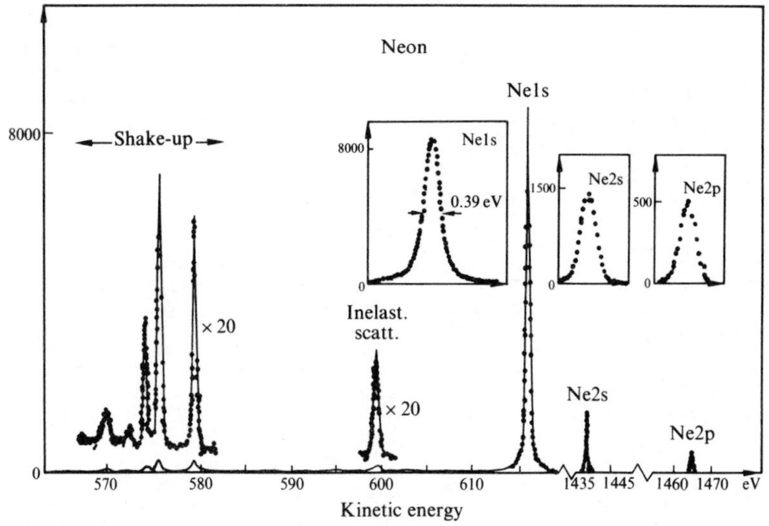

Figure 5 - ESCA spectrum of neon.

The strong line is the Ne 1*s* and the line seen is the one corresponding to *complete relaxation around the hole*, i.e. corresponding to the shifted line in the schematic illustration in figure 4.

To the left in figure 5 is a group of levels which corresponds to the other group of levels in figure 4. They correspond to excited states of the (N - 1)-particle system and are generally called *shake-up levels*. They occur with much smaller intensities.

It is of interest to compare this type of spectrum which is characteristic of an atom with the case of simple metals where the first study of the spectrum was given by B. Lundqvist [*6*], using a simple approximation for the self-energy with dynamically screened interaction. In that case one had a continuous spectrum with the level of complete relaxation around the hole as an edge. Near the edge there are contributions from particle-hole excitations and further away one finds plasmon resonances. Thus the shake up lines in an atom are in a metal replaced by electron-hole excitations and plasmons, and in both cases the state of complete relaxation gives the strong peak in the spectrum.

According to the simple model in figure 4 all discrete levels shift away from the independent particle levels ε_K, and in particular the strongest peak in the spectrum shifts by several eV from its value in a Hartree-Fock theory. The interpretation of the spectrum seems to have caused some confusion, because of worries about the time scale for the process, energy conservation etc. The key to this lies in the observation that only the full spectrum $A_c(\omega)$ should be used in these arguments. It was first found by B. Lundqvist [*6*] that the *average energy* in the spectrum, i.e. the centroid of the spectrum, coincides with the energy calculated for a frozen core energy. Thus the one-electron energy of the electron which was sitting in the core is never seen in the spectrum, but it corresponds to the average energy of the photo-electron spectrum. This makes it possible to express the polarization shift relative to the Hartree-Fock energy in terms of the shake up levels and their intensities. The polarization shifts imply that the ionization potentials will differ appreciably from the Hartree-Fock values for core levels. We refer to a paper by L. Hedin and A. Johansson [*7*] for a further discussion of this point.

We have to leave out most of the theoretical question but would like to comment briefly on the problem about the physical rearrangement, which was raised already by Pauli [*8*] in a famous review. We have assumed that the energy is sufficiently high and that we can neglect final states interaction, or in other terms that the process takes place so fast that the sudden approximation is applicable. When we lower the energy, the time of removal will be larger and if we could remove the electron very slowly we would finally approach the adiabatic limit. It is intuitively clear how the spectrum would depend on removal time. By increasing the removal time we would start cutting down the probability of shake-up levels with highest energies and when approaching the adiabatic limit, we would only see the strong line corresponding to complete relaxation and the whole shake-up spectrum should have disappeared. A theoretical discussion using finite removal time has been given by H.W. Meldner and J.D. Perez [9]. T.A. Carlson and his group at Oak Ridge has made extensive experimental studies of this effect. In the case of Ne their results [10] show that the satellite lines are independent of the energy until the region around 100 eV above threshold, where the shake-up spectrum drops to about 50% of its maximum value. It should be added that the method to simulate a slow outgoing electron by switching on the hole gradually rather than instantly has its obvious limitation. This problem seems to require the full machinery of many-body theory including final state interactions (vertex effects) and self-energy terms for the outgoing electrons should also be included. The problems arising in this context are similar for atoms and solids and are on the whole not very well understood at the present time, although one sees qualitatively how high energy satellites tend to be suppressed when the energy of the emitted electron is decreased.

5. INTRODUCTORY REMARKS ABOUT THE PHOTOABSORPTION SPECTRUM

We refer to the review by Fano and Cooper [11] for a comprehensive discussion of the theory of photoabsorption up to 1968. Here we want to emphasize some particular aspects of the many-electron interactions. In order to avoid complications with degeneracies we shall only discuss atoms having closed shells. It was mentioned in an earlier section that correlation effects on ground state properties are generally rather small. Let us use the experimental data as a guide and compare the experimental cross-section $\sigma(\omega)$ with that in the single-particle approximation, $\sigma_0(\omega)$. The ratio between the two gives a measure of the many-electron contribution. The ratio can also be given an intuitive physical meaning as follows. In an RPA description the field acting on given electron is not the external field but an effective field screened by the induced motion. Let us introduce for the moment a screening function $\sigma(\omega)$, or "dielectric function", defined by the formula

$$\sigma(\omega) = \frac{\sigma_0(\omega)}{|\mathcal{H}(\omega)|^2} . \qquad (5.1)$$

The function $|\mathcal{H}(\omega)|^2$ gives a measure of the importance of the energy-electron interactions if both $\sigma(\omega)$ and $\sigma_0(\omega)$ are known.

The function $|\mathcal{H}(\omega)|^2$ in the region of the p^6 shell absorption in Ne, Ar, Kr and Xe are shown in figure 6, taken from a paper by Wendin [12]. The interpretation of $\mathcal{H}(\omega)$ as a dielectric function should not be taken too seriously. Nevertheless the curves show a significant deviation from the single particle behaviour and exhibit a typical region of strong enhancement of the absorption, reminiscent of a tendency to produce a resonant state without really succeeding. I believe these plots should convince everyone that the interactions do indeed produce strong effects in the absorption of these shells. For the heavier

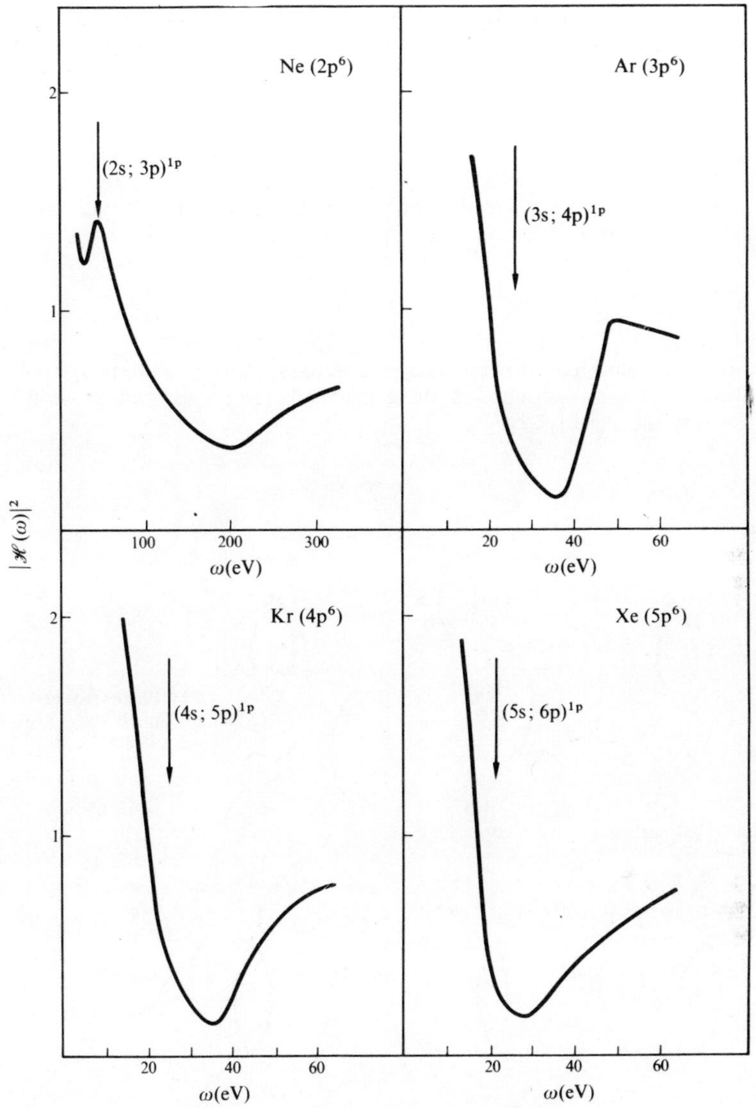

Figure 6 - The dielectric function $|\mathcal{H}(\omega)|^2$ for Ne, Ar, Kr and Xe.

atoms there is a region where the single particle absorption account for only about 20% of the total cross sections and with such strong effects it is evident that the more powerful methods of many-body theory need to be used.

Amusia [*13*] and collaborators have used a front approach to solve the integral equation numerically for the response function and obtained good agreement. Starace [*14*] has developed an approximate many-body scheme which seems to work rather well. The recent work by Wendin [12] carries the theoretical discussion considerable further and our discussion will present the main arguments and results in his thesis.

6. A FORMULA FOR THE PHOTOABSORPTION CROSS SECTION

For photons of long wave-lengths the oscillator strength distribution is given by

$$g(\omega) = \frac{\omega}{\pi} \operatorname{Im}\alpha(\omega), \tag{6.1}$$

where $\alpha(\omega)$ is the polarizability of the system. The perturbation series can be expressed as a sum of diagrams and some of them are shown in figure 7.

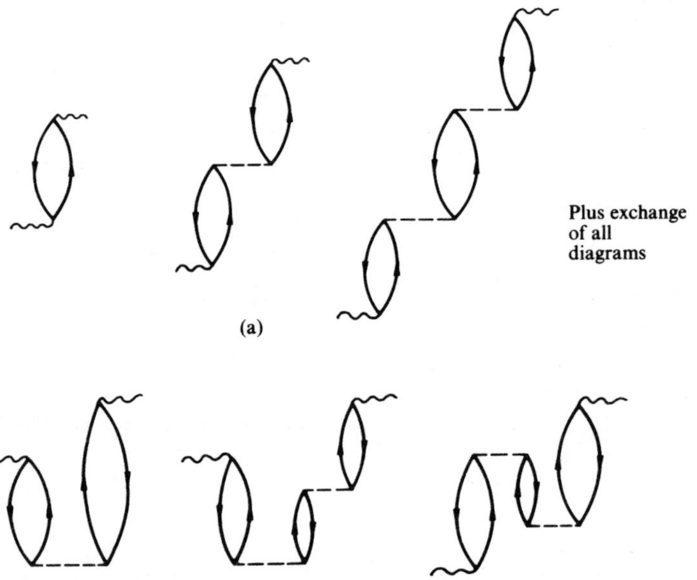

Figure 7 - The diagram expansion corresponding to the random phase approximation (RPA).

The notation is on the hole standard. The wavy lines at the bottom and top of each diagram represent the coupling of the external photon. The single particle states are calculated with a modified Hartree-Fock scheme, where the occupied (hole) states are calculated with the usual self-consistent procedure. The excited states are then calculated with the Coulomb and exchange potential of the frozen core minus one electron in the shell under consideration.

For simplicity we consider here only excitations from a single shell of N equivalent electrons of angular momentum ℓ into a single particle-hole channel. We obtain the expansion for the spectral function $\mathrm{Im}\alpha(\omega)$ by taking the imaginary part of every diagram for the polarizability $\alpha(\omega)$.

We start by separating the real and imaginary parts of each unperturbed pair propagator. For simplicity we consider here only RPA diagrams; the extension to the full expansion can be made later.

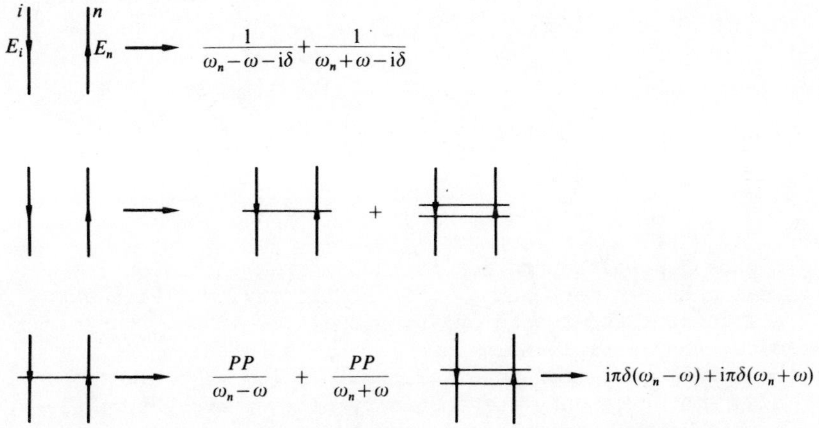

Figure 8 - Notations used in the treatment of the unperturbed particle-hole pair propagator. One bar represents the real part, two bars the imaginary part. $\omega_n = E_n - E_i$; ω, photon energy; δ, infinitesimal and positive.

In figure 9 we show the decomposition of the diagram expansion. In figure 9 we have extracted all the diagrams containing only imaginary parts in all propagators. The δ-functions lock all these contributions to the frequency of the external field so that these terms only give contributions *on the energy shell*. These terms form a geometrical series which can be summed. The remaining diagrams in figure 9a can be generated from the diagrams of figure 9b if the external coupling p_ω and internal matrix element $V_{\omega\omega}$ are replaced by the renormalized couplings $p_\omega(\omega)$ and $V_{\omega\omega}(\omega)$, defined by the diagram expansion in figure 10. The two series of diagrams can be written in closed form as integral equations. We have to refer to the thesis by Wendin for lots of technical details and give only the resulting sets of equations.

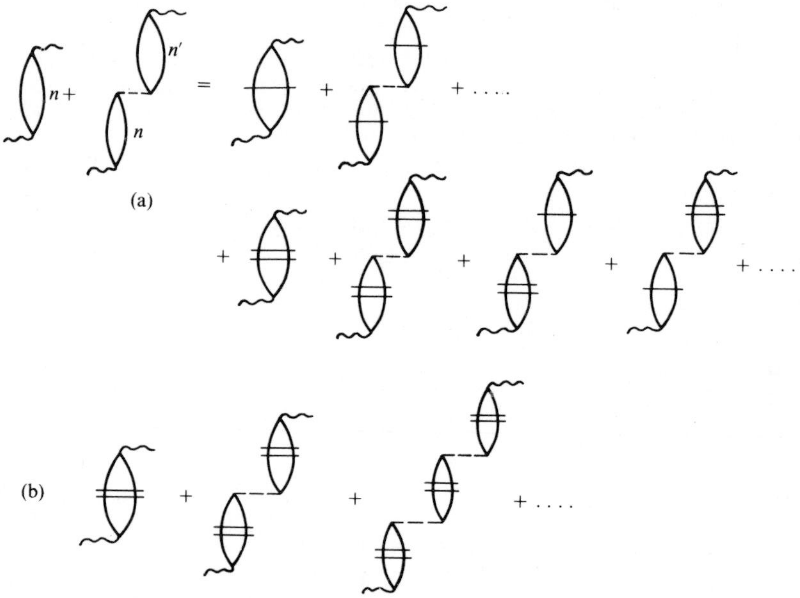

Figure 9 - (a) Separation of the diagram expansion into real and imaginary parts. For simplicity, we leave out all indices and arrows on the propagator lines. Furthermore, the particle-hole interaction is taken to include both RPA and ladder interactions. In (b) we collect the diagrams where only the imaginary part of every particle-hole pair propagator has been retained.

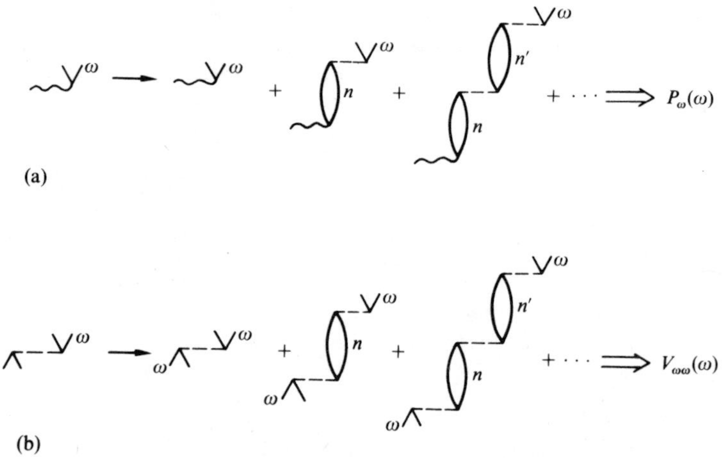

Figure 10 - Diagrams representing: (a) screening of the external perturbation, (b) screening of the internal particle-hole pair interaction.

$$\mathrm{Im}\alpha(\omega) = \frac{1}{3}\,\frac{\ell + 1}{2\ell + 1}\,\frac{N\pi P_{\omega}^{2}(\omega)}{1 + \pi(N - 1)V_{\omega\omega}(\omega)}\,,$$

$$P_{\omega}(\omega) = P_{\omega} - (N - 1)\mathop{\sum\!\!\!\!\!\!\int}_{n}\left[\frac{P_{n}V_{n\omega}(\omega)}{\omega_{n} - \omega} + \frac{P_{n}V_{n\omega}(\omega)}{\omega_{n} + \omega}\right], \qquad (6.2)$$

$$V_{n\omega}(\omega) = V_{n\omega} - (N - 1)\mathop{\sum\!\!\!\!\!\!\int}_{n'}\left[\frac{V_{nn'}V_{n'\omega}(\omega)}{\omega_{n'} - \omega} + \frac{V_{nn'}V_{n'\omega}(\omega)}{\omega_{n'} + \omega}\right].$$

$\sum\!\!\!\!\!\!\int$ indicates summation over discrete labels and integrations over continuous variables and that the principal value should be taken. Note that the effective interactions $V_{n\omega}(\omega)$ and coupling $P_{\omega}(\omega)$ to the photon both are *completely real quantities*. $V_{\omega\omega}(\omega)$ contains the screening of the bare particle-hole interactions due to the virtual fluctuations in the system and $P_{\omega}(\omega)$ includes the analog screening of the coupling to the photon.

Some remarks should be made about these formulas which permit a possibility of a more detailed analysis.

(1) Infinite order methods are necessary in many cases because the original perturbation expansion is often *strongly divergent*.

(2) The damping of excitations provides no difficulty, because the procedure is a direct calculation of the absorption.

(3) We notice that only interactions on the energy shell, $V_{\omega\omega}(\omega)$, occur in $\mathrm{Im}\alpha(\omega)$. This makes it possible to define an effective dielectric function $\mathcal{H} = \bar{\mathcal{H}} + i\bar{\bar{\mathcal{H}}}$, where $\bar{\mathcal{H}}(\omega) = V_{\omega\omega}/V_{\omega\omega}(\omega)$ (and $\bar{\bar{\mathcal{H}}} = \pi(N - 1)V_{\omega\omega}(\omega)$). Thus $\bar{\mathcal{H}}(\omega)$ describes the real screening of the particle-hole interactions, and is a measure of the enhancement due to possible phase coherence within the shell at frequency ω.

(4) The proceeding remark gives a possibility to discuss the atomic resonances. The condition for a resonance in the interaction is that $\bar{\mathcal{H}}(\omega) = 0$ and the imaginary part $\bar{\bar{\mathcal{H}}}(\omega)$ determines the damping of the resonance.

(5) The formulation includes as a special case the ordinary solution for undamped collective resonances, and gives e.g. the standard RPA result for plasmons, if applied to an electron gas.

(6) One should note that we are dealing with two infinite summations in this formulation. One is the energy shell contribution obtained by the first summation. The second is the summation leading to the effective internal

interaction $V_{\omega\omega}(\omega)$. As was just discussed it is a divergence in the second quantity which gives rise to resonances (more or less damped) in the spectrum. It may however often happen that we have no resonance but that the first summation on the energy shell is divergent. In this case we still have a situation where infinite order theory is needed; the absorption goes into a many-electron excitation of the system but without producing a resonance. The system behaves in analogy to an overdamped oscillator.

7. BEYOND THE RPA APPROXIMATION

The theory just sketched can be extended far beyond the RPA and we refer to the papers by Wendin for an extensive discussion. Here we want to make just some remarks to bring the treatment on line with the discussion of the ESCA spectrum. For excitations into the continuum it is of considerable importance to include the self-energy correction to the hole energy in a similar way as we discussed for the ESCA spectrum. This means shifting the core energy to a somewhat *higher* energy and this has a tendency of shifting the absorption spectrum to a somewhat *lower* energy. In experiments in the metallic state this shift is of key importance to include but it is always important when the hole state is well localized. As an example we can take the $4d \to nf$ excitations in Xe. When creating a hole in the $4d^{10}$ shell we will have a relaxation in the $5p^6$ shell. The first two orders are illustrated in figure 11.

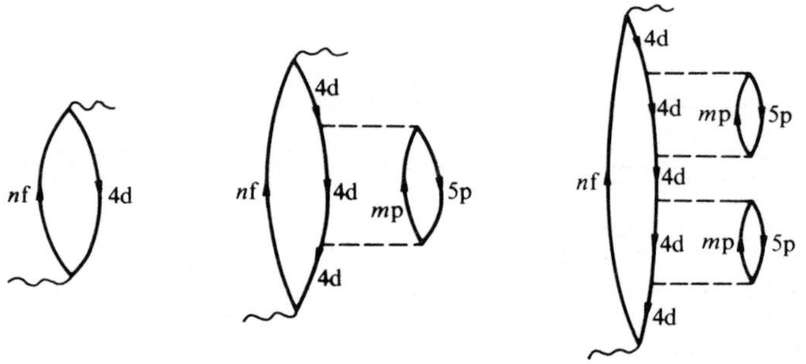

Figure 11 - Corrections describing relaxation effects.

In the rare earth series one has to consider the transitions $4d \to 4f$. In this case the self-energy correction also to the particle line should be included. However, the screening due to a hole in the d and the f level may be sufficiently equal that the difference may be neglected to a first approximation.

The diagrams shown in figure 11 should of course be summed to infinite order. If we only include diagrams connected with the hole line, this is equivalent to replacing the unperturbed hole propagator by the propagator including interactions. This brings in the full hole spectrum which we discussed in connection with the photoemission spectrum. In particular it contains the resonances which appear as shake up lines in the photoemission spectrum. This implies that these resonances will also cause characteristic effects in the photoabsorption spectrum where they show up as *autoionizating resonances*. Thus there is a one to one correspondence between the shake-up lines in the ESCA spectra and the autoionizating resonances in photoabsorption. For energies closer to threshold it is necessary to include vertex corrections as well as self-energy corrections for the outgoing electron. As in the case of the ESCA spectrum the qualitative effect of the vertex correction will be to introduce interference effects which will reduce the strength of the satellite structure. For a discussion of the effects of vertex corrections particularly in photoemission we refer to a recent paper by Langreth [*18*].

One should also remark that one channel approximation called OCA by Wendin is only an approximation. For numerical agreement it is necessary to go beyond this treatment and superimpose the spectrum of other channels contributing in the same frequency region.

8. THE SPECTRUM OF THE RARE GASES

The theory was first applied to the case of *He* where the result shows that some strength is taken from the discrete line spectrum and shifted into the continuum. In the continuum there is a shift of the oscillator strength towards higher frequencies. Thus already the lightest rare gas atom shows some of the effects, which show up more pronounced for the heavier atoms. For He of course the perturbation expansion is convergent and the corrections are only in the 10-20% range.

We skip the medium light rare gases Ar and Kr. The next interesting case is the $5p^6$ shell in Xe. Many-electron effects are here so important that perturbation theory diverges in an energy interval above the ionization threshold.

Infinite order methods have to be used, but the real part of the effective dielectric function never goes through zero. In a way the $5p^6$ shell behaves as an overdamped system. In figure 12 we show some results for the photoabsorption cross section, where the experimental curve gives the total cross section, whereas the theoretical calculations refer to the $5p$-nd channel. The many-electron properties of the absorption of the $5p^6$ shell in Xe can be discussed in terms of the effective screening constant. We defined earlier a screening function which described the internal screening of the pair interactions. We could also define a screening function which represents the screening of the dipolar coupling to the external field through the relation

$$P_\omega(\omega) = P_\omega/\mathcal{H}'(\omega). \tag{8.1}$$

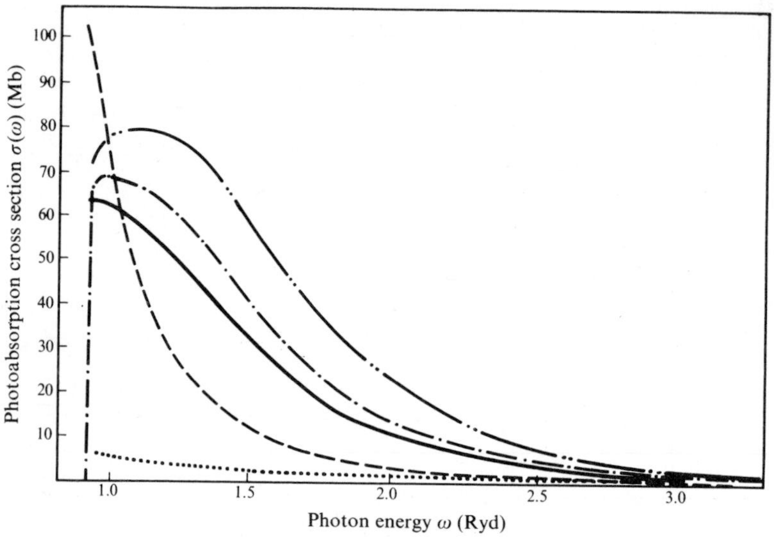

Figure 12 - Photoabsorption cross sections for the $5p \rightarrow nd$ channel in Xe. Single particle $5p \rightarrow ns$: ••••; Single particle $5p \rightarrow nd$: — — — — ; Theory: — • — • — • — •; Approximate theory (ground state) Experiments:, ——

That these functions differ relative to each other and from the full dielectric matrix is obvious from the way they have been defined. In figure 13 we give the results for the real and imaginary parts $\bar{\mathcal{H}}(\omega)$ and $\bar{\bar{\mathcal{H}}}(\omega)$ the absolute magnitude $|\mathcal{H}(\omega)|^2 = \bar{\mathcal{H}}(\omega)^2 + \bar{\bar{\mathcal{H}}}(\omega)^2$ and $\bar{\mathcal{H}}'(\omega)$, defined by equation (8.1).

In this case the many-body absorption is very strong so that the imaginary part of the perturbation expansion is divergent. However, as is shown by figure 13 the field due to the fluctuations gives rise to an enhancement but not to a resonance. Thus the many-body character of the absorption is demonstrated but there is no resonance behaviour. We note that the screening of the internal interactions described by $\bar{\mathcal{H}}(\omega)$ and the one representing screening of the external perturbation, $\mathcal{H}'(\omega)$, behave differently. The enhancement of the external coupling is numerically stronger than that of the internal interactions. No explanation has been given for this effect.

The most interesting case is the absorption of the $4d^{10}$ shell of Xe, which was mentioned in section 2. This problem has been earlier treated in the work earlier referred to [*3*,*12*-*14*]. An application of the theory outlined in section 6 using the extended RPA theory gave the results given in figure 14. The agreement is rather satisfactory considering the approximations involved. In this case both the on the energy shell contributions and the off the energy shell contributions diverge. The behaviour is illustrated in figure 15 which gives the quantities $\bar{\mathcal{H}}(\omega)$ and $\bar{\bar{\mathcal{H}}}(\omega)$

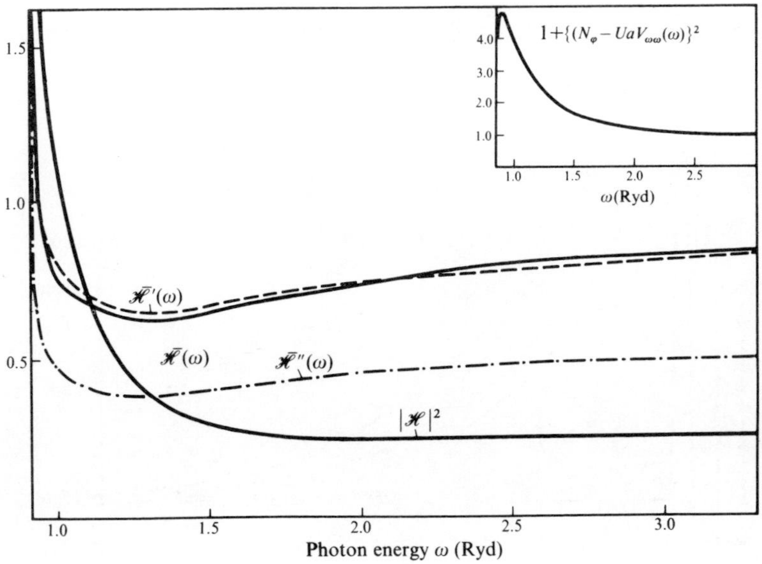

Figure 13 - Various screening functions representing the response of the $5p^6$ shell in Xe.

At the first zero of $\bar{\mathcal{H}}(\omega)$ the imaginary part is very large. At the second zero of $\bar{\mathcal{H}}(\omega)$ the damping is much smaller but nevertheless broad enough that we see only a rather broad peak which we might interpret as a broad giant resonance. This corresponds to a dipolar vibrational mode of the system but it is fairly strongly damped. We should note that the resonance has a total oscillator strength close to 10 electrons.

The strength of the theoretical methods developed by Wendin is that the calculations can be carried quite a step beyond the RPA. We have mentioned briefly the importance of accounting for the relaxation around the hole in the $4d^{10}$ shell, an effect which we found to be quite important in the ESCA spectrum. Furthermore he has shown how to relax the one channel approximation to include also the $4d \to np$ transitions.

Due to purely technological difficulties, to deal with self-energy corrections etc., the higher order calculations by Wendin did not cover the entire range of frequencies. The interesting part is the tail of the profile and in particular the low energy tail where there is a marked discrepancy between the RPA and the experimental results. In figure 16 we reproduce a comparision between experiments and theory in different approximations.

These results demonstrate the power of the theoretical method and show that the physically important modifications beyond the extended RPA can be treated in a systematic way. In particular this calculations shows the necessity to account for the relaxation around the hole in the $4d$ shell in order to

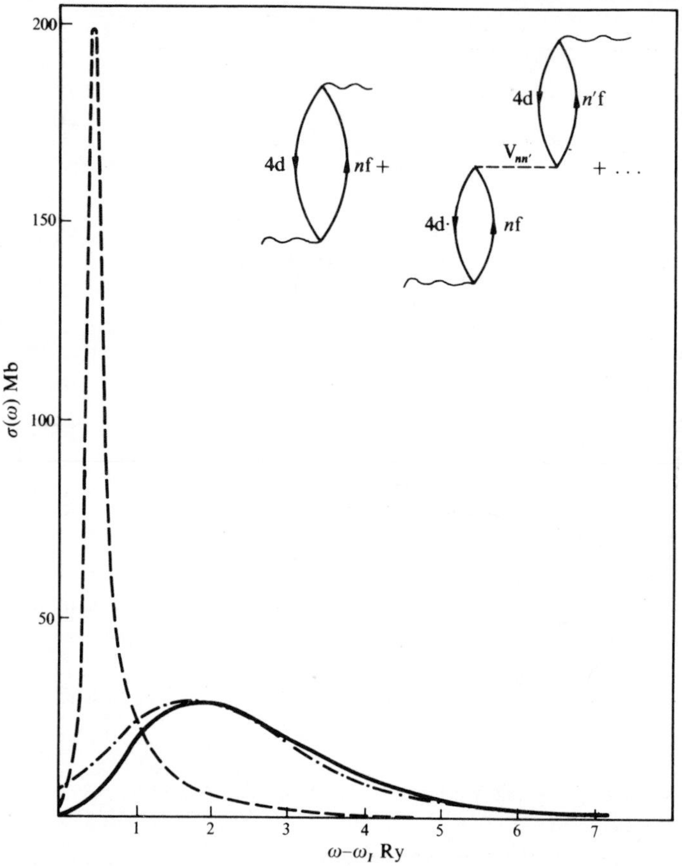

Figure 14 - Photoabsorption cross section for the $4d^{10}$ shell in Xe. Single particle approximation (first diagram): — — — — ; RPAE (sum of diagrams to infinite order): ——; Experiment: — • — • — • — • —.

obtain a good agreement below the absorption maximum.

In conclusion it seems that the theory seems to explain the properties of the rare gas atoms quite well. For He, Ar and Kr the extended RPA seems adequate; there are moderate to strong many-electron effects and the contributions on the energy shell converge for the lighter atoms but diverge for the heavier ones, thus emphasizing the need to use infinite order methods. For the $4d \rightarrow nf$ transitions also the fluctuations give rise to divergences and the effective dielectric function goes through zero, however with a rather strong damping part. The extended RPA accounts well for the gross features. However, in order to get good agreement it is necessary to go beyond the RPA and in particular the relaxation around the $4d$ hole seems to be quite important to get agreement with the experimental data.

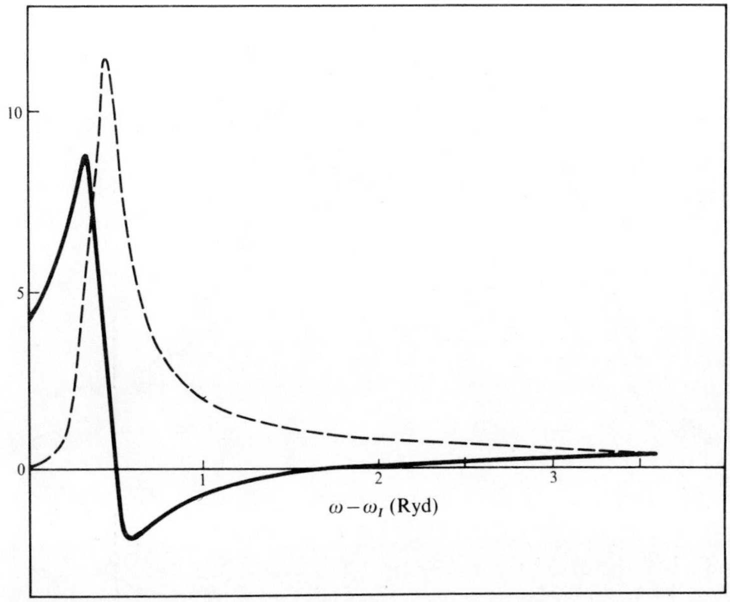

Figure 15 - Real and imaginary parts of the screening function.

9. MANY-ELECTRON RESONANCES IN THE RARE EARTH METALS

Recent experiments show that the photoabsorption of the $4d^{10}$ shell in the rare earth metals strongly peaks about 15-20 eV above the $N_{\mathrm{IV,V}}$ threshold, with a width in the range 10-20 eV. A survey over the data is shown in figure 17 [*17*].

In the light of our discussion of the results for Xe, we can physically understand the key features of the spectrum. In the case of Xe we studied the $4d \to nf$ transitions where the f levels form a virtual state in the continuum such that the zero order spectrum is strongly peaked about 80 eV excitation. When we increase the nuclear charge and go into the rare earth series the f states become bound and gradually filled with electrons. Thus the dominating process is now the $4d \to 4f$ between *discrete* states. However, as in the case of Xe, the interaction shifts the spectrum to higher energies and pushes the transition up into the continuum. The gradual decrease in the total oscillator strength is due to the gradual filling of the f-level which reduces the available transition strength.

Recent yet unpublished calculations by Wendin for Ba and La give excellent confirmation that these peaks should be interpreted as giant resonances, corresponding to the vibrations of the entire d shell. As in the case of Xe the perturbation

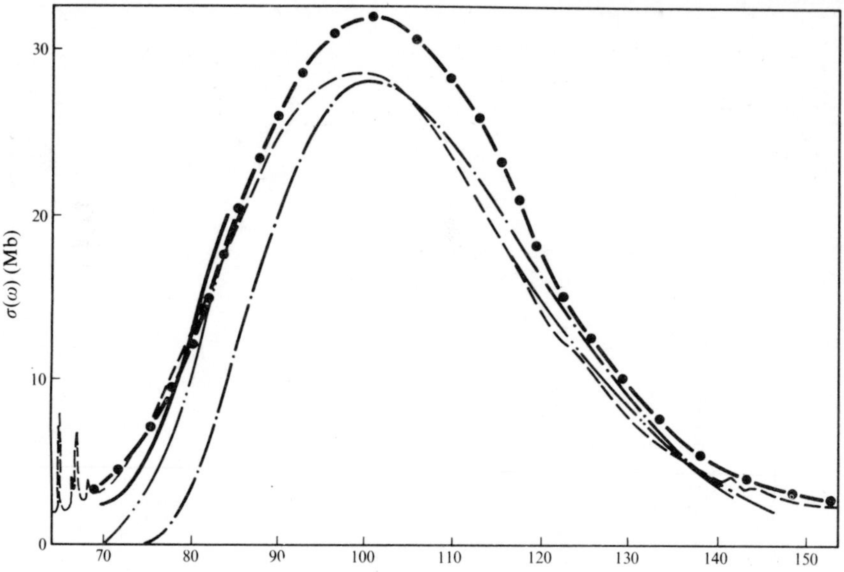

Figure 16 - Absorption of the $4d^{10}$ shell in Xe. Experimental result by Haensel *et al.*[15]: — — — —; Experimental results by El-Sherbini and van der Wiel: — × — × — × —; One channel approximations (OCA): — • — • — • —; OCA with core relaxation: — •• — •• — •• —; OCA with core relaxation on a back-ground of pair excitations $4d \rightarrow np$: ———.

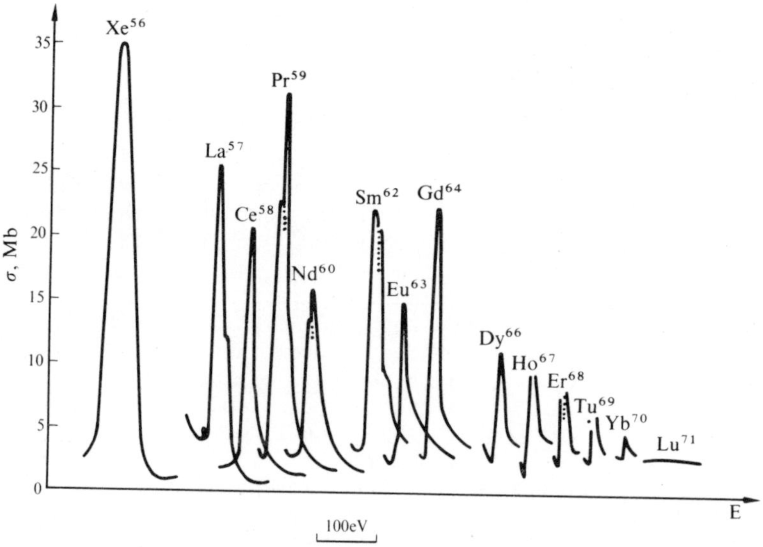

Figure 17 - Absorption of the d-shell in rare earths.

expansion is strongly divergent, the matrix for the effective interaction has a pole and the dielectric function goes through zero. The width is considerably smaller for the rare earth metals than for the case of Xe. The results for Ba and La are in excellent agreement with the experimental data.

The experiments are done in the metallic state, and therefore the calculation is not entirely atomic. The atomic 6*s* (5*d*) wave function becomes strongly compressed when forming the metallic conduction bands and therefore the core levels will shift upwards, and the conduction electrons are more effective in relaxing around the 4*d* than the atomic 6*s* (5*d*) electrons. Because of this it would be of great interest to have the absorption cross sections for some metal vapours. This should unambiguously show what are the true atomic effects and should show what features are true solid state effects.

It should be mentioned that Wendin has also studied the absorption of the $5p^6$ shell in Ba. The results indicate a strong collective effect and predict a very pronounced structure in the wavelength region 550-650 Å.

10. FINAL REMARKS

We should again mention that the ESCA and photoabsorption spectra are related for processes where an electron comes out with high enough energy and we neglect inelastic processes on its way out. In the case of the ESCA spectrum it is absolutely necessary to renormalize the hole propagator to account for the relaxation around the hole. However, the calculations by Wendin show that it is equally necessary to renormalize the hole propagator to get agreement for the absorption spectrum. Therefore in terms of diagrams the XPS and photoabsorption are closely related. Physically this means that structure revealed in XPS could also show up in the photoabsorption. This is particularly true for the shake-up levels in the ESCA spectrum which come from the additional peaks in the hole spectral function. In the photoabsorption each shake-up line should show up as an *autoionizing resonance*. Experimentally this has not been observed for technical reasons. The shake-up spectrum is quite weak and is seen most easily in the spectrum of rather deep core levels, where the intensity is high. For the Ne 1*s* spectrum the corresponding effects should occur around 500 eV, which is a fairly high energy at which the photoabsorption spectrum has not yet been observed.

Finally we should mention the solid state effects. For Xe the absorption of the 4*d* shell has been studied both in the atomic and the solid state by Haensel *et al.* [*15*]. The results show that the gross behaviour in the solid is the same as the atomic but there is a characteristic structure with oscillations, additional peaks, etc. which is due to the solid state environment. In the case of the rare earth metals mentioned earlier, the metallic environment seems to give rise to some interesting effects.

The experimental and theoretical work for the rare gases and rare earths seem to indicate an area where further work is needed both to understand the dynamical properties of atoms better in the high frequency regime and also to understand how these properties are modified in a solid state environment.

The final question to be raised is: Where is the atomic plasmon? We mentioned in the introductory remarks that the applications of the electron gas theories have been unsuccessful so far to treat the oscillations of an atom. A major feature is certainly the nature of the one-electron spectrum and we have seen in the examples that only when sufficient one electron oscillator strength bunches together in a narrow region, the interaction will be effective in producing a considerable shift and a resonance in the system. Qualitatively we have three major ingredients: (a) the characteristic single particle energies involved (b) the many-electron shift which is a measure of the collective behaviour and finally (c) the damping of the resonance. It is only the shift (b) which could be related to a plasma frequency and, indeed, numerically the shift is of the order of an average plasma frequency for the absorbing shell. This idea could be developed further along similar lines as discussed by Professor March in his lectures. In fact if one writes down the equation for motion for the second time derivative for small vibration around the equilibrium density one finds two characteristic contributions in a linear approximation: (a) one instantaneous response which is determined by the local plasma frequency; this term is the classical contribution; (b) one retarded response determined by the energy spectrum and wave functions for the system. Such a formulation might be used to look further into the problem of the atomic plasmon and how the single-particle and collective aspects are related.

REFERENCES

1. Bloch, F. (1933). *Z. Phys.*, **81**, 363.
2. Jensen, H. (1937). *Z. Phys.*, **106**, 620.
3. Brandt, W., Ederer, D. L. and Lundqvist, S. (1967). *J. Quant, Spectrosc. Radiat. Transfer*, **7**, 185.
4. Cooper, J.W. (1964). *Phys. Rev. Lett.*, **13**, 762.
5. Siegbahn, K. et al. (1969). *ESCA — Applied to Free Molecules*, (North-Holland, Amsterdam).
6. Lundqvist, B.I. (1969). *Phys. Kondens. Mater.*, **9**, 236.
7. Hedin, L. and Johansson, A. (1969). *J. Phys.*, **B2**, 1336.
8. Pauli, W. (1933). *Handbuch der Physik, Vol. 24.* (Springer, Berlin). p. 1.
9. Meldner, H.W. and Perez, J.D. (1971). *Phys. Rev.*, **A4**, 1388.
10. Carlson, T.A. and Krause, M.O. (1965). *Phys. Rev.*, **A140**, 1057. There are many other papers by this group of great interest for these problems.
11. Fano, V. and Cooper, J.W. (1968). *Rev. Mod. Phys.*, **40**, 441.
12. Wendin, G. (1973). (Dissertation), (Chalmers University of Technology), which includes the following papers, (1970).

J. Phys., **B3**, 455, 466; (1971). *J. Phys.*, **B4**, 1080; (1972). *J. Phys.*, **B5**, 110; (1973). *J. Phys.*, **B6**, 42.

13. Amusia, M.Ya., Cherepkov, N.A. and Chernysheva, L.V. (1970). *Phys. Lett.*, **A31**, 553; (1971). *Sov. Phys., JETP*, **33**, 90.
14. Starace, A. (1970). *Phys. Rev.*, **A2**, 118.
15. Haensel, R., Keitel, G. and Schreiber, P. (1969). *Phys. Rev.*, **188**, 1375.
16. El-Sherbini, Th.M. and Van der Wiel, M.J. (1972). *Physica*, **62**, 119.
17. Zimkina, T.M., Fomichev, V.A., Gribovskii, S.A. and Zhukova, I.I. (1967). *Sov. Phys. Solid State*, **9**, 1128.
18. Langreth, D. *Proceedings of Nobel Symposium XXIV on Collective Properties of Physical Systems*, (Almqvist and Wiksell, Uppsala), (in press).

SOFT X-RAY SPECTRA OF MOLECULES†

A. BARRY KUNZ

Department of Physics and Materials Research Laboratory, University of Illinois, Urbana, Illinois 61801

1. THE THEORETICAL MODELS

The basic starting point for our discussion is to be the Spin-Polarized-Hartree-Fock method [*1*] (SPHF). In this model, the electronic wavefunction is assumed to be a single Slater determinant of one-particle wavefunctions. The Hamiltonian of the system is given to be

$$\mathcal{H} = -\sum_i \nabla_i^2 - \sum_{i,I} \frac{2Z_I}{|\underset{\sim}{r}_i - \underset{\sim}{R}_I|} + \tfrac{1}{2}\sum_{i,j}{}' \frac{2}{|\underset{\sim}{r}_i - \underset{\sim}{r}_j|} + \tfrac{1}{2}\sum_{I,J}{}' \frac{2Z_I Z_J}{|\underset{\sim}{R}_I - \underset{\sim}{R}_J|}. \quad (1)$$

In obtaining this Hamiltonian, one must make one of two approximations. Either one must separate out the nuclear motion by the Born-Oppenheimer approximation or one must assume the nuclei to be infinitely heavy [*2*]. The corrections to the theory due to the kinetic energy of the nuclei are beyond the scope of the present lecture. In equation (1), an atomic system of units is used. Here $e = \sqrt{2}$, $\hbar = 1$, $m_e = 0.5$ and the unit of energy is the Rydberg (1 Ry ≈ 13.6 eV), the unit of the length is the atomic unit (1 a.u. = 1 bohr radius ≈ 0.53 Å). The upper case letters refer to nuclear coordinates and properties, while the lower case letters refer to electronic coordinates and properties.

The usual starting point is with the variational thecrem. That is one assumes,

---

† Work supported in part by the National Science Foundation under Grant GH-33634 and by the Aerospace Research Laboratory, Air Force Systems Command, USAF, Wright-Patterson AFB, Ohio, Contract No. F33615-72-C-1506.

$$\delta \frac{\int \psi^* \mathcal{H} \psi d\tau}{\int \psi^* \psi d\tau} = 0. \tag{2}$$

If the choice of ψ is allowed to be completely free, that is for a system of n electrons if one permits

$$\psi \to \psi(\vec{r}_1, \vec{r}_2, \ldots, \vec{r}_n), \tag{3}$$

on has the usual Schrödinger equation

$$\mathcal{H}\psi = E\psi. \tag{4}$$

This equation is unfortunately too difficult to solve exactly except for a few simple cases. Therefore, approximations are to be employed. One particularly simple approximation which takes into account the Pauli exclusion principle is to assume

$$\psi(\vec{r}_1, \ldots, \vec{r}_n) \cong (n!)^{-\frac{1}{2}} \det|\phi_i(\vec{r}_j)|. \tag{5}$$

The ϕ_i are one-particle functions and the coordinate $\vec{r}_j$ is assumed to include the spin dependence. If one requires the ϕ_i to be orthonormal and if one doesn't require the spatial part of a spin-up, spin-down pair of orbitals to be identical and if one inserts the approximate wavefunction, equation (5), into the variational theorem, equation (2), one obtains the spin-polarized Hartree-Fock equation for defining the ϕ_i's.

$$\left[-\nabla^2 - \sum_{I=1}^{N} \frac{2Z_I}{|\underset{\sim}{r} - \underset{\sim}{R}_I|} + V_C + V_E \right] \phi_i(\vec{r}) = \varepsilon_i \phi_i(\vec{r}). \tag{6}$$

Here one has

$$V_C = \sum_{j=1}^{n} \int \frac{\phi_j^*(\underset{\sim}{r}')\phi_j(\underset{\sim}{r}')}{|\underset{\sim}{r} - \underset{\sim}{r}'|} d\underset{\sim}{r}', \tag{7}$$

and also

$$V_E \phi_i(\vec{r}) = \sum_{j=1}^{n} \phi_j(\underset{\sim}{r}) \int \frac{\phi_j^*(\underset{\sim}{r}')\phi_i(\underset{\sim}{r}')}{|\underset{\sim}{r} - \underset{\sim}{r}'|} d\underset{\sim}{r}'. \tag{8}$$

Thus one finds that equation (6) represents a set of n coupled second order, non-homogeneous, integro-partial differential equations for the ϕ's. It is important to recognize that one has two general Hermitian operators arising from equation (6), one for the spin-up orbitals and a second for the spin-down orbitals. Thus the solutions to equation (6) include orbitals which occur in the approximate ψ and other which don't. Those orbitals which occur in ψ are termed occupied orbitals, whereas those which don't are termed virtual orbitals. We will employ quantum numbers a,b,c, etc. for the virtual states and i,j,k, etc. for the occupied states. Clearly, here we have assumed

$$\phi(\vec{r}) = \tilde{\phi}(\vec{r})\alpha, \qquad \text{or} \qquad \phi(\vec{r}) = \tilde{\phi}(\vec{r})\beta, \tag{9}$$

where α and β are the usual Pauli spin-up or spin-down matrices. It is often possible and sometimes desirable to impose further approximations onto the wavefunction of equation (5). In principle, one might require the spatial part of a spin-up, spin-down pair of wavefunctions be identical. This is in essence the Restricted Hartree-Fock approximation (RHF). If one has all closed subshells present, the RHF and the SPHF method are the same. There are a number of available prescriptions for obtaining RHF wavefunctions for open shell cases. The one used in this report is due to Nesbet [3] and essentially involves using an average operator for the spin-up spin-down system to represent both systems. This is the method of symmetry and equivalence restrictions. It has the disadvantage that the defining equation for the ϕ's is not obtained from a variational principle. Nonetheless, the total energy obtained using these wavefunctions is an upper bound and is very close to that obtained using the SPHF equations.

If one works with the SPHF wavefunction, there are two important theorems which apply. To see these effects most clearly, let us define a set of creation and annihilation operators for the electrons. Let α_a or α_i be the annihilation operators and $\alpha_a^\dagger$ or $\alpha_i^\dagger$ be the creation operators. Let us then say that

$$|\ \rangle = (n!)^{-\frac{1}{2}}\det|\phi_i(\vec{r}_j)|. \tag{10}$$

The first important theorem is Koopman's theorem [4]. This says that

$$\varepsilon_i = \langle\ |\mathcal{H}|\ \rangle - \langle\ |\alpha_i^\dagger\mathcal{H}\alpha_i|\ \rangle. \tag{11}$$

Thus we see the physical significance of the SPHF eigenvalues. These are the energies needed to remove the electron in orbital i from the crystal if all other orbitals are frozen. The second theorem is Brillouin's theorem [4]. This theorem says

$$\langle\ |\mathcal{H}\alpha_a^\dagger\alpha_i|\ \rangle = \langle a|F|i\rangle \equiv 0. \tag{12}$$

Here F is the Hartree-Fock operator in equation (6). Thus there are only zero Hamiltonian matrix elements between a Slater determinant and another Slater determinant which differs only by a single one-electron orbital. In addition, there are some other useful properties. These are

$$\langle\,|\mathcal{H}\alpha_a^\dagger\alpha_b^\dagger\alpha_c^\dagger\alpha_i\alpha_j\alpha_k|\,\rangle \equiv 0, \tag{13}$$

and

$$\langle\,|\mathcal{H}\alpha_a^\dagger\alpha_b^\dagger\alpha_i\alpha_j|\,\rangle = \iint\frac{\phi_i^*(\underset{\sim}{r})\phi_a(\underset{\sim}{r})\phi_j^*(\underset{\sim}{r}')\phi_b(\underset{\sim}{r}')}{|\underset{\sim}{r}-\underset{\sim}{r}'|}\,d\underset{\sim}{r}d\underset{\sim}{r}'$$

$$-\iint\frac{\phi_i^*(\underset{\sim}{r})\phi_b(\underset{\sim}{r})\phi_j^*(\underset{\sim}{r}')\phi_a(\underset{\sim}{r}')}{|\underset{\sim}{r}-\underset{\sim}{r}'|}\,d\underset{\sim}{r}d\underset{\sim}{r}'. \tag{14}$$

Thus one finds that the Hamiltonian connects a given Slater determinant only with ones differing by two orbitals provided the orbitals are eigenstates of the SPHF set of equations.

Finally, there are some useful expressions which one can use in calculating the total system energy using the SPHF wavefunction. Let us define

$$W = \tfrac{1}{2}\sum_{I,J}{}' \frac{2Z_I Z_J}{|\underset{\sim}{R}_I - \underset{\sim}{R}_J|}\,, \tag{15}$$

$$f_{ii} = \int\phi_i^*\left(-\nabla^2 - \sum_I \frac{2Z_I}{|\underset{\sim}{r}_i - \underset{\sim}{R}_I|}\right)\phi_i d\vec{r}, \tag{16}$$

$$v_{iijj} = \int\frac{e^2\phi_i^{\,2}(\underset{\sim}{r})\phi_j^{\,2}(\underset{\sim}{r}')}{|\underset{\sim}{r}-\underset{\sim}{r}'|}\,d\underset{\sim}{r}d\underset{\sim}{r}', \tag{17}$$

$$v_{ijij} = \int\frac{e^2\phi_i^*(\underset{\sim}{r})\phi_j(\underset{\sim}{r})\phi_j^*(\underset{\sim}{r}')\phi_i^*(\underset{\sim}{r}')}{|\underset{\sim}{r}-\underset{\sim}{r}'|}\,d\underset{\sim}{r}d\underset{\sim}{r}'. \tag{18}$$

The total energy E is given as

$$E = \sum_{i=1}^{n} f_{ii} + \sum_{i,j=1}^{n} \tfrac{1}{2}(v_{iijj} - v_{ijij}). \tag{19}$$

Alternately one has

$$E = \tfrac{1}{2}\sum_i (\varepsilon_i + f_{ii}). \tag{20}$$

There are many variants on the Hartree-Fock scheme of greater or lesser utility. The interested person is directed to the book by Schaefer [1] or the review by Löwdin [5].

Finally, it is worth noting that one clear advantage of the SPHF or for that matter any of the Hartree-Fock methods is that it is improvable in a formally direct and simple way. The basic technique is that of configuration interaction (CI). Here one recognizes the ϕ's and hence the Slater determinants form a complete set and thus

$$\psi(\underset{\sim}{r}_1,\underset{\sim}{r}_2,\ldots,\underset{\sim}{r}_n) = \beta|\ \rangle + \sum_{a=n+1}^{\infty}\sum_{i=1}^{n} \beta_i^a \alpha_a^{\dagger}\alpha_i|\ \rangle + \sum_{a,b=n+1}^{\infty}\sum_{i,j=1}^{n} \beta_{ij}^{ab}\alpha_a^{\dagger}\alpha_b^{\dagger}\alpha_i\alpha_j|\ \rangle + \ldots \quad (21)$$

Here the coefficients, β, are determined by the variational method. Clearly in practice the sums over virtual orbitals are restricted to a finite set and hence the energy obtained using the wavefunction is an upper bound. It is also clear that these techniques apply to states other than the ground state if proper case is used in insuring orthogonality of a given state to those of lower energy. We do not detail the notational changes to the above prescriptions for states other than the ground state since they are quite simple and are freely available in the literature. We observe that the above form of CI is only one of a large number of types available.

It is worthwhile discussing briefly one useful and simple variant on the above theory of CI. This is motivated by realizing that the virtual orbitals of the Hartree-Fock equation are very diffuse in practice whereas the occupied orbitals are quite compact. Therefore the expansion, equation (21), converges very slowly. The second important realization is that the virtual orbitals in the Hartree-Fock theory are quite arbitrary [6]. It is quite advantageous to exploit this arbitrariness. Consider, if one defines

$$\rho(\underset{\sim}{r},\underset{\sim}{r}') = \sum_{i=1}^{n} \phi_i(\underset{\sim}{r})\phi_i^{\dagger}(\underset{\sim}{r}'). \quad (22)$$

The quantity $\rho(\underset{\sim}{r},\underset{\sim}{r}')$ is the first order density matrix. Let us define an operator A^{1-F} as,

$$A^{1-F} = (1 - \rho)A(1 - \rho), \quad (23)$$

with A being any Hermitian operator. One finds that for the occupied orbitals of the SPHF operator

$$\left[-\nabla^2 - \sum_I \frac{2Z_I}{|\underset{\sim}{r}_i - \underset{\sim}{R}_I|} + V_c + V_E\right]\phi_i = \quad \text{(Contd)}$$

$$\text{(Contd)} \qquad = \left[-\nabla^2 - \sum_I \frac{2Z_I}{\underset{\sim}{r}_i - \underset{\sim}{R}_I} + V_C + E_E + A^{1-F} \right] \phi_i. \qquad (24)$$

Thus the edition of A^{1-F} to the SPHF equation leaves the occupied orbitals unchanged and alters only the virtual space. Thus it should be possible to choose an A which accelerates the convergence of the expansion, equation (21). Huzinaga has suggested that the choice of $- e^2/|\underset{\sim}{r}|$ would be an appropriate choice for atomic studies in that the virtual orbitals of equation (6) sees the repulsion of all the occupied orbitals, whereas the occupied states see the repulsion of only $n - 1$ electrons [6]. Thus, this choice tends to cancel the repulsion of one of the occupied electrons.

The author has developed a simple test of this idea [6]. In the author's test, A was defined to be $A = D$, $r \leqslant R_M$, $A = 0$, $r > R_M$. The system chosen was the He atom. C and A are varied and two approximations to $\Psi(\underset{\sim}{r}_1 \underset{\sim}{r}_2)$ are used. In the first case only two configurations were used, the $1s^2$ and the $2s^2$, in the second, three configurations were used, the $1s^2$, $2s^2$, $3s^2$. The results of these tests are seen in figures 1 and 2. The value of $R_M = 0$ corresponds to using orbitals for the SPHF equation

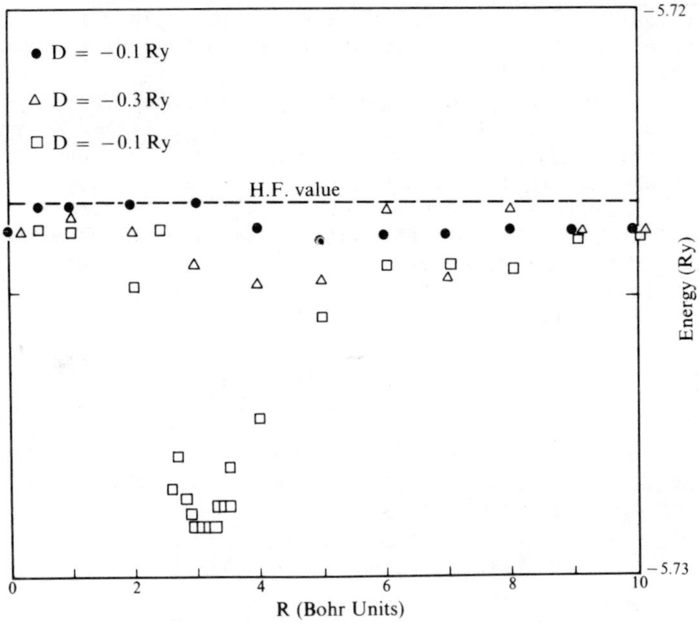

Figure 1 - The total energy of He is seen for a two configuration calculation, choosing the configuration as described in the text.

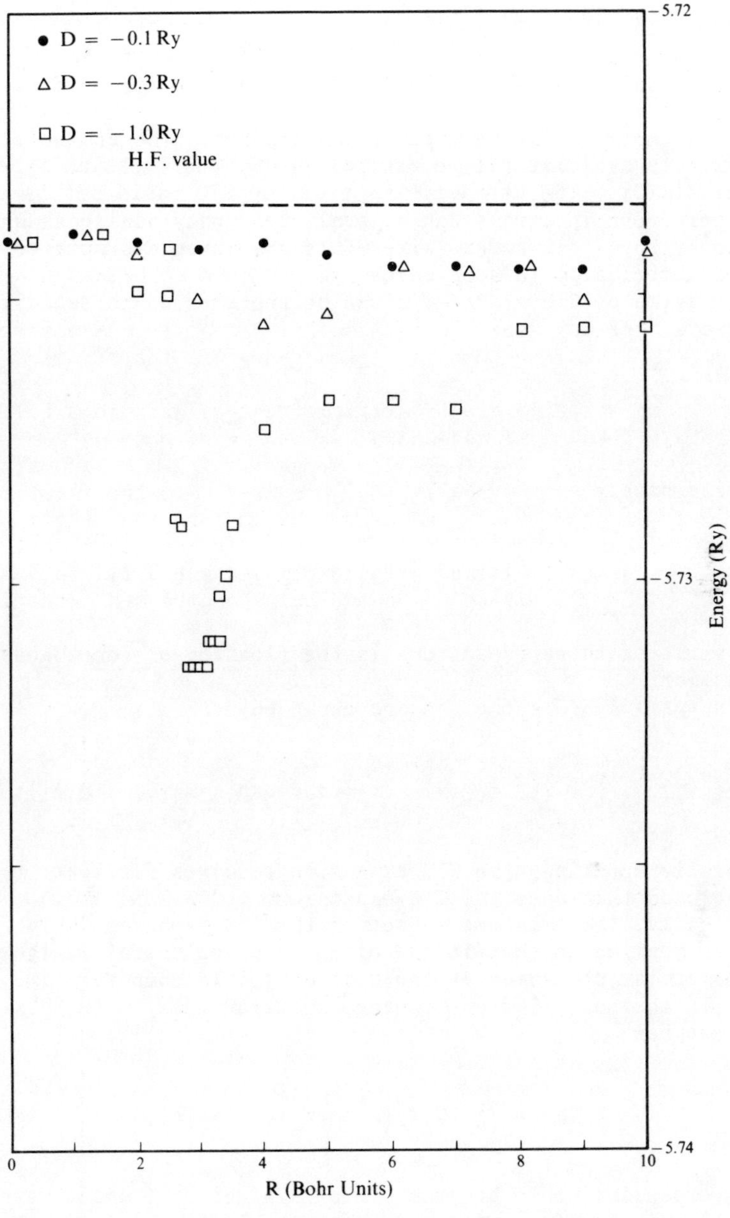

Figure 2 - The total energy of He is seen for a three configuration calculation, choosing the configuration as described in the text.

(6). It is clearly seen that the use of A^{1-F} has clearly accelerated the convergence of the CI expansion.

2. METHOD OF SOLUTION

In the case of molecules, it is usually found to be impractical to solve for the eigenstates of equation (6) by direct numerical integration. It is usual to solve for the ϕ's by expanding the ϕ's in some type of a basis set. The commonly used basis sets are Slater type orbitals (STO) and Gaussian type orbitals (GTO). The attractiveness of the GTO basis set is that the pertinent integrals can be evaluated analytically as was shown by Boys [7] whereas with STO's the integrals must be evaluated numerically in some cases.

In terms of GTO's, the ϕ's can be represented in two different ways. We may use

$$\phi_k(\underset{\sim}{r}) = \sum_{j=1}^{m} C_j^k \exp[i\alpha_j(\underset{\sim}{r} - \underset{\sim}{r}_j)^2], \tag{25}$$

or alternately

$$\phi_k(\underset{\sim}{r}) = \sum_{j=1}^{m} C_j^k Y_{\ell_j}^{m_j}(\theta,\phi)\exp[-\alpha_j(\underset{\sim}{r} - \underset{\sim}{r}_j)^2]. \tag{26}$$

The first of these expansions is the floating or lobe Gaussian expansion.

In terms of STO's the ϕ's are expanded as

$$\phi_k(\underset{\sim}{r}) = \sum_{j=1}^{m} C_j^k(\underset{\sim}{r} - \underset{\sim}{r}_j)^{\ell+A_j} Y_{\ell_j}^{m_j}(\theta,\phi)\exp[-\alpha_j|(\underset{\sim}{r} - \underset{\sim}{r}_j)|]. \tag{27}$$

Generally speaking, the STO expansion requires far fewer terms to be used than does the GTO expansion.

Finally, there is one subset of the STO expansion which deserves mention in that it has often provided useful insights into molecular processes at the cost of little computer time. This is the so-called one center expansion (OCE). In this case, one assumes

$$\phi_k(\underset{\sim}{r}) = \sum_{j=1}^{m} C_j^k Y_{\ell_j}^{m_j} r^{\ell_j+A_j}\exp[-\alpha_j|\underset{\sim}{r}|]. \tag{28}$$

This expansion would be exact if all values of ℓ and m were used as well as a sufficient number of ϕ's. In principle, only a few values of ℓ are used and thus this expansion is chiefly used for studies on molecules with a large central atom surrounded by smaller atoms, e.g., NH_4^+ or GeH_4. In the case of soft X-ray spectroscopy, where the transition is from a core state localized on the large atom, this may be useful since the bonding orbitals are not directly involved in the physical process and thus the limitations of the OCE method are minimized. This method has been given a good review by Parr [*8*].

3. SOFT X-RAY SPECTROSCOPY

In recent years, the advent of good synchrotron radiation sources have prompted interest in soft X-ray spectra of molecules. This is a new field and will be briefly reviewed in this section. The experimental references are not meant to be complete, but are sited simply to illustrate the theoretical ideas presented.

A very steep photon absorption threshold is found for SiH at an energy close to a similar threshold found in solid Si films [*4*,*9*]. The $Li_{II,III}$ spectra of Si almost certainly arises from the electric dipole transition of a single 2*p* electron to the conduction band. By analogy, the $L_{II,III}$ spectra in SiH_4 is believed to arise from the excitation of a single 2*p* electron to a molecular anti-bonding state. It is found that there exists a Rydberg-like series of such states converging to the continuum limit. In fact, for SiH_4 several series of such states are observed. The experimental situation is seen in figure 3.

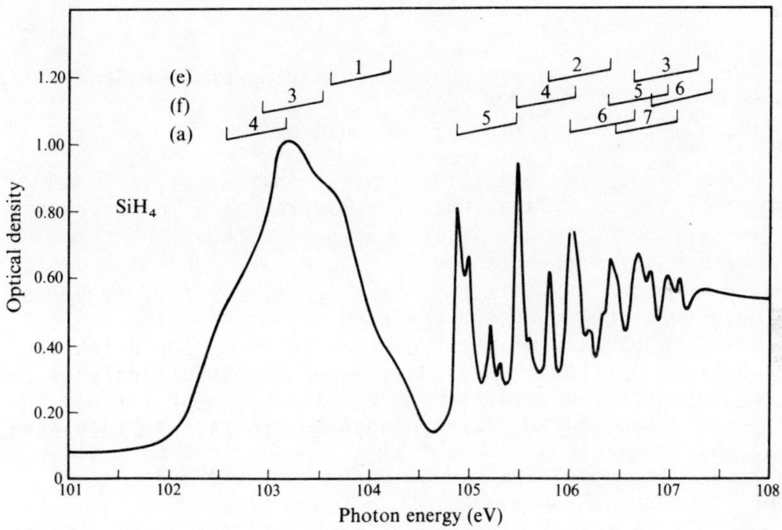

Figure 3 - The soft X-ray spectra of SiH_4 is seen (reference [*9*]) and theoretical levels are indicated for a set of 6 (spin included) Rydberg series.

A series of Hartree-Fock self-consistent calculations were performed for SiH_4, CH_4, and NH_4^+. Matrix techniques standard for Hartree-Fock calculations were used [*1*]. The basis functions were centered on the dominant atom. They consisted of standard STO's. A minimal angular set of just one type of spherical harmonic for each one-electron orbital was used. Radial functions were taken from standard basis sets for the main atomic constituents. Diffuse basis functions were then added to allow for the existence of charge density in the bonding regions, on the hydrogen atoms, and — for the case of the excited Hartree-Fock orbitals — in the regions far beyond the molecular skeleton.

The point symmetry of SiH_4 and CH_4 in its ground state is tetrahedral or T_d. Here we use the point group notation of Bethe. In this symmetry *s* orbitals transform like the Γ_1 representation, *p* orbitals like Γ_5 and *d* orbitals split into a triply degenerate Γ_5 and a doubly degenerate Γ_3 representation. The ground configuration of SiH_4 is $1\Gamma_1{}^2 2\Gamma_1{}^2 3\Gamma_1{}^2 1\Gamma_5{}^6 2\Gamma_5{}^6$ (or ${}^1\Gamma_1$ overall) and allowed electric dipole transitions occur to the underlined states from the $1\Gamma_5$ level. For brevity, we list only the *2p* shell and the excited one-electron orbital. The overall molecular symmetries are given in parentheses:

$$1\Gamma_5^5 n\Gamma_1({}^{1,3}\underline{\Gamma_5})n \geqslant 4$$

$$1\Gamma_5^5 n\Gamma_5({}^{1,3}\underline{\Gamma_5}, {}^{1,3}\Gamma_3, {}^{1,3}\Gamma_1)n \geqslant 3$$

$$1\Gamma_5^5 n\Gamma_3({}^{1,3}\underline{\Gamma_5}, {}^{1,3}\Gamma_4)n \geqslant 1.$$

The index *n* labels the molecular orbital energetically from lowest to highest. *N* is one for the lowest energy occurrence of a given symmetry type. It is hypothesized that the three excited state symmetries listed form three independent Rydberg series.

Calculations were done for the ground state of SiH_4 and for the lowest members of the three excited series. In addition, configuration coordinate diagrams in the breathing mode were obtained both for the Hartree-Fock ground state and for the lowest ${}^{1,3}\Gamma_5$ excited state. The exchange interaction between the *2p* hole and the excited $4\Gamma_1$ orbital in the excited molecule produced in the $1\Gamma_5 \rightarrow 4\Gamma_1$ transition was calculated and found to be 0.003 Ry. Being found small, the correction was ignored for this and subsequent calculations. Spin-orbit effects were not calculated but were estimated from atomic silicon calculations of Herman and Skillman [*10*] to be about 0.60 eV.

In both potential surface calculations, the silicon-hydrogen bonding distance was stepped in 0.1 Bohr unit intervals from 2.5 through 3.5 Bohr units. The molecule was kept in tetrahedral coordination and the basis set was kept fixed for all bond lengths.

The intent for CH_4 was to obtain an estimate for the threshold of the *K* edge absorption. The onset should correspond to a $1s \rightarrow 3p$ atomic transition. The ground configuration of methane is $1\Gamma_1{}^2 2\Gamma_1{}^2 1\Gamma_5{}^6$. So the excited electronic configuration most closely approximating the threshold should be: $1\Gamma_1 2\Gamma_1{}^2$ $1\Gamma_5{}^6 2\Gamma_5{}^1$. Again we calculated breathing coordinate diagrams to find the mechanical positions of minimum total energy for both the ground and excited states, and again basis functions were kept fixed as the coordinate was changed. Neither spin orbit nor core-hole interactions were included, being deemed quite small.

Finally, a ground state calculation on the tetrahedral molecule NH_4^+ was performed as a check on code accuracy. The experimental bonding length was used.

The results of ground state calculations for all three molecules are presented in Table I. In all cases our results were

TABLE I

Computed Binding Energies and ⟨X-H⟩

	Molecule	CH_4	SiH_4	NH_4^+
This Calculation	Eg (Ry)	-79.039	-580.647	-56.006
	⟨X-H⟩ Bohr	2.050	2.90	2.07
Previous Similar Calculation	Eg (Ry)	-79.00†	-580.16†	-56.218†
	⟨X-H⟩ Bohr	2.0655	2.79	1.99
Experiment	⟨X-H⟩ Bohr	2.07	2.87	1.96

† From Parr (1964).

in reasonable agreement with previous ones. The bond lengths are in good agreement with experiment. A minimum angular basis was used in our calculations. So what discrepancies exist between our calculations and previous ones could easily arise from the use of this angular truncation. We therefore conclude that no serious errors arise in our computer coding.

In the case of SiH_4, calculations of peak positions generally support the experimental ordering and positioning of the molecular levels. The atomic estimates of the spin-orbit splittings are borne out. The calculated molecular field splittings are off by a factor or two, but this may be due to the approximate nature of the calculated charge densities near the hydrogen atoms. The results are presented in Tables I and II, and in figure 4. The configuration calculations indicate a very

TABLE II

Some Soft X-Ray Transitions for SiH_4

Transition	J	Theory (eV)	Experiment (eV)
$2\Gamma_5^6 \rightarrow 1\Gamma_5^5 4\Gamma_1$	3/2	100.2	102.59
$1\Gamma_5^6 \rightarrow 1\Gamma_5^5 4\Gamma_1$	1/2		103.18
$1\Gamma_5^6 \rightarrow 1\Gamma_5^5 5\Gamma_1$	3/2	103.2	104.92
$1\Gamma_5^6 \rightarrow 1\Gamma_5^5 5\Gamma_1$	1/2		105.52
$1\Gamma_5^6 \rightarrow 1\Gamma_5^5 3\Gamma_5$	3/2	101.0	102.95
$1\Gamma_5^6 \rightarrow 1\Gamma_5^5 3\Gamma_5$	1/2		103.55
$1\Gamma_5^6 \rightarrow 1\Gamma_5^5 4\Gamma_5$	3/2	103.4	105.52
$1\Gamma_5^6 \rightarrow 1\Gamma_5^5 4\Gamma_5$	1/2		106.10
$1\Gamma_5^6 \rightarrow 1\Gamma_5^5 1\Gamma_3$	3/2	102.2	103.65
$1\Gamma_5^6 \rightarrow 1\Gamma_5^5 1\Gamma_3$	1/2		104.25
$1\Gamma_5^6 \rightarrow 1\Gamma_5^5 2\Gamma_3$	3/2	104.0	105.84
$1\Gamma_5^6 \rightarrow 1\Gamma_5^5 2\Gamma_3$	1/2		106.45

slight shift in the mechanical equilibrium position of the molecule in the first soft X-ray excited state, so that the transition should occur vertically with little mechanical readjustment in the excited state. When this is assumed, the excitation energy is 100.3 ± 0.2 eV. This value falls short of the center of gravity of the measured absorption peak which is 102.8 eV.

Table II shows the general agreement between calculated and experimental peak positions. The main discrepancy is that the

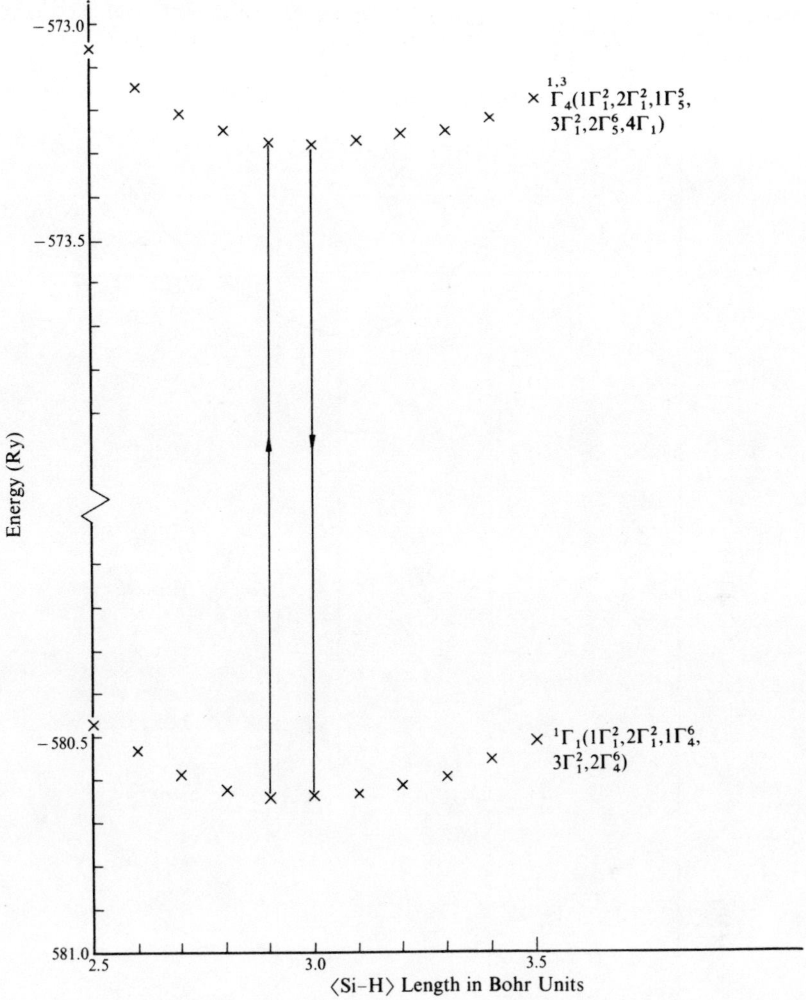

Figure 4 - The SiH_4 energy levels are given as a function of the breathing mode.

molecular field splittings are in error. The splitting happens because the T_d symmetry of the molecule splits the excited d-orbital into a Γ_3 and a Γ_5 component.

Here the splittings are about one-half of the calculated values. This could easily by the fault of the approximate nature of the one-center charge distribution in the region of the hydrogen atom.

The ground and first excited states resulting from a $1s \rightarrow 3p$ dipole transition were calculated for CH_4. Potential energy curves for the breathing coordinate were obtained for both the ground and excited states. (See figure 5). The mechanical

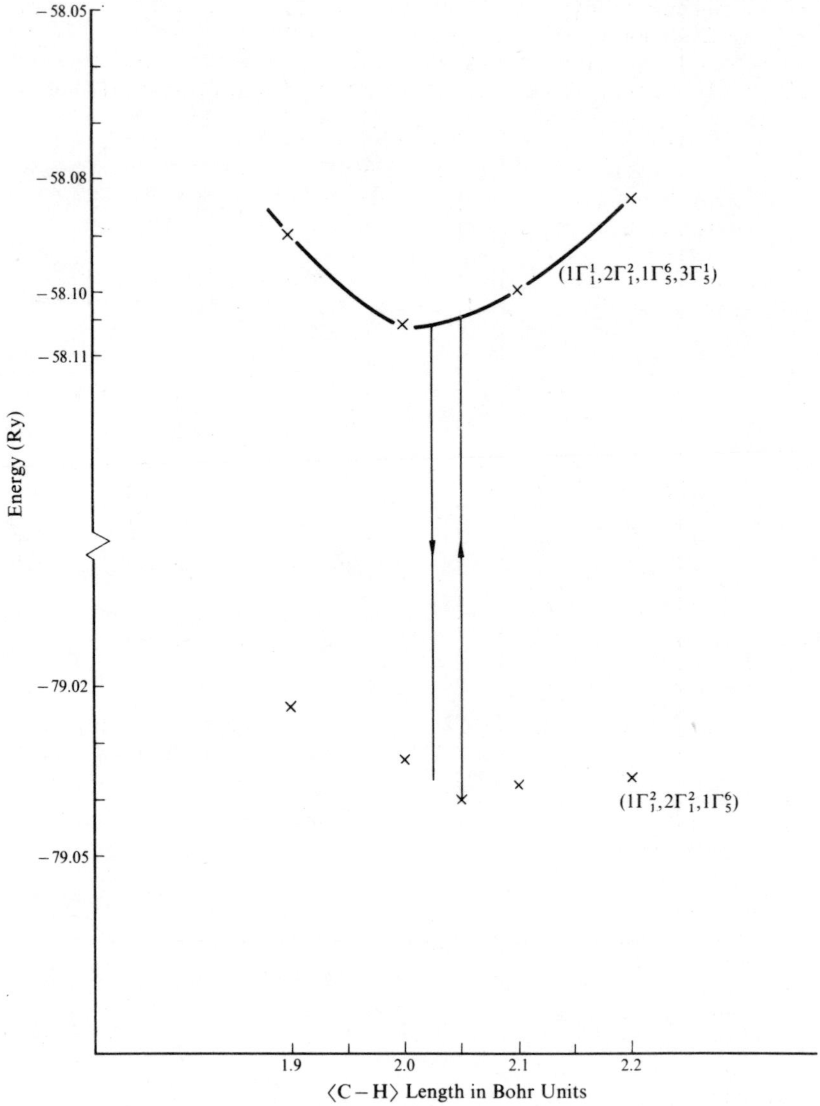

Figure 5 - The CH_4 energy levels are given as a function of the breathing mode.

equilibria for both states are nearly the same, and one is encouraged that the transition is vertical. When the Franck-Condon principle is applied, the onset of absorption occurs at about 284.7 ± 0.3 eV. The experiment of H.U. Chun [*11*] indicates an onset between 286.7 and 287 eV.

It is clear from studies of figure 3 that the dominant features are gotten in terms of Rydberg series calculated using

Hartree-Fock theory for the molecules. It is also clear from the great breadth of the initial absorption structure and the presence of small features on the high energy side of the main absorption peaks that the interaction of electronic states with vibronic motion plays a role in the proper interpretation of such spectra. No detailed theory of these effects in the soft X-ray spectra of molecules has been developed. It is also clear that many body effects play at best a small part in determining the absorption edge positions in these systems. Just what role matrix element effects and many body corrections to these matrix elements will play in the detailed understanding of the soft X-ray spectra of molecules is as yet unclear.

ACKNOWLEDGMENT

The author wishes to thank Mr. Peter Deutsch, Dr. W. Hayes, and Professor F.C. Brown for their discussions and for the use of their work in preparing this report.

REFERENCES

1. Hartree, D.R. (1957). *The Calculation of Atomic Structure*, (John Wiley and Sons, Inc., New York); Slater, J.C. (1960). *Quantum Theory of Atomic Structure, Vol.s I, II*, (McGraw-Hill, New York); Schaefer, Henry F., III. (1972). *The Electronic Structure of Atoms and Molecules*, (Addison Wesley Company, Reading, Massachusetts).
2. Slater, J.C. (1963). *Quantum Theory of Molecules and Solids, Vol. I*, (McGraw-Hill, New York).
3. Nesbet, R.K. (1961). *Rev. Mod. Phys.*, **33**, 28.
4. Berthier, G. (1966). In *Molecular Orbitals in Chemistry, Physics, and Biology*, (eds. Löwdin, P.O. and Pullman, B.), (Academic Press, New York).
5. Löwdin, P.O. (1966). In *Quantum Theory of Atoms, Molecules, and The Solid State*, (ed. Löwdin, P.O.), (Academic Press, New York).
6. Kunz, A.B. (1971). *Phys. Stat. Sol.*, **46**, 697; Huzinaga, S. and Arnau, C. (1971). *J. Chem. Phys.*, **54**, 1948.
7. Boys, S.F. (1966). *Proc. Roy. Soc.*, **A200**, 542, and references contained therein.
8. Parr, R.G. (1964). In *Molecular Orbitals in Chemistry, Physics, and Biology*, (eds. Löwdin, P.O. and Pullman, B.), (Academic Press, New York).
9. Hayes, W. and Brown, F.C. (1972). *Phys. Rev.*, **A6**, 21; Hayes, W., Brown, F.C. and Kunz, A.B. (1971). *Phys. Rev. Lett.*, **27**, 774; Gahwiller, C. and Brown, F.C. (1970). *Phys. Rev.*, **B2**, 1918.
10. Herman, F. and Skillman, S. (1963). *Atomic Structure Calculations*, (Prentice-Hall, Englewood Cliffs, New Jersey).
11. Chun, H.U. (1969). *Phys. Rev. Lett.*, **30A**, 445.

COLLECTIVE EXCITATIONS OF DIPOLAR LIQUIDS†

ROBERTO LOBO¶, JOHN E. ROBINSON§ and SERGIO RODRIGUEZ

Department of Physics, Purdue University
West Lafayette, Indiana 47907, *U.S.A.*

1. INTRODUCTION

Collective excitations in condensed matter have received considerable attention in the literature. The most thoroughly studied among such elementary excitations are the long-wavelength plasma oscillations in charged plasmas and the phonons in all states of aggregation of matter. The collective oscillations of a system of interacting neutral dipoles has not received the same attention.

In an electron plasma in a solid, for example, the electron density can experience self-sustained oscillations of frequency $(4\pi ne^2/m)^{\frac{1}{2}}$ where n is the electron concentration, and e and m their charge and mass. Similarly if we have dipoles of electric dipole moment μ and moment of inertia I, we anticipate fluctuations of the polarization of such a medium to experience self-sustained oscillations of frequency $(4\pi n\mu^2/I)^{\frac{1}{2}}$.

The authors have studied the collective excitations of such a system in some detail [*1*]. The purpose of this work is to review the derivation of the dielectric function $\varepsilon(\omega)$ for dipolar liquids and exhibit the form in which the collective

---

† Supported at Purdue University by the National Science Foundation (Grant GH 33774) and based in part on work performed at Argonne National Laboratory under the auspices of the U.S. Atomic Energy Commission.

¶ Permanent Address: Escola de Engenharia de São Carlos, Universidade de São Paulo, S. Paulo, Brazil.

§ On leave of absence from Argonne National Laboratory, Argonne, Illinois 60439 dipolar oscillations appear in such liquids. Section 2 gives a summary of the theories of the static dielectric constant of liquids. Section 3 gives a derivation of $\varepsilon(\omega)$ following the work of Nee and Zwanzig [2], except that we include the inertial effects and the effects of the correlation time of the random torque acting on the dipoles of the system. Section 4 contains a discussion of the properties of the dielectric function.

2. STATIC DIELECTRIC CONSTANT OF A POLAR LIQUID

All theories of the dielectric constant of a polar liquid depend on the determination of the mutual interactions among the molecules of the liquid. The oldest and simplest theory is that of Debye in which the local field at the position of a molecule is assumed to be the Maxwell macroscopic field $\vec{E}$ in the medium plus the Lorentz field $(4\pi/3)\vec{P}$ where $\vec{P}$ is the macroscopic polarization of the medium assumed to be uniform. This result is derived by imagining that we cut out a sphere of the material centered at the molecule in question and suppose that the net field produced by other molecules within this sphere is zero. With this assumption we can write

$$P = n\alpha_e E_\ell + n\mu\mathcal{L}(\beta\mu E_\ell), \tag{1}$$

where n is the number of molecules per unit volume, α the electronic polarizability of a molecule, $E_\ell = E + (4\pi/3)P$, μ the permanent dipole moment of each molecule, $T = (k_B\beta)^{-1}$ the absolute temperature and $\mathcal{L}(x) = \coth x - x^{-1}$, the Langevin function. At ordinary temperatures in most circumstances $\beta\mu E_\ell \ll 1$ and we have for the dielectric constant

$$\frac{\varepsilon_0 - 1}{\varepsilon_0 + 2} = \frac{4\pi}{3} n\alpha \equiv \frac{4\pi}{3} n\left(\alpha_e + \frac{\mu^2}{3k_BT}\right) . \tag{2}$$

An attempt to account for the mutual interactions of the molecules within the Lorentz cavity was made by Debye [3] in 1935. However, all such theories involve the difficulty that the left hand side of equation (2) can never exceed unity but the right hand side can, in principle, attain larger values. This is, of course, partly due to the fact that this equation is derived assuming that the local field E_ℓ is small compared to k_BT/μ. Furthermore, according to this theory all dipolar liquids should exhibit a spontaneous polarization below a temperature which is typically a few hundred degrees K, a phenomenon which is not observed.

In order to avoid these difficulties Onsager [4] considered a spherical cavity (called the Onsager cavity) containing a single molecule. The field acting on the molecule is

$$\vec{E}_\ell = \vec{G} + \vec{R}, \tag{3}$$

where $\vec{G}$, the Onsager cavity field, is the field which would exist within the cavity if the permanent dipole were absent, while $\vec{R}$, the reaction field, is the field in the cavity arising from the polarization induced in the surrounding medium by the permanent dipole. We shall suppose that the electronic polarizability of the molecules is isotropic and can be described by the optical dielectric constant ε_∞. We can obtain the cavity field G by solving the appropriate boundary value problem. Let V be the potential. Then, the field inside the cavity is

$$V_i = - Gr\cos\theta, \tag{4}$$

while that outside is

$$V_O = - Er\cos\theta - \frac{\mu}{\varepsilon r^2}\cos\theta, \tag{5}$$

where we have selected spherical polar coordinates with the origin of the coordinate system at the center of the Onsager cavity whose radius we take to be a (roughly the molecular radius). Now the boundary conditions require the continuity of V and of the normal component of $\vec{D}$, i.e. $- \varepsilon_\infty(\partial V_i/\partial r)_a = - \varepsilon(\partial V_O/\partial r)_a$. This yields

$$\vec{G} = \frac{3\varepsilon}{2\varepsilon + \varepsilon_\infty}\vec{E}, \tag{6}$$

The reaction field is obtained in a similar fashion. We take

$$V_i = \frac{\mu}{\varepsilon_\infty r^2}\cos\theta - Rr\cos\theta, \tag{7}$$

and

$$V_O = \frac{\mu'}{\varepsilon r^2}\cos\theta, \tag{8}$$

and find

$$\vec{R} = \frac{8\pi}{3} n \frac{\varepsilon - \varepsilon_\infty}{\varepsilon_\infty(2\varepsilon + \varepsilon_\infty)}\vec{\mu}, \tag{9}$$

with $(4\pi/3)a^3 n = 1$, and

$$\mu = \frac{2\varepsilon + \varepsilon_\infty}{3\varepsilon}\mu'. \tag{10}$$

We notice, in particular, that in all expressions concerning the dynamics of the dipole moment, the dipole moment in the medium is given by

$$\mu = \frac{1}{3}(\varepsilon_\infty + 2)\mu_V, \tag{11}$$

where μ_V is the dipole moment in the vacuum. The dielectric constant ε is then obtained by recognizing that the reaction field $\vec{R}$, being parallel to $\vec{\mu}$ exerts no torque on the molecule. At sufficiently high temperature (i.e., when $x \equiv \mu G/k_B T \ll 1$) we find

$$\frac{(\varepsilon - \varepsilon_\infty)(2\varepsilon + \varepsilon_\infty)}{3\varepsilon} = \frac{4\pi n\mu^2}{3k_B T}. \tag{12}$$

A more general theory by Kirkwood [5] yields similar results except that the right hand side of this equation is multiplied by a factor $g = \left\langle \sum_{ij} \vec{\mu}_i \cdot \vec{\mu}_j / N\mu^2 \right\rangle$ where the average is assumed to be carried out over the equilibrium ensemble (in the absence of a field) and the sums extend over all N molecules. Long range correlations among dipoles are already accounted for as in the Onsager theory, with the exception that the sphere is macroscopic rather than of molecular size. Only short range correlations contribute to g, and hence the sum need be carried only to a few molecular diameters.

3. THE DIELECTRIC CONSTANT AT FINITE FREQUENCIES

We now turn to our study of the dielectric function at finite frequencies. We shall assume that we can always neglect the displacement current in the Maxwell equations. This limits our discussion to frequencies below those in the visible region of the electromagnetic spectrum. The first attempt at such a theory is that due to Debye in which it is assumed that the response to a field varying as $e^{-i\omega t}$ gives rise to an orientational part of the polarizability of the form

$$\alpha = \frac{\alpha_0}{1 - i\omega\tau_D}, \tag{13}$$

where α_0 is the static value. For liquids τ_D is of the order of

$$\tau_D = \frac{4\pi\eta a^3}{k_B T}, \tag{14}$$

where η is the viscosity of the liquid. These times are of the order of 10^{-11} sec for water at room temperature ($a \simeq 10^{-8}$ cm, $\eta = 0.01$ poise). This theory leads to the result

$$\varepsilon(\omega) - \varepsilon_\infty = \frac{\varepsilon_0 - \varepsilon_\infty}{1 - i\omega\tau_D} \qquad (15)$$

or, for $\omega\tau_D \ll 1$

$$\varepsilon(\omega) - \varepsilon_0 = (\varepsilon_0 - \varepsilon_\infty)i\omega\tau_D. \qquad (16)$$

We now sketch a theory of the dielectric function of a dipolar liquid at finite frequencies and long wavelengths. We assume rigid dipoles, and since translational motion is of secondary interest here, concentrate on the rotational motion. Within the framework of a dynamical Onsager model the extension of equation (12) to finite frequencies is

$$\frac{[\varepsilon(\omega) - \varepsilon_\infty][2\varepsilon(\omega) + \varepsilon_\infty]}{3\varepsilon(\omega)} = 4\pi\alpha(\omega), \qquad (17)$$

where Kubo's formulation of linear response theory gives [6]

$$\alpha(\omega) = -\frac{n}{3k_BT}\int_0^\infty dt e^{i\omega t}\langle\vec{\mu}(0)\cdot\dot{\vec{\mu}}(t)\rangle. \qquad (18)$$

Statistical averages are to be taken in the equilibrium ensemble in the absence of an applied field. Here and in the following we denote one-sided Fourier (or Laplace) transforms by

$$f(\omega) = \mathcal{L}f(t) = \int_0^\infty dt e^{i\omega t} f(t). \qquad (19)$$

Our task is to calculate the transform of the dipole autocorrelation function $\langle\vec{\mu}(0)\cdot\vec{\mu}(t)\rangle$. The kinematic equation

$$\frac{d\vec{\mu}(t)}{dt} = \vec{\Omega}(t)\times\vec{\mu}(t), \qquad (20)$$

where $\vec{\Omega}(t)$ is the angular velocity of the dipole at time t, enables us to express the distribution function $f(\vec{\mu},t)$ of dipolar orientations in terms of correlation functions of $\vec{\Omega}(t)$. The angular velocity components obey the equation of motion

$$I_{ij}\dot{\Omega}_j(t) = -\int_0^t ds\Gamma_{ij}(s)\Omega_j(t-s) + N_i(t). \qquad (21)$$

Here $\underset{\sim}{I} = (I_{ij})$ is the inertia tensor of the molecule, $\underset{\sim}{\Gamma}(s)$ is a tensor describing the systematic torque on the rotating molecule, and $\vec{N}(t)$ is the randomly fluctuating torque on the dipole. Summation over repeated indices is implied. Equation (21) is a generalization of the Langevin equation familar in the theory of Brownian motion. We multiply equation (21) by $\Omega_k(0)$ and average to obtain

$$I_{ij} \frac{d}{dt} \langle \Omega_j(t)\Omega_k(0) \rangle = - \int_0^t ds \Gamma_{ij}(s) \langle \Omega_j(t - s)\Omega_k(0) \rangle . \quad (22)$$

We solve this equation by using the one-sided Fourier transform and find

$$(\Gamma_{ij}(\omega) - i\omega I_{ij}) \mathcal{L} \langle \Omega_j(t)\Omega_k(0) \rangle = I_{ij} \langle \Omega_j(0)\Omega_k(0) \rangle . \quad (23)$$

By virtue of the equipartition theorem, the right hand side equals $k_B T \delta_{ik}$ so that

$$(\Gamma_{ij}(\omega) - i\omega I_{ij}) \mathcal{L} \langle \Omega_j(t)\Omega_k(0) \rangle = k_B T \delta_{ik} . \quad (24)$$

The distribution function $f(\vec{\mu},t)$ satisfies the Liouville equation

$$\frac{\partial f}{\partial t} + \dot{\vec{\mu}} \cdot \frac{\partial f}{\partial \vec{\mu}} = 0, \quad (25)$$

or using equation (20),

$$\frac{\partial f}{\partial t} = - [\vec{\Omega}(t) \times \vec{\mu}] \cdot \frac{\partial f}{\partial \vec{\mu}} = - \vec{\Omega}(t) \cdot \vec{\mu} \times \frac{\partial f}{\partial \vec{\mu}} \equiv \Lambda(t) f . \quad (26)$$

A formal solution of this equation, obtained by iteration, is

$$f(\vec{\mu},t) = [1 - \int_0^t \Lambda(s) ds + \int_0^t ds_1 \int_0^{s_1} ds_2 \Lambda(s_1)\Lambda(s_2) \mp \ldots] f(\vec{\mu},0), \quad (27)$$

which could be written more compactly as a time ordered exponential. We now take an average over the ensemble of fluctuations and assume that $\Omega(s)$ and thus $\Lambda(s)$ is Gaussian. We thus find for the coarse grained distribution function $\bar{f}(\vec{\mu},t)$,

$$\bar{f}(\vec{\mu},t) = \exp[\int_0^t ds_1 \int_0^{s_1} ds_2 \langle \Lambda(s_1)\Lambda(s_2) \rangle] \bar{f}(\vec{\mu},0). \quad (28)$$

Here

$$\langle \Lambda(s_1)\Lambda(s_2)\rangle = \left(\vec{\mu}\times\frac{\partial}{\partial\vec{\mu}}\right)\cdot\langle\vec{\Omega}(s_1)\vec{\Omega}(s_2)\rangle\cdot\left(\vec{\mu}\times\frac{\partial}{\partial\vec{\mu}}\right), \tag{29}$$

and

$$\langle\vec{\Omega}(s_1)\vec{\Omega}(s_2)\rangle = \langle\vec{\Omega}(s_1 - s_2)\vec{\Omega}(0)\rangle.$$

The time evolution of $\bar{f}(\vec{\mu},t)$ is thus governed by a rotational diffusion coefficient dyadic defined by

$$\underset{\sim}{D}(\omega) \equiv \mathcal{L}\langle\vec{\Omega}(t)\vec{\Omega}(0)\rangle \equiv \mathcal{L}\underset{\sim}{D}(t). \tag{30}$$

For simplicity we assume inertial isotropy for the molecules so that $\underset{\sim}{D}$ is a scalar, and equation (29) becomes

$$\langle \Lambda(s_1)\Lambda(s_2)\rangle = D(s_1 - s_2)\left(\vec{\mu}\times\frac{\partial}{\partial\vec{\mu}}\right)^2. \tag{31}$$

We put (31) into (28) and differentiate with respect to t, obtaining

$$\frac{\partial}{\partial t}\bar{f}(\vec{\mu},t) = \int_0^t ds D(t - s)\left(\vec{\mu}\times\frac{\partial}{\partial\mu}\right)^2\bar{f}(\vec{\mu},s). \tag{32}$$

In spherical coordinates, with $\vec{\mu}$ having polar and azimuthal angles θ and ϕ.

$$\left(\vec{\mu}\times\frac{\partial}{\partial\vec{\mu}}\right)^2 = \frac{1}{\sin\theta}\frac{\partial}{\partial\theta}\left(\sin\theta\frac{\partial}{\partial\theta}\right) + \frac{1}{\sin^2\theta}\frac{\partial^2}{\partial\phi^2}.$$

We obtain the equation of motion for the average moment

$$\vec{\mu}(t) \equiv \int\sin\theta d\theta\int d\phi\vec{\mu}\bar{f}(\vec{\mu},t), \tag{33}$$

namely

$$\frac{\partial\vec{\mu}(t)}{\partial s} = -2\int_0^t ds D(t - s)\vec{\mu}(s). \tag{34}$$

Taking the scalar product of equation (34) with $\vec{\mu}(0)$ and averaging over a uniform distribution of initial orientations gives for the correlation function itself

$$\frac{\partial}{\partial t}\langle\vec{\mu}(0)\cdot\vec{\mu}(t)\rangle = -2\int_0^t ds D(t-s)\langle\vec{\mu}(0)\cdot\vec{\mu}(s)\rangle . \tag{35}$$

We define a normalized autocorrelation by

$$\phi(t) \equiv \frac{1}{\mu^2}\langle\vec{\mu}(0)\cdot\vec{\mu}(t)\rangle , \tag{36}$$

and obtain

$$\mathcal{L}\phi(t) = [2D(\omega) - i\omega]^{-1}. \tag{37}$$

In equation (18) we require

$$\mathcal{L}\left(-\frac{d\phi(t)}{dt}\right) = \left(1 - \frac{i\omega}{2D(\omega)}\right)^{-1}. \tag{38}$$

In the isotropic case equation (24) and equation (30) give

$$D(\omega) = k_B T[\Gamma(\omega) - i\omega I]^{-1}. \tag{39}$$

In the low frequency limit the inertial term $i\omega I$ is negligible, and $\Gamma(\omega)$ approaches a constant $2\tau_0 k_B T$ where τ_0 is of the order of the macroscopic Debye relaxation time τ_D. The relation between τ_0 and τ_D will be given later. In the high frequency limit, $\Gamma(\omega)$ approaches zero, since at sufficiently high frequencies, corresponding to very short times, the system cannot react back on the molecule, which then rotates freely. In general we may distinguish two kinds of contributions to $\Gamma(\omega)$. The first is an ordinary Stokes friction term which we write as

$$\Gamma_S(\omega) = 2k_B T \frac{\tau_0}{1 - i\omega\tau} \tag{40}$$

which is the simplest form satisfying the limiting conditions just stated. In equation (40) τ has the significance of the correlation time of the random torque; if the random torque correlation function is assumed to decay exponentially then $\tau \approx 2(k_B T)^2\tau_0/\langle N^2\rangle$. One expects $(\tau/\tau_0) \sim O(0.1)$.

The other contribution comes from the long range reaction field. Because the molecules of the medium have inertia they cannot respond instantaneously to the central dipole (i.e., the dipole in the Onsager cavity), the reaction field lags the instantaneous orientation of that dipole and hence exerts a finite torque on it. When this torque is linearized in the angular velocity, Nee and Zwanzig [2] show that the Fourier Transorm

$$\vec{N}(\omega) = -\Gamma_R(\omega)\vec{\Omega}_\perp(\omega) \tag{41}$$

where $\vec{\Omega}_\perp(\omega)$ is the transform of the component of $\vec{\Omega}(t)$ which is perpendicular to $\vec{\mu}(t)$, and

$$\Gamma_R(\omega) = (-)\,\frac{2k_BT}{i\omega}\,\frac{(\varepsilon_0 - \varepsilon_\infty)(\varepsilon_0 - \varepsilon(\omega))}{\varepsilon_0(2\varepsilon(\omega) + \varepsilon_\infty)}\,. \tag{42}$$

The presence of the factor i signals that Γ_R is not simply a purely dissipative term.

Collecting the results of equations (38)-(42) and inserting them in equation (18) we obtain

$$\alpha(\omega) = \frac{n\mu^2}{3k_BT}\left[1 - \frac{\omega^2 I}{2k_BT} - \frac{i\omega\tau_0}{1 - i\omega\tau} + \frac{(\varepsilon_0 - \varepsilon_\infty)(\varepsilon_0 - \varepsilon(\omega))}{\varepsilon_0(2\varepsilon(\omega) + \varepsilon_\infty)}\right]^{-1}. \tag{43}$$

The properties of the dielectric function obtained from equations (43) and (17) are studied in the next section.

For low frequencies equation (43) in conjunction with equations (17) and (16) permit us to recover a simple Debye relaxation and relate the microscopic (τ_0) to the macroscopic (τ_D) Debye relaxation time by expanding $\varepsilon(\omega)$, obtaining

$$\tau_0 = \frac{\varepsilon_0 + 2\varepsilon_\infty}{2\varepsilon_0 + \varepsilon_\infty}\,\tau_D.$$

4. PROPERTIES OF THE DIELECTRIC FUNCTION

We first introduce some reduced variables to condense our formal expressions, writing

$$\bar{\varepsilon}(\omega) = \frac{\varepsilon(\omega)}{\varepsilon_\infty}\,, \tag{44}$$

$$\omega_0^2 = \frac{4\pi n\mu^2}{I\varepsilon_\infty} = \frac{4\pi n\mu_V^2(\varepsilon_\infty + 2)^2}{9I\varepsilon_\infty}\,, \tag{45}$$

$$\omega_T^2 = \frac{2k_BT}{I}\,, \tag{46}$$

$$\bar{\omega} = \frac{\omega}{\omega_T}\,, \tag{47}$$

$$\bar{\tau} = \omega_T\tau, \tag{48}$$

$$z^2(\omega) = \bar{\omega}^2 + \frac{i\bar{\omega}\tau_0}{(1 - i\omega\tau)} = \bar{\omega}^2 + \frac{i\bar{\omega}\bar{\tau}_0}{(1 - i\bar{\omega}\bar{\tau})} . \tag{49}$$

In what follows it is necessary to assume that ω has a positive infinitesimal imaginary part δ. In these units

$$\bar{\varepsilon}(\omega) - 1 = \frac{2\bar{\varepsilon}(\omega)}{2\bar{\varepsilon}(\omega) + 1}\,\omega_0^2\left\{1 - z^2(\omega) - \left(\frac{1}{\bar{\varepsilon}_0} - 1\right)\frac{\bar{\varepsilon}_0 - \bar{\varepsilon}(\omega)}{2\bar{\varepsilon}(\omega) + 1}\right\}^{-1} . \tag{50}$$

The polarizability is immediately seen to have poles of two kinds: at $\varepsilon = 0$, $|\alpha| = \infty$ for $\omega = \omega_p$ given by

$$z^2(\omega_p) \equiv z_p^2 \equiv \bar{\varepsilon}_0 = \frac{\varepsilon_0}{\varepsilon_\infty} , \tag{51}$$

and at $|\varepsilon(\omega)| = \infty$, $|\alpha| = \infty$ for ω_t given by

$$z^2(\omega_t) \equiv z_t^2 = \tfrac{1}{2}\left(1 + \frac{1}{\bar{\varepsilon}_0}\right) = \tfrac{1}{2}\left(1 + \frac{\varepsilon_\infty}{\varepsilon_0}\right) = z_p^2 - \bar{\omega}_0^2 . \tag{52}$$

The first of these is the dipolar plasmon resonance, while the second corresponds to complete screening, $\varepsilon^{-1} = 0$. For the electron plasma, the complete screening occurs at $\omega = 0$, corresponding, of course, to the exclusion of static fields from a conductor. Solving for $\bar{\varepsilon}(\omega)$ we have

$$\bar{\varepsilon}(\omega) = \frac{1}{4}\,\frac{z_p^2 - z^2(\omega)}{z_t^2 - z^2(\omega)} + \frac{3}{4}\left\{\frac{[z_1^2 - z^2(\omega)][z_p^2 - z^2(\omega)]}{[z_t^2 - z^2(\omega)]^2}\right\}^{\frac{1}{2}} . \tag{53}$$

and

$$\frac{1}{\bar{\varepsilon}(\omega)} = -\frac{1}{2} + \frac{3}{2}\left\{\frac{z_1^2 - z^2(\omega)}{z_p^2 - z^2(\omega)}\right\}^{\frac{1}{2}} , \tag{54}$$

where

$$z_1^2 \equiv \frac{1}{9}\,(z_p^2 + 8z_t^2) = z_p^2 - \frac{8}{9}\,\bar{\omega}_0^2 = z_t^2 + \frac{1}{9}\,\bar{\omega}_0^2 . \tag{55}$$

We shall also need $\omega = \omega_1$ given by

$$z^2(\omega_1) = z_1^2 . \tag{56}$$

Due to the form of the cavity field factor and of the linearized reaction field torque, the plasmon singularity is a branch point rather than a simple zero (pole) of $\varepsilon(\omega)$ $(\varepsilon^{-1}(\omega))$.

4.a Positions of the Singularities (Values of ω_t, ω_l, ω_p)

If τ_0 were negligible, then we would have $z^2(\omega) = (\omega + i\delta)^2$ and all singularities lie infinitesimally below the real axis, and the branch cuts $z_l^2 < \bar{\omega}^2 < z_p^2$ would be the only regions of true absorption. In this case $\bar{\omega}_t = \pm z_t$ $\bar{\omega}_l = \pm z_l$, $\omega_p = \pm z_p$. In the collisionless regime, $\omega\tau_0 > \omega\tau \gg 1$:

$$z^2(\omega) \approx \bar{\omega}^2 - \frac{\tau_0}{\tau}$$

which again implies that the singularities lie close to the real axis, with real parts $\bar{\omega}_t^2 = (\tau_0/\tau) + z_t^2$, $\bar{\omega}_l^2 = (\tau_0/\tau) + z_l^2$, and $\bar{\omega}_p^2 = (\tau_0/\tau) + z_p^2$. The damping, of course, displaces the singularities into the lower half of the complex ω-plane, and in all cases we have investigated, each of the cubic equations for ω_l, ω_t, and ω_p give three solutions of the form $\omega = -i/\tau''$, and $\omega = (-i/\tau') \pm \omega'$. We find that in all cases in which $(\tau\ /\tau) \approx 10$ the real parts of the complex roots are very much larger than the imaginary parts, implying the existence of weakly damped oscillations. For cases of interest, $\varepsilon_0/\varepsilon_\infty$ is large and the positions of the singularities are relatively insensitive to variations in ε_∞ since then $z_p^2 \approx \bar{\omega}_0^2$ and the factor $(\varepsilon_\infty + 2)^2/9\varepsilon_\infty$ is slowly varying (it has a minimum of 8/9 at $\varepsilon_\infty = 2$ increasing to 1 at $\varepsilon_\infty = 1$, or 4). Accordingly, we see, again for $(\varepsilon_0/\varepsilon_\infty) \gg 1$, that the real parts of ω_l and ω_p will be dominated by the reaction field while the real part of ω_t is dominated by the dissipative relaxation, just as would be expected on physical grounds.

4.b Collective Mode Charge Density

The conventional formal hallmark of a significant collective mode (or elementary excitation) is that the analytic continuation of the relevant response function (or propagator across the real axis in the complex frequency plane) has an *isolated* pole with real part much larger than imaginary part. The variable of concern — density, magnetization, etc. — in consequence has a component which decays slowly but remains coherent throughout the decay interval because no other components are close enough in frequency to produce additional dephasing. All familiar response functions known to the authors which have been shown to lead to well-defined collective modes or quasi-particle excitations have at least one such pole, whatever other complexities they exhibit.

The response function relevant to longitudinal polarization waves is $(\varepsilon^{-1} - 1)$, which is the Fourier time transform of the charge density induced by an infinitesimal probe and hence of collective oscillations which can be self-sustained by the system [7]. Since we do not treat the electronic polarizability dynamically but only as background, the appropriate object for examination is

$$\frac{1}{\varepsilon(\omega)} - \frac{1}{\varepsilon_\infty} = \frac{1}{\varepsilon_\infty}\left(\frac{1}{\bar{\varepsilon}(\omega)} - 1\right) .$$

Accordingly we wish to analyze the quantity

$$f(t) \equiv \int_{-\infty}^{\infty} \frac{d\omega}{2\pi} e^{-i\omega t}\left(\frac{1}{\bar{\varepsilon}(\omega)} - 1\right) , \qquad t > 0. \tag{57}$$

As we have seen, the complex singularities of ε and ε^{-1} would, taken individually, fulfil the frequency >> damping rate criterion. However, not only is ω_p a branch point, so that there is a continuum of possibly interfering frequencies, but the infinity in ε^{-1} at ω_p is significantly weaker than for, e.g., the electron plasma. The extent to which dipolar plasmons resemble more familiar and simpler collective modes can only be settled by analysis of $f(t)$ in some detail.

By the theory of residues, the integral for $f(t)$ is equal to the sum of integrals in the lower half-plane around the branch cuts of $\bar{\varepsilon}^{-1}(\omega)$. The cut along the negative imaginary axis makes a purely decaying contribution to the charge density, and we concentrate on the finite frequency contributions from the cuts in the third and fourth quadrants. Expansion of $\bar{\varepsilon}^{-1}(\omega)$ near the branch points shows that the ratio of the coefficient of $|\omega - \omega_p|^{-\frac{1}{2}}$ near ω_p to the coefficient of $|\omega - \omega_1|^{\frac{1}{2}}$ near ω_1 is of order unity, being a bit greater than unity for $\varepsilon_0/\varepsilon_\infty \gg 1$. Accordingly the branch cut integrals should be dominated by the region close to the infinity at ω_p. Since $|\mathrm{Re}\omega| \gg |\mathrm{Im}\omega|$ everywhere on these cuts, it is adequate to our immediate purpose to ignore $\mathrm{Im}\omega$ and take the cuts on the real axis. In this case, $f(t)$ can be evaluated analytically (with some travail). The complete expression is:

$$\text{for} \quad \left\{ \begin{array}{l} z_1{}^2 \to (\mathrm{Re}\bar{\omega}_1)^2 \equiv (\bar{\omega}_1')^2 \\[2ex] z_p{}^2 \to (\mathrm{Re}\bar{\omega}_p)^2 \equiv (\bar{\omega}_p')^2 \end{array} \right.$$

$$f(t) = -\,\omega_p' \,\frac{3}{2}\left(1 - \frac{\omega_1'^2}{\omega_p'^2}\right) J_1(\omega_p' t) + \omega_p' \,\frac{\omega_1'^2}{\omega_p'^2}\left(1 - \frac{\bar{\omega}_1'^2}{\bar{\omega}_p'^2}\right) J_3(\omega_p' t)$$

$$+ \,\omega_p' \sum_{n \geqslant 1} \frac{1}{2n + 4}\left\{P_n\left(1 - 2\,\frac{\omega_1'^2}{\omega_p'^2}\right) - \right. \qquad \text{(Contd)}$$

$$- \left(1 - 2\frac{\omega_1'^2}{\omega_p'^2}\right)P_{n+1}\left(1 - 2\frac{\omega_1'^2}{\omega_p'^2}\right)\Bigg\}J_{2n+1}(\omega_p't), \qquad (58)$$

where J_n and P_n are ordinary Bessel functions and Legendre polynomials respectively. We note, first of all, that at very long times we would have

$$f(t) \sim \text{constant} \times \frac{1}{\sqrt{\omega_p t}} \cos\omega_p t.$$

At moderately long times there would be some dephasing and apparent decay even without the Debye dissipative damping. Numerical estimates, suggest that in all cases of interest the s series from J_5 on may be neglected and that even the J_3 term is substantially smaller than the leading term. For an isotropic Onsager model for H_2O at 300° K (with $\tau_0/\tau = 10$),

$$\frac{f(t)}{\omega_p} = -\ 0.836\ J_1 + 0.297\ J_3 + 0.111\ J_5$$

$$-\ 0.060\ J_7 - 0.008\ J_9 + \ldots \qquad (59)$$

the argument of the Bessel functions being $\omega_p't$. Had we neglected the Debye relaxation entirely (i.e. neglected the effect of τ_0/τ on ω_1,ω_p) only the leading term would be significant at all.

With sufficient art and persistence the collective mode charge density function $f(t)$ perhaps can be evaluated analytically for the general case, but we have not tried to do so. The main points are in any event by now clear. The dominant contributions come from the parts of the branch cuts nearest the ω_p branch points. The purely decaying contribution, from the cut on the imaginary axis, is a superposition over a *continuum* of decay times with the *shorter* times substantially more heavily weighted (we find $|\text{Im}\omega_p| > |\text{Im}\omega_1|$ for this cut). In strongly dipolar systems the oscillatory part of $f(t)$ should be dominated by a single term of form

$$\text{constant} \times e^{-\lambda t} J_1(\omega_p't),$$

for some real, positive λ. We note that modes of J_1 after its first maximum exhibit 2π periodicity within 2%, so that the polarization charge density oscillations for not-too-long times can closely imitate damped but strictly periodic vibrations. We note also that the dominance of the frequencies near ω'_p for finite t likely is greater than our numerical estimates of the last paragraph would indicate, since we find $|\text{Im}\omega_1| > |\text{Im}\omega_p|$ for oscillatory branch cuts.

4.c Response to Transverse Probes

A pole in a dielectric function commonly signals an anomaly in the response to a transverse probe. For example, the dielectric function at $k = 0$ for a polar crystal has a pole at the TO mode frequency as well as a zero at the LO frequency. We have so far calculated a longitudinal dielectric function at infinite wavelength, but may safely use it to discuss propagation of electromagnetic waves because (i) in the long wavelength limit the longitudinal and transverse dielectric functions become equal, and (ii) the k-dependence of the dielectric function can only become significant at wavelengths as short as a few times the interparticle spacing while electromagnetic wavelengths are orders of magnitude larger.

If, for the moment, we neglect the Debye dissipation completely, then the dispersion relation for propagation of transverse electromagnetic waves is

$$\omega^2 = \frac{c^2k^2}{\varepsilon(k,\omega)} \approx \frac{c^2k^2}{\varepsilon(\omega)} . \tag{60}$$

We then find undamped propagation for $0 < \omega < \omega_t$ and for $\omega > \omega_p$, total reflection for $\omega_t < \omega < \omega_l$, and true absorption for $\omega_l < \omega < \omega_p$. Asymptotically far from the anomalies we have

$$\text{for } \omega \to 0, \qquad \omega = \frac{ck}{\sqrt{\varepsilon_0}} ,$$

$$\text{for } \omega \to \infty, \qquad \omega = \frac{ck}{\sqrt{\varepsilon_\infty}} ,$$

and just above ω_p we find

$$\omega \approx \omega_p\left\{1 + \left(\frac{\omega_0}{\varepsilon_\infty\omega_p}\right)^2\left(\frac{ck}{\omega_p}\right)^4\right\} . \tag{61}$$

In a metallic conductor, the electron distribution remains degenerate right up to the melting point, thermal excitation of the plasma is negligible, and there is no propagation at *any* frequency below ω_p. Here, however, in the classical Boltzmann distribution, purely thermal noise spoils the dipoles' capacity to respond fully and screen out field having frequencies lower than thermal rotational frequencies for dipoles in equilibrium. Recall that when the Debye terms are ignored,

$$\omega_t^2 = \left(\frac{k_BT}{I}\right)\left(1 + \frac{\varepsilon_\infty}{\varepsilon_0}\right) . \tag{62}$$

One also notes that the ratio of longitudinal to transverse frequency, still ignoring Debye terms,

$$\frac{{\omega_p}^2}{{\omega_t}^2} = \frac{\varepsilon_0}{\varepsilon_\infty}\frac{2\varepsilon_0}{\varepsilon_0 + \varepsilon_\infty} \tag{63}$$

resembles the Lyddane-Sachs-Teller relationship (LST)

$$\frac{\omega_{LO}^2}{\omega_{TO}^2} = \frac{\varepsilon_0}{\varepsilon_\infty} \tag{64}$$

for the optical dispersion and longitudinal mode frequencies in polar crystals. If, for the *transverse* response, we were to neglect the long-range reaction field in calculating the polarizability, then we would have $\omega_t = \omega_T$ and would find exactly the LST relation. Since in simple physical interpretations of the LST relation it is common to assert that only longitudinal vibrations are affected by the macroscopic polarization this result is, on reflection, no surprise. However, the assertion does not hold if the massive constituents have substantial permanent dipole moments rather than the dominant monopole charge distributions. Equation (63) represents an extension to the present Onsager model of the LST relation. One must bear in mind, however, that frequency shifts due to Debye damping are not allowed for in that equation.

The restoration of the dissipative Debye terms of course modifies details, but the primary features listed above survive. We represent the complex amplitude for an electromagnetic wave travelling in the x- direction by

$$\exp\{i[k'(\omega) + ik''(\omega)]x - i\omega t\}$$

where, as usual,

$$k'(\omega) = \frac{\omega}{c}\, n(\omega)$$

$$k''(\omega) = \frac{\omega}{c}\, \kappa(\omega) \tag{65}$$

$$[n(\omega) + i\kappa(\omega)]^2 = \varepsilon(\omega).$$

The resulting dispersion relation between frequency ω and propagation constant resembles that for the undamped case for $\omega > \omega_p$ and $\omega < \omega_t$ but has also a branch for $\omega_t < \omega < \omega_p$ corresponding to highly damped propagation. A detailed discussion of the response probes will be given elsewhere.

5. APPLICATIONS

We expect our results to be applicable to a large number of dipolar liquids. In particular they should provide an adequate description of the longitudinal polarization charge density oscillations since, as we have discussed, the short range correlations responsible for the Kirkwood enhancement of the static dielectric constant are ineffective at large frequencies. In the case of liquid H_2O at 300° K our results would indicate an absorption in the range 30 μm to 15 μm, which is consistent with what little is known from measurements of the refractive index. The so-called 'hindered rotator' peak observed in incoherent neutron scattering also fits into our picture, since ω_p is also a resonance of the single-particle motion. However, the clearest and most direct support for our picture of dipolar plasmons is provided by molecular dynamics calculations for their model of liquid water by Rahman and Stillinger [*8*]. Using an improved pair interaction, they have now calculated [*9*] the correlation function for the longitudinal polarization charge density and find very clear oscillations with angular frequency $\omega \sim 1.6 \times 10^{14}$ sec^{-1}, while we estimate about 1.2×10^{14} sec^{-1}. Moreover, the very weak dispersion they find is in accord with our predictions [*1*].

REFERENCES

1. Lobo, R., Rodriguez, S. and Robinson, J.E. (1967). *Phys. Rev.*, **161**, 513; Lobo, R. (1968). *Phys. Rev. Lett.*, **21**, 145.
2. Nee, T.W. and Zwanzig, R. (1970). *J. Chem. Phys.*, **52**, 6353.
3. Debye, P. (1935). *Phys. Z.*, **36**, 100, 193.
4. Onsager, L. (1936). *J. Amer. Chem. Soc.*, **58**, 1886.
5. Kirkwood, J.G. (1939). *J. Chem. Phys.*, **7**, 911.
6. Kubo, R. (1966). *Rep. Prog. Phys.*, **29**, (Part I), 255.
7. Nozières, P. and Pines, D. (1958). *Nuovo Cim.*, **9**, 470.
8. Rahman, A. and Stillinger, F. (1971). *J. Chem. Phys.*, **55**, 3336.
9. Rahman, A. and Stillinger, F. (Private Communication), (to be published).

DYNAMICAL IMAGE CHARGE INTERACTIONS

J. HEINRICHS†

Université de Liège, Institut de Physique,
Sart-Tilman, Liège, Belgium.

1. INTRODUCTION

In this lecture a discussion is given of generalized image charge interactions for particles which are moving at some distance from a metal surface. The importance of this problem lies in the fact that many investigations of surface phenomena involve the scattering of charged particles by the surface (low energy electron diffraction, fast electron spectroscopy, etc.) or the ejection of charged particles from the surface (photoemission, field emission). These dynamical interactions are dependent on the nature of the elementary excitations at the surface.

In the studies of the dynamical charge-metal interaction it is generally assumed that the particle is moving uniformly along a classical trajectory. This involves, in particular, the assumption that the dynamics of the external particle and of the free electrons in the metal may be treated separately. Such a decoupling is valid if the external charge is sufficiently heavy (Born-Oppenheimer approximation) or if it is located sufficiently far away from the electron screening cloud at the surface, in a region where the interaction is weak. On the other hand, the acceleration of the particle is negligible as long as its kinetic energy is large compared with the image potential which causes it to accelerate. Recent treatments of the dynamical

† Chercheur Qualifié au Fonds National de la Recherche Scientifique, Bruxelles. Part of this work was performed during the author's stay at Kernforschungsanlage, Institut für Festkörperforschung, 517 Jülich, West Germany.

image potential by Sunjic, Toulouse and Lucas [1], Ray and Mahan [2], and Harris and Jones [3] have focussed attention on the case of a particle approaching the surface along the z-direction (normal to the surface) with constant velocity v. Using a different method, Heinrichs [4] has studied various other cases as well and, in particular, the case of a particle ejected from the surface, where additional real excitation effects come into play. Since a detailed review of image potential interactions is already given in Mahan's lectures we shall emphasize more recent developments [4,5].

2. GENERAL EXPRESSIONS FOR INTERACTION ENERGY

We consider a classical jellium model with a step function variation for the unperturbed electron density at the surface. Furthermore, we assume that the potential barrier associated with the step in the electron density is infinitely high so that the electrons are reflected specularly. The surface is placed normally to the z-axis at $z = 0$ and the metal occupies the region $z > 0$. The moving external charge of magnitude q is located in the vacuum region ($z < 0$) at $r(t) = (0,0,a(t) < 0)$. For the time being the nature of the trajectory $a(t)$ and the initial conditions are left unspecified. The instantaneous charge-metal interaction energy is given by

$$W[a(t)] = \frac{q}{2} V_i(x = 0, y = 0, z = a(t)), \tag{1}$$

where V_i denotes the induced potential (total potential minus free space potential of the external charge) created in the vacuum by the charge density fluctuation which is induced in the surface. The factor $\frac{1}{2}$ arises from the definition of W as the energy lowering which results when the charge q is brought from infinity to a distance $a(t)$ from the surface. The induced potential is obtained by matching boundary conditions in the usual way. To apply these conditions one needs the solutions for the electrostatic potential and for the displacement field in the metal interior. In the present model these quantities may be expressed in terms of the bulk dielectric function $\varepsilon(k,\omega)$, using a straightforward extension of a method due to Ritchie and Marusak [6]. The calculations are performed explicitly in reference [4] to which we refer the reader for details. The final result for W is [4]

$$W[a(t)] = \frac{q^2}{4\pi}\int_{-\infty}^{\infty} d\omega e^{-i\omega t}\int_0^{\infty} dk_{\|}$$

$$\times \int_{-\infty}^{\infty} dt' e^{i\omega t'} e^{k_{\|}(a(t)+a(t'))} \frac{1 - \varepsilon_s(k_{\|},\omega)}{1 + \varepsilon_s(k_{\|},\omega)} \tag{2}$$

$$a(t) < 0.$$

The function $\varepsilon_S(k_{||},\omega)$ depending on wave vector parallel to the surface and on frequency is defined by

$$\varepsilon_S(k_{||},\omega) = \left[\frac{k_{||}}{\pi}\int_{-\infty}^{\infty} \frac{dk_z}{k^2\varepsilon(k,\omega)}\right]^{-1} . \tag{3}$$

It is referred to as a surface dielectric function because from the point of view of the screening it represents the surface analog of the bulk dielectric function. One may verify explicitly that ε_S and ε coincide in the limit of a local Drude dielectric constant $\varepsilon(k,\omega) \equiv \varepsilon(\omega)$. As shown below the image potential may be significantly affected by frequencies corresponding to the solutions of

$$\varepsilon_S(k_{||},\omega) + 1 = 0. \tag{4}$$

This is the standard dispersion relation for the surface plasmon frequency in the present model [6].

The expression for W simplifies considerably in the case of a uniformly moving particle (of velocity v) which is approaching the surface from $-\infty$, so that $a(t) = vt$, $t \in [-\infty,0]$. In this case the t'-integral is

$$\int_{-\infty}^{0} dt' \exp(i\omega + k_{|}v)t' = (i\omega + k_{|}v)^{-1},$$

and recalling that as a result of causality requirements the singularities of $\varepsilon_S(k_{||},\omega)$ and $\varepsilon(k,\omega)$ are restricted to the lower half of the complex ω-plane one gets

$$W[a(t)] = \frac{q^2}{2}\int_0^{\infty} dk_{||} e^{2k_{||}a(t)} \frac{1 - \varepsilon_S(k_{||},ik_{||}v)}{1 + \varepsilon_S(k_{||},ik_{||}v)} . \tag{5}$$

We emphasize that since (5) is just the contribution from the pole $\omega = ik_{|}v$ the interaction is not affected by real surface plasmon (or single particle) excitation in this case. It arises purely from a polarization of the electron gas by the external charge through virtual excitations.

In the case of a static point charge located at $a(t) = a$, the interaction energy (2) takes a form which differs from (5) simply by the replacement of $\varepsilon_S(k_{||},ik_{|}v)$ by the static dielectric constant $\varepsilon_S(k_{||},0)$.

3. DETAILED APPLICATIONS

Some applications of equation (2) have been discussed in reference [4] for the case of a Drude dielectric constant $\varepsilon(k,\omega) \equiv \varepsilon(\omega)$, as well as for the case of stationary charges. Subsequently, the dynamical results have been generalized to include the effect of single-particle excitations [5], by using a dielectric constant $\varepsilon(k,\omega)$ which depends on wave vector as well as on frequency. To keep the problem simple, we have used the expression

$$\varepsilon(k,\omega) = 1 - \frac{\omega_p^2}{\omega^2 - \beta^2 k^2 + i0^+} , \tag{6}$$

which results from a hydrodynamic treatment of the electron gas [7]. Here ω_p is the plasma frequency and the parameter β^2 may be chosen to be either $\beta^2 = \beta_1{}^2 = (3/5)v_F{}^2$ (v_F = Fermi velocity), a value which yields the correct RPA dispersion relation for bulk plasmons, or, alternatively, $\beta^2 = \beta_2{}^2 = v_F{}^2/3$, which is such that (6) reduces to the Thomas-Fermi limit for $\omega = 0$. It is seen that (6) simulates the actual single-particle spectrum of an electron gas via an average excitation frequency $\omega = \beta k$. By inserting (6) in (3) and performing the integral we obtain

$$\varepsilon_s(k_{||},\omega) = \frac{(\omega + i0^+)^2 - \omega_p^2}{(\omega + i0^+)^2 \mp \gamma^{-1} k_{||}\omega_p^2} , \tag{7}$$

$$\gamma = \frac{1}{\beta} [\omega_p^2 + \beta^2 k_{||}^2 - (\omega + i0^+)^2]^{\frac{1}{2}} \tag{8}$$

where the expression with the upper sign is valid for Re $\omega <$ $(\omega_p^2 + \beta^2 k_{||}^2)$ and the one with the lower sign for Re $\omega > (\omega_p^2 + \beta^2 k_{||}^2)^{\frac{1}{2}}$.

Before discussing the case of moving charges in more detail it is useful to consider the limit of a static point charge placed at a distance a from the surface. In this case we substitute the static limit of (7) (for $\beta = \beta_2$) in (5) and obtain [4,8]

$$W(a) = - \frac{q^2 k_{FT}}{2} \int_0^\infty dx e^{-2k_{FT}|a|x} [x - (1 + x^2)^{\frac{1}{2}}]^2 , \tag{9}$$

which for $k_{FT}|a| >> 1$ reduces to

$$W(a) = - \frac{q^2}{4|a|} \left[1 - \frac{1}{k_{FT}|a|} + \frac{1}{k_{FT}{}^2|a|^2} + \ldots \right] , \tag{10}$$

where $k_{FT} = \omega_p/\beta_2 = (6\pi ne^2/m)^{\frac{1}{2}}$ is the Thomas-Fermi wave vector. This shows that the classic expression $-q^2/4|a|$ for the image potential is valid for distances large compared with the electron screening length $k_{FT}{}^{-1}$. Furthermore, as a consequence of a non-zero screening length (9) is finite for $|a| \to 0$). These results are discussed in more detail in reference [4] which also describes an alternative treatement avoiding the use of the step density assumption.

We now examine the dynamical image potential interaction for two specific cases of practical interest.

Case 1: Particle Approaching the Surface with a Uniform Velocity v

We first substitute the local limit $\varepsilon_S(k_{\|} = 0,\omega) \equiv \varepsilon(\omega)$ in (5) and obtain the exact expression [2,4]

$$W[a(t)] = -\frac{q^2\omega_S{}^0}{2v} f\left(\frac{2\omega_S{}^0|a(t)|}{v}\right) , \tag{11}$$

where $\omega_S{}^0 = \omega_p/\sqrt{2}$ is the surface plasmon frequency and $f(x)$ is the tabulated function [9]

$$f(x) = \mathrm{Ci}(x)\sin x - \mathrm{si}(x)\cos x. \tag{12}$$

This expression is finite for $|a(t)| \to 0$ and vanishes for $v \to \infty$ ($|a| \neq 0$), in agreement with the fact that a sufficiently fast external charge cannot induce any charge fluctuation in the metal surface. For large separations equation (11) becomes

$$W[a(t)] = -\frac{q^2}{4|a|}\left[1 - \frac{v^2}{2\omega_S{}^{02}|a|^2} + \frac{3v^4}{2\omega_S{}^{04}|a|^4} + \ldots\right] , \tag{13}$$

where the reduction of the interaction compared with the classic result $-q^2/4|a|$ corresponds to the time delay with which the electrons respond to the dynamical perturbation. The fact that the interaction energy is finite for $|a| \to 0$, unlike the expression $-q^2/4|a|$, is a consequence of deviations from perfect static screening in the case of (9) and from perfect dynamic screening in the case of (11) [4].

In the case where the $k_{\|}$-dependence of (7) is included it is possible to evaluate (5) for large distances by expanding the coefficient of $\exp[2k_{\|}a(t)]$ in powers of $k_{\|}$. The result for $\beta = \beta_2$ is

$$W[a(t)] = -\frac{q^2}{4|a|}\left[1 - \frac{1}{k_{FT}|a|} + \frac{1}{k_{FT}{}^2}\left(1 - \frac{v^2}{\beta_2{}^2}\right)\frac{1}{|a|^2} - \frac{3}{4k_{FT}{}^3}\left(1 - \frac{7v^2}{\beta_2{}^2}\right)\frac{1}{|a|^3} + \ldots\right] . \tag{14}$$

The comparison with equations (10) and (13) indicates that the first dynamical single-particle correction to the sum of static and local results is of order $v^2|a|^{-4}$. This term differs from the corresponding result of Sunjič, Toulouse and Lucas [1] who chose to use an *ad hoc* model expression for $\varepsilon_S(k_{||},\omega)$ instead of deriving it from the bulk dielectric function as done here. Equation (14) shows that for $v^2 \ll \beta_2{}^2 = v_F{}^2/3$ the static treatment is valid, while for $v^2 \gg \beta_2{}^2/7$ single-particle dispersion effects are unimportant and the local Drude treatment becomes adequate.

We note that the present study does not lead to an asymptotic single-particle correction proportional to v of the type found by Harris and Jones [3], in their improved treatment of the infinite potential barrier model. However, their treatment does reduce to the present one in the static and local cases [3]. On the other hand, the existence of other dynamical corrections, due to quantum mechanical recoil effects and to effects non-linear in q^2, has been demonstrated by Sunjič, Toulouse and Lucas [1].

Case 2: Particle Ejected from the Surface with a Uniform Velocity v

In this case the trajectory is given by $a(t) = -vt$, $t \in [0,\infty]$ and by performing the t'-integral in (2) we obtain

$$W[a(t)] = \frac{iq^2}{4\pi}\int_0^\infty dk_{||}e^{-k_{||}|a(t)|}\int_{-\infty}^{\infty} d\omega\, \frac{e^{-i\omega t}(1-\varepsilon_S(k_{||},\omega))}{(\omega + ik_{||}v)(1+\varepsilon_S(k_{||},\omega))} \, . \quad (15)$$

Since $t > 0$ it follows that W has two contributions in the local case [4]. The first one is a polarization effect similar to (5), which arises from the pole $\omega = -ik_{||}v$. The second one is the contribution from real surface plasmons whose frequencies are given by (4). As discussed in reference [4] this latter contribution is associated with the sudden creation of the charge with initial velocity v at the surface at $t = 0$. The evaluation of (15) in the local approximation yields [4]

$$W[a(t)] = -\frac{q^2\omega_S{}^0}{2v}\left[f\left(\frac{2\omega_S{}^0|a|}{v}\right) - f\left(\frac{\omega_S{}^0|a|}{v}\right)\cos\left(\frac{\omega_S{}^0|a|}{v}\right) + g\left(\frac{\omega_S{}^0|a|}{v}\right)\sin\left(\frac{\omega_S{}^0|a|}{v}\right)\right] \, . \quad (16)$$

where

$$g(x) = -\,\mathrm{Ci}(x)\cos x - \mathrm{si}(x)\sin x. \quad (17)$$

We now discuss the extension of these results in the more general case where the $k_{||}$-dependence of $\varepsilon_S(k_{||},\omega)$ given by (7) is included [5]. We use the residue theorem to perform the ω-integration in (15). The integrand has poles at $\omega = -ik_{\|}v$ and at the surface plasmon frequencies [10]

$$\omega_s^{\pm}(k_{\|}) = \pm \tfrac{1}{2}[(2\omega_p^2 + \beta^2 k_{\|}^2)^{\frac{1}{2}} + \beta k_{\|}] - i0^+ \tag{18}$$

(exact solutions of (4)), and single-particle branch points at $\omega = \omega_{\pm}(k_{\|}) = \pm(\omega_p^2 + \beta^2 k_{\|}^2)^{\frac{1}{2}} - i0^+$. Since $t > 0$ the residue theorem is applied with the contour formed by the real axis, a semi-circle of infinite radius in the lower half ω-plane and two branch cuts $C_{\pm}$ parallel to the real axis and extending from $\omega_+(k_{\|})$ to $\infty - i0^+$ and from $\omega_-(k_{\|})$ to $-\infty - i0^+$. We have thus (schematically)

$$W[a(t)] + \int_{C_+} + \int_{C_-} = -2\pi i \sum \mathrm{Res}[-ik_{\|}v, \omega_s^{\pm}(k_{\|})],$$

and from the explicit evaluation of the residues and the expression of the branch cut contributions we obtain

$$W[a(t)] = W_o + W_{spo} + W_{sp}, \tag{19}$$

where

$$W_o = \frac{q^2}{2}\int_0^{\infty} dk_{\|} e^{-2\kappa_{\|}|a|} \frac{1 - \varepsilon_s(k_{\|}, -ik_{\|}v)}{1 + \varepsilon_s(k_{\|}, -ik_{\|}v)}, \tag{20}$$

$$W_{spo} = \frac{q^2}{4}\int_0^{\infty} dk_{\|} \frac{e^{-k_{\|}|a|}}{\omega_s^2 + k_{\|}^2 v^2} \frac{(\omega_0 - \beta k_{\|})^2(\omega_0 - 3\beta k_{\|})}{(2\omega_p^2 - 3\beta k_{\|}\omega_0 + \beta^2 k_{\|}^2)} \times \left[\omega_s \cos\left(\frac{\omega_s|a|}{v}\right) - k_{\|}v \sin\left(\frac{\omega_s|a|}{v}\right)\right], \tag{21}$$

$$W_{sp} = -\frac{2q^2\omega_p^2}{\pi}\int_0^{\infty} dk_{\|} e^{-k_{\|}|a|} I(k_{\|}, |a|), \tag{22}$$

$$I(k_{\|}, |a|) = -k_{\|}\mathrm{Im}\int_{\omega_+(k_{\|})}^{\infty - i0^+} d\omega \frac{e^{-i\omega|a|/v}}{\omega + ik_{\|}v} \times \frac{\gamma(\omega_p^2 - \omega^2)}{\gamma^2(2\omega^2 - \omega_p^2)^2 - k_{\|}^2\omega_p^4}, \tag{23}$$

where $\omega_0 = (2\omega_p^2 + \beta^2 k_{||}^2)^{\frac{1}{2}}$ and $\omega_s = |\omega_s^{\pm}(k_{||})|$. W_{sp} gives the contribution from the cuts which is a purely dynamical single-particle effect, while W_{spo} describes the effect of the surface plasmon poles (18). In the limit $\beta \to 0$, $W_o + W_{spo}$ reduces of course to equation (16). For large separations $|a|$ the contributions (20) and (21) may be evaluated analytically by expanding the coefficients of the exponentials and of the argument of the circular functions in power series in $k_{||}$. W_o is then given by (14) and the result for W_{spo} is calculated to be ($\lambda = \beta/2v$)

$$W_{spo} \simeq \frac{q^2}{2|a|}\frac{1}{1+\lambda^2}\left\{1 - \lambda\tan\left(\frac{\omega_s^0|a|}{v}\right) - \frac{v}{\omega_s^0|a|}\frac{1}{1+\lambda^2}\right.$$

$$\times\left[\lambda(5 - 3\lambda^2) + (1 - 7\lambda^2)\tan\left(\frac{\omega_s^0|a|}{v}\right)\right]$$

$$\left. + \theta(|a|^{-2})\right\}\cos\left(\frac{\omega_s^0|a|}{v}\right). \quad (24)$$

This expression involves a minor approximation, namely the neglect at successive orders in $|a|^{-1}$ of the non-linear dependence on $k_{||}$ of the argument of the circular functions. On the other hand one may easily verify that W_{sp} is of relative order $|a|^{-3}$ and that with a crude estimate of (23) one obtains for large $|a|$,

$$W_{sp} \simeq \frac{13}{16\pi}\frac{q^2\beta v}{\omega_p^2|a|^3}\sin\left(\frac{\omega_p|a|}{v}\right) + \theta(|a|^{-4}). \quad (25)$$

Thus W_{sp} is only one order lower in $v/\omega_p|a|$ than the first velocity-dependent single-particle correction in (14).

For $\lambda = 0$ ($\beta = 0$) the above results reduce to the local treatment which is thus valid for $v \gg v_F$. On the other hand, for $\lambda \gg 1$, i.e. for $v \ll v_F$, we obtain from (24)

$$W_{spo} \simeq -\frac{q^2}{|a|}\frac{v}{\beta}\left[\sin\left(\frac{\omega_s^0|a|}{v}\right) - \frac{3v}{\omega_s^0|a|}\cos\left(\frac{\omega_s^0|a|}{v}\right) + \dots\right], \quad (26)$$

which shows that the surface plasmon effect decreases gradually with decreasing velocity. It follows, therefore, that $W(a)$ reduces to the static expression (10) for $v \to 0$. Such a result is not expected from the local treatment for an ejected particle [4]. This demonstrates the importance of including single-particle dispersion effects for the purpose of reconciling the dynamical treatment with the quasistatic one in the limit of low velocities. We also note that for low velocities and large distances the velocity-dependent terms in $W(a)$ are of lower order in the present case than in the case of a particle approaching the surface. Finally, the smallness of the dynamical single-particle contribution (25) compared with the surface plasmon effect (24) is not surprising since one expects the coupling to real single-particle excitations to be much weaker than the coupling to charge density waves.

REFERENCES

1. Šunjić, M., Toulouse, G. and Lucas, A.A. (1972). *Sol. State Commun.*, **11**, 1629.
2. Ray, R. and Mahan, G.D. (1972). *Phys. Lett.*, **42A**, 301.
3. Harris, J. and Jones, R.O. (1973). *J. Phys.*, C, (Solid State Physics), (to be published).
4. Heinrichs, J. (1973). *Phys. Rev.*, **B**, (in press).
5. Heinrichs, J. (to be published).
6. Ritchie, R.H. and Marusak, A.L. (1966). *Surface Sci.*, **4**, 234.
7. Pines, D. and Nozières, P. (1966). *The Theory of Quantum Liquids*, (Benjamin, New York), chapter 4.
8. Newns, D.M. (1969). *J. Chem. Phys.*, **50**, 4572.
9. Abramowitz, M. and Stegun, I.A. (eds.). (1965). *Handbook of Mathematical Functions*, (National Bureau of Standards), p. 228.
10. See, e.g., Heinrichs, J. (1973). *Phys. Rev.*, **B7**, 3487.

FINE STRUCTURE CORRECTION TO THE EDGE SINGULARITY IN THE X-RAY SPECTRA OF METALS

PIERRE LONGE†

Institut de Physique, Univerité de Liège,
Sart Tilman, B-4000 Liège, Belgium

In 1969, an expression was proposed by Nozières and de Dominicis [1] (ND) and by other authors [2] to describe the edge singularity of X-ray band spectra of metals. This expression is

$$I_\ell(\varepsilon) = I_\ell{}^0(\varepsilon)\left(\frac{\xi_0}{|\varepsilon|}\right)^{\alpha_\ell}, \tag{1}$$

where the frequency ε is measured from the Fermi edge and where $I_\ell{}^0(\varepsilon)$ is the one electron transition intensity. The exponent depends only on the electron phase shifts δ_ℓ at the Fermi level and is given by

$$\alpha_\ell = \frac{2}{\pi}\,\delta_\ell - 2\sum_{\ell'=0}^{\infty}(2\ell'+1)\left(\frac{\delta_{\ell'}}{\pi}\right)^2. \tag{2}$$

This singularity, already pointed out by Mahan [3] in 1967, is related to an infra-red catastrophe, involving the electrons close to the Fermi surface. The sudden variation of the ion potential at the moment of the X-ray absorption (or emission) gives rise to the creation of a large number of weak energy electron-hole pairs.

In a recent paper [4] where we extended expression (1) beyond the close neigborhood of the Fermi edge, we noticed that another factor $A(\varepsilon)$ appeared in (1). This factor is particularly

† Chercheur I.I.S.N.

interesting because it presents *a singularity in its slope* for $\varepsilon = 0$ and thus brings a contribution in the shape of the edge singularity. This factor comes out when the *fine structure* of the zeroth order propagator is treated exactly, or more precisely when the fine structure is not averaged, as in reference [1, 2]. Such an exact treatment becomes important when the Dyson equation applies to times when the potential varies suddenly.

Of course our factor $A(\varepsilon)$ does not modify the 'sign' of the singularity, i.e. the sign of α_ℓ, discussed by Ausman and Glick [5], and also in reference [4]. However, it may modify the shape of the singularity sufficiently to make questionable an experimental determination of α_ℓ from the edge shape. The effect of this factor may be larger than the smearing effects due for instance to the temperature, hole recoil, self absorption, already discussed in various papers [6-8].

In reference [4], $A(\varepsilon)$ appears in the calculation of the open line part of the ND problem. In that paper, the problem was treated using a realistic (non separable) potential, which requires a certain number of approximations. At first sight, $A(\varepsilon)$ could appear as due to these approximations. It is the reason why we would show in the present paper that such a factor having a steep slope at $\varepsilon = 0$ is indeed contained in the ND model and was missed up to now because of the approximation of the averaged fine structure.

First let us summarize the method used to solve the ND problem. The X-ray response function is obtained from a propagator describing an electron in absorption (or a hole in emission) moving close to the Fermi surface between times t and t'. During the interval $t' - t$, this electron (or hole) and all the other electrons interact with a potential switched on suddenly at time t and switched off in the same way at t'. This transient potential is the potential of the ion where-from comes the extra electron (we will only consider the absorption case) and the infra-red catastrophe is essentially due to its sudden switching, on and off. The interaction of the emitted electron with the transient potential is described by the open line part of the propagator and the interaction of the other conduction electrons by a closed loop (vacuum)part. These two parts of the dressed propagator give rise respectively to the first and to the second term of (2).

The calculations were made in ND by using the approximation of the separable core-hole potential

$$V(\mathbf{k},\mathbf{k}') = -\sum_{\ell} V_\ell u_\ell(\varepsilon_k) u_\ell(\varepsilon_{k'}), \tag{3}$$

with $\varepsilon_k = k^2/2m - \mu$ and

$$u(\varepsilon) = e^{-|\varepsilon|/2\xi_0}, \tag{4}$$

where ξ_0 is a cutoff of the order of the Fermi energy μ. Likewise, the X-ray matrix element was given by

$$W(\mathbf{k}) = \sum_{\ell} W_\ell u_\ell(\varepsilon_k). \tag{5}$$

We will only consider the open line contribution, (giving the factor $A(\varepsilon)$), and in this contribution only the ℓ-th harmonic. Index ℓ will be dropped for simplicity.

When there is no transient core-hole potential, the one electron propagator is

$$G(\tau' - \tau) = -i\nu_0\left[P\,\frac{1}{\tau' - \tau} + \pi\tan\theta\;\delta(\tau' - \tau)\right], \tag{6}$$

where ν_0 is the electron density at the Fermi level; P stands for 'principal part'. The last term of (6) represents the total weight of the fine structure; it will be discussed below. On the other hand, when the transient potential is present, i.e. when $t < (\tau,\tau') < t'$, (6) has to be replaced, according to ND, by

$$\phi(\tau',\tau;t',t) = G'(\tau' - \tau)\left[\frac{(t' - \tau)(\tau' - t)}{(t' - \tau')(\tau - t)}\right]^{\delta/\pi}, \tag{7}$$

where $G'(\tau' - \tau)$, slightly different from (6), is given by

$$G'(\tau' - \tau) = -\frac{i\nu_0}{\beta}\left[P\,\frac{1}{\tau' - \tau} + \pi\tan\theta'\;\delta(\tau' - \tau)\right]. \tag{8}$$

It is the one-electron propagator with a permanent core-hole potential. The open line contribution is then obtained by letting $\tau' = t'$ and $\tau = t$ in (7). This requires the introduction in (7) of the cutoff ξ_0. One obtains in that way

$$\phi(t',t;t',t) = -\frac{i\nu_0}{\beta(t' - t)}\left[i\xi_0(t' - t)\right]^{2\delta/\pi}. \tag{9}$$

Taking the real part of the Fourier transform of this latter expression, one gets the open line part of (1), which is

$$z(\varepsilon) = \frac{\nu_0(\xi/\varepsilon)^{2\delta/\pi}}{\beta\Gamma(1 - \delta/\pi)}. \tag{10}$$

Here we have to pay attention to two points. First there is no reason to use the same μ function, or more exactly the same ξ_0 in (3) and (5), secondly the introduction of this ξ_0 in (9) is a little bit heuristic.

Let us go back to (3) and let us write this potential in the form

$$V(\varepsilon,\varepsilon') = -Vu(\varepsilon)u(\varepsilon'), \tag{11}$$

with now

$$u(\varepsilon) = e^{-a\varepsilon/2} \qquad \text{for } \varepsilon > 0,$$

$$= e^{b\varepsilon/2} \qquad \text{for } \varepsilon < 0.$$

The one-electron propagator, established from expression (6)

$$G(\tau' - \tau) = -\int_{-\infty}^{0} d\varepsilon \nu_0 u^2(\varepsilon) e^{-i\varepsilon(\tau'-\tau)}$$

$$+ \int_{0}^{\infty} d\varepsilon \nu_0 u^2(\varepsilon) e^{-i\varepsilon(\tau'-\tau)}.$$

is then

$$G(\tau' - \tau) = -i\nu_0 \left[\frac{\eta(\tau' - \tau)}{\tau' - \tau - ia} \quad \frac{\eta(\tau - \tau')}{\tau' - \tau + ib}\right], \qquad (12)$$

η being the step function. This latter expression becomes equivalent to (6) if the fine structure (at $\tau \sim \tau'$) is averaged by letting

$$\pi \tan\theta = (-i\nu_0)^{-1} \int_{\tau-\alpha}^{\tau+\alpha} d\tau' G(\tau' - \tau)$$

$$= \ln \frac{b}{a} . \qquad (13)$$

This requires, however, $\alpha \gg a$ or b. Similarly, in the presence of a permanent core-hole potential, we can write

$$G'(\tau' - \tau) = -\frac{i\nu_0}{\beta}\left[\frac{\eta(\tau' - \tau)}{\tau' - \tau - ia'} + \frac{\eta(\tau - \tau')}{\tau' - \tau + ib'}\right], \qquad (14)$$

which, when compared to (8), gives

$$\pi \tan\theta' = \ln \frac{b'}{a'} .$$

All this shows that the propagator (7) with G' given by (8) is strictly correct only if $t' - \tau$ or $\tau' - t$ is much larger than a' or b'. We can however improve (7), for $\tau' \sim t'$ and $\tau \sim t$, by considering the Dyson equation leading to (7). This equation is

$$\phi(\tau',\tau;t',t) = G(\tau' - \tau) + \qquad \text{(Contd)}$$

$$+ iV\int_t^{t'} d\tau'' G(\tau' - \tau'')\phi(\tau'',\tau;t',t). \qquad (15)$$

If G is given by (6), (7) will be the exact solution. However, when $\tau' \sim t'$ or $\tau \sim t$, expression (12) is more correct than (6), as shown by the integration in (13), and (12) has to be substituted in (15). For the limit $\tau' = t'$ *or* $\tau = t$, only the first term of (12) appears. We will write it in the form

$$\mathfrak{G}(\tau' - \tau) = \frac{-i\nu_0}{\tau' - \tau - ic}\,. \qquad (16)$$

Here we replace the relaxation time a by another value c. Indeed $\mathfrak{G}$ connects a core-hole vertex (11) to an X-ray vertex (5) where, as mentioned before, we use another u function, whose cutoff is much larger than a^{-1} or b^{-1}. Indeed the states involved in an X-ray transition are not restricted to the neighborhood of the Fermi surface by the momentum conservation. Thus, in (16), we can estimate that $2c \sim a$ or b. In the same way, if $\tau' = t'$ *and* $\tau = t$, (16) will be replaced by

$$\mathfrak{G}^*(t' - t) = \frac{-i\nu_0}{t' - t - ic^*}\,, \qquad (17)$$

where $c \ll a$ or b since we now connect two X-ray vertices.

We can now replace the Dyson equation (15) by another equation which applies to the case $\tau' = t'$ and $\tau = t$. This equation is

$$\phi(t',t;t',t) = \mathfrak{G}^*(t' - t) + iV\int_t^{t'} d\tau\,\mathfrak{G}(t' - \tau)\mathfrak{G}(\tau - t)$$

$$+ (iV)^2\int_t^{t'} d\tau'\int_t^{t'} d\tau\,\mathfrak{G}(t' - \tau')\phi(\tau',\tau;t',t)\mathfrak{G}(\tau - t). \qquad (18)$$

Using an iteration method by substitution of (7) and (8) in the second member, one obtain the following expression

$$\phi(t',t;t',t) = \frac{\nu_0}{is}\left(\frac{is}{c}\right)^{2g} - \frac{gb}{(is)^2}\ln\frac{(is)^2}{c(c - b)}\,, \qquad (19)$$

with $s = t' - t$ and $g = \delta/\pi$.

The second order terms in g are dropped. The details of the calculations will be published elsewhere. The Fourier transform of (19) is

$$z(\varepsilon) = (\varepsilon c)^{-2g} + gb\varepsilon\ln\frac{\varepsilon^2}{c(b - c)}\,,$$

or, writing $b \approx 2c$,

$$z(\varepsilon) = \left[1 + \frac{4g}{1 + 2g} (c\varepsilon)^{1+2g} \ln(c\varepsilon)^{1+2g}\right] (c\varepsilon)^{-2g}.$$

The expression in the square brackets is the announced function $A(\varepsilon)$ which can be obtained by that method in the logarithmic accuracy. The slope tends to present a logarithmic divergence when δ is small (K-bands). An estimation of the term in the bracket may be obtained. For Na-L absorption edge, one has [*4*, *5*] $\delta/\pi = 0.3$. On the other hand, writing $C \approx \frac{1}{2}\xi_0^{-1}$ and using the value $\xi_0 \approx 0.4\ \mu$ proposed in reference [*4*] for various metals, one obtains for this function,

$$A(\varepsilon) = 1 + 0.3\ (\varepsilon/\mu)^{1.6} \ln(\varepsilon/\mu),$$

which is the factor multiplying the power divergence (1) (with $\alpha_\ell = 0.4$ for Na L-band). This result is in agreement with the result obtained in reference [*4*].

REFERENCES

1. Nozières, P. and de Dominicis, C.T. (1969). *Phys. Rev.*, **178**, 1097.
2. Schotte, K.D. and Schotte, U. (1969). *Phys. Rev.*, **185**, 509.
3. Mahan, G.D. (1967). *Phys. Rev.*, **163**, 612.
4. Longe, P. (1973). *Phys. Rev.*, **B8**.
5. Ausman, G.A. and Glick, A.J. (1969). *Phys. Rev.*, **183**, 687.
6. Ferrell, R.A. (1969). *Phys. Rev.*, **186**, 399.
7. Bergersen, B., McMullen, T. and Carbotte, J.P. (1971). *Can. J. Phys.*, **49**, 3155.
8. Yuval, G. (1971). *Phys. Rev.*, **B4**, 4315.

AUTHOR INDEX TO PART A

SUBJECT INDEX TO PART A